Agent-Based Modeling of Environmental Conflict and Cooperation

Agent-Based Modeling of Environmental Conflict and Cooperation

Todd K. BenDor, PhD and Jürgen Scheffran, PhD

CRC Press
Taylor & Francis Group
Boca Raton London New York

CRC Press is an imprint of the
Taylor & Francis Group, an **informa** business

CRC Press
Taylor & Francis Group
6000 Broken Sound Parkway NW, Suite 300
Boca Raton, FL 33487-2742

First issued in paperback 2020

ISBN 13: 978-0-367-57082-8 (pbk)
ISBN 13: 978-1-138-47603-5 (hbk)

Library of Congress Cataloging-in-Publication Data

Names: BenDor, Todd K., author. | Scheffran, Jèurgen, author.
Title: Agent-based modeling of environmental conflict and cooperation / Todd K.
BenDor and Jèurgen Scheffran.
Description: Boca Raton, FL : Taylor & Francis, 2018. | Includes bibliographical references.
Identifiers: LCCN 2018017475 | ISBN 9781138476035 (hardback: alk. paper)
Subjects: LCSH: Environmental sciences—Mathematical models. | Multiagent
systems. | Conflict management.
Classification: LCC GE45.M37 B46 2018 | DDC 363.7/05250113—dc23
LC record available at https://lccn.loc.gov/2018017475

To Amanda, Marianne, Astrid, and all those who endeavor each day to solve

problems, resolve disputes, and leave our world healthier, safer, and better off.

About the cover

About 20 km from Cape Town, South Africa, lies the picturesque gated suburb of Lake Michelle. With easy access to surfing and other amenities, real estate prices can reach several million South African Rand (~USD $100,000s). Beyond Lake Michelle's electrified fence, and across a narrow strip of wetlands, is the neighboring township of Masiphumelele ("let us succeed" in the Xhosa language). The site of frequent protests over public service delivery and government corruption, there is no police station, only one small day clinic, and 35% of Masiphumelele's 38,000 residents are believed to be infected with HIV or tuberculosis. Fires frequently sweep through the thicket of small tin shacks, sometimes displacing residents by the hundreds. In 2011, a fire burned down nearly 1,500 formal and informal homes, killing one and displacing approximately 5,000 residents.

This photograph depicts these places in stark contrast. The photographer, Johnny Miller, wrote: "I see the wetlands between them as a sort of no-man's land; an area too scary to venture into from either side. I imagine both sides peer across at their neighbors with distrust and suspicion." (https://www.millefoto.com, https://www.unequalscenes.com/)

Authors

Todd K. BenDor is a professor of City and Regional Planning at the University of North Carolina at Chapel Hill. His research and teaching aim to better understand and reduce the impacts that human activities and development can have on sensitive ecosystems. Dr. BenDor's work uses qualitative and quantitative methods to improve environmental policy, particularly ecosystem service markets, such as those used domestically to improve water quality and restore wetlands. His work also focuses on finding better and more user-friendly models of urban growth and change, as well as improving techniques for resolving environmental conflicts. He received a Ph.D. in Regional Planning from the University of Illinois, an M.S. in Environmental Science from Washington State University, and a B.S. in System Dynamics from the Worcester Polytechnic Institute.

Jürgen Scheffran is a professor of integrative geography at the University of Hamburg and head of the Research Group Climate Change and Security (CLISEC) in the CliSAP Cluster of Excellence and the Center for Earth System Research and Sustainability (CEN). He had positions at University of Marburg, Technical University of Darmstadt, University of Illinois, and Potsdam Institute for Climate Impact Research. His research fields include security risks and conflicts of climate change and climate engineering, energy security and energy landscapes, the water–food–land nexus, human migration and rural–urban relations, modeling of human–environment interactions, complex systems analysis and sustainability science, and technology assessment and international security.

Acknowledgments

This book has been a long time coming. The early foundations of this work were laid over a decade ago when one of the authors (BenDor) was a graduate student of the other (Scheffran). Over the years, we have had a lot of help in making this work a reality. We would first like to thank Jim Westervelt, Nikhil Kaza, and Mostafa Shaaban for their generous assistance in helping us to develop and improve the models presented in Part III of this text. Kyle Vangel was superb in helping to collect much of the literature that we review. Shanwen Liu was key in assisting with the figure design throughout this text, and Ellen Emeric generously helped with the cover design. We would also like to thank Amanda BenDor, Ashton Verdery, Larry Susskind, Jean-Pierre Aubin, Michael Link, Thomas Dietz, Asim Zia, Alexey Voinov, the participants of the 2015 System Dynamics Winter Camp, and a half-dozen anonymous reviewers for their helpful input into this book. Bruce Hannon's thoughtful wisdom and guidance were crucial to setting the early stage for this work, while the gracious sabbatical hospitality of Denise Pumain (*Université Paris 1 Panthéon-Sorbonne*) and the late Judy Layzer (MIT) was key in helping this work become the final text that lays before you.

Acknowledgments

Pour ce qui est de l'avenir, il ne s'agit pas de le prévoir mais de le rendre possible.
As for the future, your task is not to foresee it, but to enable it.

—*Antoine de Saint-Exupéry, Citadelle [The Wisdom of the Sands] (1948)*

Contents

Part II Modeling Environmental Conflict

Part III   Applications of the VIABLE Model Framework

Preface: Reframing Conflict

Peace is not absence of conflict, it is the ability to handle conflict by peaceful means.
—**U.S. President Ronald Reagan** *(1982), Eureka College commencement address*

Introduction

Conflicts are an essential part of human life. In fact, much of human history can be described as a history of conflicts. Although no single definition of conflict has become standard, definitions commonly include themes like interactions, interdependence, and perception of incompatibility between parties. For example, Kenneth Thomas (1976, p. 891) concisely defined conflict as "...the process which begins when one party perceives that another has frustrated, or is about the frustrate, some concern of his." Barki and Hartwick (2004, p. 234) elaborated by defining conflict as "...a dynamic process that occurs between interdependent parties as they experience negative emotional reactions to perceived disagreements and interference with the attainment of their goals."

However, Dean Tjosvold (2006) convincingly argues that any definition of conflict needs to take account of the reality of scarcity, competition, and trade-offs, acknowledging the roles that the parties involved have in escalating their own conflicts. Therefore, we should think of conflict in terms of opposing goals and incompatible activities, challenging parties to confront and make the most of their conflicts.

Environmental conflicts occur around the world every day. Disputes over land and energy resources pit neighbors against each other (Nolon, Ferguson, and Field 2013). Water, food, and land scarcities affect billions around the world, raising concerns about violence, mass migrations, and prolonged regional and international disputes (Maxwell and Reuveny 2005). These persistent conflicts are also becoming costlier, fragmenting into complex dispute constellations. All of this transpires as climate change looms on an uncertain horizon, hinting at drastic changes in our livelihoods and access to many of the basic resources and opportunities that make us free and bring quality to our lives.

The words "*dispute*" and "*conflict*" have separate origins and represent different meanings of similar phenomena in modern English (Dietz 2001). They can be taken to technically represent two points along a continuum of disagreement. In the negotiation literature, a *dispute* represents a short-term incident that can be resolved through negotiation processes (Burton 1990). In contrast, conflicts occur over a longer term, with deeply rooted issues that add extra complexity to any efforts at resolution. The principal idea is that disputes can turn into conflicts if left unchecked and unaddressed. However, given the wide range of conflict and dispute definitions out there, it can be hard to exactly distinguish between disputes and conflicts. Therefore, while we readily acknowledge—and will explore at length—the continuums describing the type and severity of conflicts, we will use these terms interchangeably throughout this book.

Environmental conflict, like many other types of environmental problems, requires thoughtful consideration as a precondition for its resolution. While conflicts can consume an enormous amount of resources, whether they play a destructive or a constructive role in social change, critically depends on the way they are handled by the actors involved (we will use the terms *stakeholder, actor, party,* and *agent* interchangeably). This book is devoted to innovative thinking around environmental conflict and creating new types of solutions. Our aim is to provide a basis for improved understanding of environmental conflict, which can help guide readers toward novel ways of viewing and resolving environmental disputes.

There is now widespread recognition of the interplay between human society and the natural systems in which we exist (Millennium Ecosystem Assessment 2005). Technology and the growth of our civilizations are transforming the way we view and manage the earth's resources, broadening and complicating the scale and goals of environmental management efforts around the world (World Water Assessment Program 2006). As in many endeavors in our increasingly complex world, a big part of this means that we must first tear down the walls that have long separated numerous professions that aim to resolve environmental conflicts.

These transformations also mean that complex negotiations now necessitate the use of models—tools for representing our complex world in simple ways that improve our understanding. Whether these are informal mental models of how companies use resources or sophisticated computer simulations of conflict dynamics and ecological change, it is clear that no single party has a perfect way of handling all the relevant information needed to make decisions (Shmueli, Kaufman, and Ozawa 2008). As a result, parties need to depend on procedures within conflict resolution processes to jointly generate needed understanding for producing viable agreements.

It is important to note that most conflict resolution processes do not employ formal models. However, environmental conflicts (and policies meant to avoid them) usually involve complex scientific and technical issues, scientific uncertainty, and a wide array of stakeholders, many of whom have major conflicts in their own values (Stave 2002). At its core, many environmental disputes epitomize decision-making complexity; for example, Dietz and Stern (1998) aptly argue that during conflicts, decision makers are forced to weigh the different types of values of their constituents while equivalently balancing the values of society as a whole.

How do we know what these values are for the various actors? A litany of work now suggests that stakeholders should be involved in conflict resolution processes to discuss trade-offs, and ultimately, to help make decisions (Gray et al. 2018). This is important as decisions to change environmental or social conditions in conflicts involve not just scientific judgments, but also value judgments about whether actions would create situations that are more desirable than they were before (Stave 2002). As William Ruckelshaus (1998), the former chief of the US Environmental Protection Agency (EPA), has notably argued, problems in complex environmental systems, such as salmon fishery management, are often not due to failures to understand salmon or its habitat, but rather an inability to focus on system stakeholders, their behavior, and their needs; the issue is not fish management, but rather managing stakeholders with different values, laws, institutions, needs, and accessibility.

Our Goals for This Book

Our view emerges from many years of studying different types of conflicts from widely varying vantages; over several decades, our work includes studies of conflict in the realm of international security and nuclear disarmament, water resources and energy, and local and regional decisions over land use change and ecological impacts. This work has shaped our approach to conflict analysis, and it is our hope that this book will effectively convey conceptual and computer modeling strategies for understanding conflict. In the end, our focus will fall on "agent-based" models, which explicitly represent the *agency* of conflicting parties, to better determine how resolution to conflicts can be affected by innovative approaches. This means considering the actors (e.g., institutions, parties, individuals, groups, coalitions, associations) involved in conflicts, their ability to know about the world and act on that knowledge (i.e., actions, functions, decision structures, and their ability to evolve and adapt), and the resources (e.g., expertise, money, arms) at their disposal.

In this book, we have two primary goals; first, we will demonstrate that a natural, interdisciplinary union is appearing between the fields of alternative dispute resolution, system dynamics modeling and systems thinking, agent-based modeling, and a variety of other relevant areas. Second, we will introduce a new conceptual and technical framework for using agent-based modeling as a tool for better understanding environmental conflicts. This framework will help us represent and understand the alternative realities of agents in a conflict system, including the alternative information available, decisions, capabilities to act, and goals of conflicting parties.

Complexity Science as a Tool for Unifying Studies of Conflict

Concerning the first goal, we will show that these fields are linked through theories of *complexity* and system viability that have emerged from physics and mathematical sciences over the last several decades. By demonstrating this convergence, we hope to introduce readers to the application of system dynamics and agent-based modeling for the analysis of environmental conflicts with many disputing parties. We intend for this technical representation of conflicting agents to complement the outstanding conflict studies in the anthropological and sociological fields that have emerged over the last 40 years. In particular, these fields have made a convincing argument for participatory approaches that recognize environmental complexity, multiple stakeholder viewpoints (the reality of multiple "realities"), scientific uncertainties, and nonlinear dynamics (Videira, Antunes, and Santos 2009).

The New England Complex Systems Institute (NECSI 2016) defines *complexity science* as "...a new field of science studying how parts of a system give rise to the collective behaviors of the system, and how the system interacts with its environment." Complexity science is composed of theories of how relationships between system parts create collective system structures and behaviors (de Roo and Silva 2010; de Roo and Hillier 2016).

A good portion of the dispute resolution literature centers on case studies of individual conflicts (e.g., examples in Tolba and Rummel-Bulska 1998; Susskind, Moomaw, and Hill 1997; Gregory 2000; Susskind, Moomaw, and Gallagher 2002), as well as literature that analyzes conflicts where standard consensus-building strategies fail (Lewicki, Gray, and Elliott 2003). Paralleling this has been expansive work on modeling conflict and conflict resolution processes. We will contend that expanding our technical capability to understand and model conflicts can leverage and enhance the insights that social science explorations have lent to our understanding of environmental disputes (Folger, Poole, and Stutman 2005). We will also argue for a reframing and convergence of conflict social science and simulation modeling efforts under the banner of complexity sciences. At the core of this argument is the notion that technical and sociological approaches are mature enough to work hand in hand toward crafting proposals for real-world environmental conflict resolution.

This book, in part, is meant as a step toward answering the call of seminal research such as that of Judith Innes and David Booher (1999), who in their landmark paper, *Consensus Building and Complex Adaptive Systems*, and later in their 2010 book, *Planning with Complexity: An Introduction to Collaborative Rationality for Public Policy*, argued forcefully for new consensus-building frameworks that explicitly deal with complexity, self-organization, and adaptive systems. Innes and Booher rightly saw these ideas as an important organizing apparatus for addressing conflicting values, political fragmentation, joint governance, and power imbalances in today's conflicted world. Following this, many researchers have recognized problems with the so-called rational model of decision making, whereby "expert" analysts, in isolation, define problems, collect and select information they deem to be objective ("scientific"), build technical models pointing the way toward problem explanations, and thereafter, unilaterally develop optional solutions for use by formal decision makers (Shmueli, Kaufman, and Ozawa 2008).

Instead, we will present a wave of research that emerges from a variety of fields, which produces a forceful argument that data collection, technical modeling work, and eventual conflict resolution should be done openly. Negotiation and collaborative efforts to solve conflict can now be perceived as a form of joint decision making. By combining new technical tools—emerging from physical, computer, and complexity sciences—with sophisticated, ethnographic descriptions of conflicts and their resolution processes, our work suggests that collaborative decision modeling should be at the core of our continually renewing efforts to improve our ability to resolve disputes.

Applying Agent-Based Modeling to Conflict and Conflict Resolution

Our second goal in this book is to introduce a new framework for thinking about how and why conflicts emerge from the complex interactions of multiple parties and the environment. This framework will use agent-based modeling as a tool for better understanding the dynamics of conflicts that emerge from the incompatible actions, values, behavioral modes, and priorities of actors who fail to reduce their differences and tensions to tolerable levels. These ideas build on computational techniques in complex adaptive systems, as well as work in social networks, institutional analysis, and coalition formation. These fields, which have each grown immensely in recent years, now each lend their own useful tools and ways of thinking to our exploration of environmental disputes. Our particular focus will be in exploring numerous examples of environmental and resource conflicts around the world, as well as cooperative approaches for conflict resolution.

Book Organization

This book consists of three parts. In Part I (Chapters 1–4), we introduce key issues concerning environmental conflicts, the role and history of simulation modeling in conflict understanding and resolution, and the central tenants of participation in creating and using these models of conflict.

In Chapter 1, we introduce central ideas concerning difficulties in defining *environmental* conflict, the evolution of alternative dispute resolution techniques, and key conceptual and background issues around enduring states of conflict and cooperation. We also present an overview of how relevant disciplines—such as geography, urban planning, business and management, economics, and environmental studies—approach environmental conflict and cooperation. In Chapter 2, we discuss the basic philosophy around modeling, models as representation of more complex systems, and the concept of "mental modeling." In particular, we address the question: How does modeling help to resolve conflict?

In Chapter 3, we review the historical evolution of efforts to model conflicts, beginning with applications of models of historical cases of arms races and armed conflicts. We discuss the history, contributions, and limitations of game theory in conflict resolution and introduce more sophisticated approaches to conflict modeling, including complex dynamic models of conflict and models that include conflict geography. This chapter is aimed at providing readers a better understanding of the progression of conflict modeling in a way that contextualizes both recent advances in the art and science of conflict simulation, as well as our own efforts to find convergence between social science and technical views of conflict.

Finally, in Chapter 4, we introduce concepts and practices around *participatory modeling*, a useful tool to bridging the gap between science and policy while supporting collaborative processes that engage nonscientists, the public, and conflict stakeholders. This chapter provides a concise overview of the literature on participatory modeling, focusing on relevant issues, such as social learning, participatory decision making, stakeholder selection issues, and a typology of modeling methods (e.g., decision analysis, fuzzy cognitive mapping) commonly used in participatory settings. We should note that while we do not directly incorporate the input or involvement of stakeholders in the models we present in Part III of this book, we instead demonstrate how stakeholders could adapt them for use in real environmental disputes. We hope that using our detailed reviews of best practices and advice from the participatory modeling literature, readers will use and build on these techniques for your own dispute resolution purposes.

Part II of this book is focused on first introducing central tenants of system dynamics and agent-based modeling, followed by a detailed presentation of the VIABLE agent-based modeling framework that we have created.

In Chapter 5, we introduce readers to the system dynamics modeling approach and emerging systems-focused techniques that have been used for intervening in conflicts. We present a concise synthesis of the last 15 years of literature in this area, highlighting group model building efforts and demonstrating their use in collaborative learning and participatory processes. In Chapter 6, we present a practical introduction to agent-based modeling, its application to conflict research to conflict modeling and resolution, and present several examples of agent-based modeling, including applications to environmental conflict. We also discuss multi-agent modeling concepts and theory (e.g., stability, chaos, emergence), comparing them with the dynamic modeling approaches more frequently used in conflict modeling.

Finally, in Chapter 7, we present the "VIABLE" modeling framework that we have created, which involves the application of agent-based modeling to environmental conflicts, the use of a mathematical concept called *viability theory* to shift our thinking concerning the sustainability of human–environment interactions (or of resolved disputes), and the use of particular interventions to understand the behavior of individual actors in conflict. We will explore the capabilities of the VIABLE framework, which simulates the <u>V</u>alues and <u>I</u>nvestments for <u>A</u>gent-<u>B</u>ased interaction and <u>L</u>earning for <u>E</u>nvironmental systems by leveraging aspects of agent-based modeling, evolutionary game theory, system dynamics modeling, and network theory.

In Part III of this book, we present three applications of our VIABLE modeling technique that are exemplary of key conflict constellations of our age. We use agent-based modeling to explore conflicts over fisheries (Chapter 8), carbon emissions (Chapter 9), and land use and biofuels (Chapter 10). These disputes concern a range of conflicting parties, conflict patterns, institutional roles, and geographies, which range from local (land use) to regional (fisheries) to global (carbon) in scale. These models are created using the NetLogo software (Wilensky 1999), a free, popular, and widely supported agent-based modeling framework. We detail the development of these models, providing further step-by-step instructions on this book's website (http://todd.bendor.org/viable). These models explore relevant aspects of conflict and cooperation, including insights into the simplicity of agent behavior that creates complex, emergent dynamics within conflict systems.

We are confident that readers of all types—including conflict resolution professionals, researchers, students, and those interested in producing better dispute outcomes—will gain an appreciation of the union that can occur between social science-based explorations of conflict—including case studies and participatory interventions—and the elegant and insightful representation of individual-level conflict behavior produced by agent-based modeling. Through the background discussions given in Part I, the modeling techniques introduced in Part II, and the applications of our framework in Part III, it is our hope that readers can draw on this book as a resource for exploring conflict resolution concepts, reframing your mental models of conflicts and conflict resolution strategies, better understanding the roles and uses of specific conflict modeling techniques, and articulating your own interesting and useful models of environmental conflicts. We wish you all the best in these efforts.

References

Barki, Henri, and Jon Hartwick. 2004. "Conceptualizing the Construct of Interpersonal Conflict." *International Journal of Conflict Management* 15 (3): 216–44.

Burton, John Wear. 1990. *Conflict: Resolution and Prevention*. New York: Macmillan.

de Roo, Gert, and Jean Hillier. 2016. *Complexity and Planning: Systems, Assemblages and Simulations*. New York: Routledge.

de Roo, Gert, and Elisabete A. Silva, eds. 2010. *A Planner's Encounter with Complexity*. Aldershot, England: Ashgate.

Dietz, Thomas. 2001. "Thinking about Environmental Conflicts." In *Celebrating Scholarship*, edited by Lisa M. Kadous, 31–54. Fairfax, VA: College of Arts and Sciences, George Mason University.

Dietz, Thomas, and P. C. Stern. 1998. "Science, Values, and Biodiversity." *Bioscience* 48 (6): 441–45.

Folger, J. P., M. S. Poole, and R. K. Stutman. 2005. *Working through Conflict: Strategies for Relationships, Groups, and Organizations*. Boston, MA: Pearson/Allyn and Bacon.

Gray, Steven, Alexey Voinov, Michael Paolisso, Rebecca Jordan, Todd BenDor, Pierre Bommel, Pierre Glynn, et al. 2018. "Purpose, Processes, Partnerships, and Products: Four Ps to Advance Participatory Socio-Environmental Modeling." Ecological Applications 28 (1): 46–61.

Gregory, Robin. 2000. "Using Stakeholder Values to Make Smarter Environmental Decisions." *Environment: Science and Policy for Sustainable Development* 42 (5): 34–44.

Innes, Judith E., and David E. Booher. 1999. "Consensus Building and Complex Adaptive Systems: A Framework for Evaluating Collaborative Planning." *Journal of the American Planning Association* 65 (4): 412–23.

Innes, Judith E., and David E. Booher. 2010. *Planning with Complexity: An Introduction to Collaborative Rationality for Public Policy.* New York: Routledge.

Lewicki, R. J., B. Gray, and M. Elliott, eds. 2003. *Making Sense of Intractable Environmental Conflicts: Concepts and Cases.* Washington, DC: Island Press.

Maxwell, J. W., and Rafael Reuveny. 2005. "Continuing Conflict." *Journal of Economic Behavior and Organization* 58: 30–52.

Millennium Ecosystem Assessment. 2005. *Millennium Ecosystem Assessment: Ecosystems and Human Wellbeing.* Washington, DC: Island Press.

NECSI. 2016. *About Complex Systems.* http://necsi.edu/guide/

Nolon, Sean, Ona Ferguson, and Pat Field. 2013. *Land in Conflict: Managing and Resolving Land Use Disputes.* Cambridge, MA: Lincoln Institute of Land Policy.

Ruckelshaus, William D. 1998. "Foreword: Managing Commons and Community: Pacific Northwest People, Salmon, Rivers, and Sea." In *Managing the Commons* (2nd Edition), edited by J. A. Baden and D. S. Noonan, ix–xiv. Bloomington: Indiana University Press.

Shmueli, Deborah F., Sanda Kaufman, and Connie Ozawa. 2008. "Mining Negotiation Theory for Planning Insights." *Journal of Planning Education and Research* 27 (3): 359–64.

Stave, Krystyna. 2002. "Using System Dynamics to Improve Public Participation in Environmental Decisions." *System Dynamics Review* 18 (2): 139–67.

Susskind, Lawrence, William Moomaw, and Kevin Gallagher. 2002. *Transboundary Environmental Negotiation: New Approaches to Global Cooperation.* San Francisco, CA: Jossey-Bass.

Susskind, Lawrence, W. Moomaw, and T. L. Hill. 1997. *Global Environment: Negotiating the Future.* Papers on International Environmental Negotiation. Cambridge, MA: Program on Negotiation (PON Books), Harvard Law School.

Thomas, Kenneth W. 1976. "Conflict and Conflict Management." In *Handbook of Industrial and Organizational Psychology,* edited by M. D. Dunnette, 889–935. Chicago, IL: Rand McNally.

Tjosvold, Dean. 2006. "Defining Conflict and Making Choices about Its Management: Lighting the Dark Side of Organizational Life." *International Journal of Conflict Management* 17 (2): 87–95.

Tolba, M. K., and I. Rummel-Bulska. 1998. *Global Environmental Diplomacy: Negotiating Environment Agreements for the World, 1973–1992.* Cambridge, MA: MIT Press.

Videira, Nuno, Paula Antunes, and Rui Santos. 2009. "Scoping River Basin Management Issues with Participatory Modelling: The Baixo Guadiana Experience." *Ecological Economics* 68 (4): 965–78.

World Water Assessment Program. 2006. *Water: A Shared Responsibility (World Water Development Report No. 2).* Paris, France: United Nations.

Part I

Conflict and the Promise of Conflict Modeling

In Part I of this book, we argue that scholars and policy makers should reframe how they approach and understand conflict. Our goal is to provide the reader with a strong foundation for understanding:

- The forces that create and sustain environmental conflicts and the array of techniques developed to resolve conflicts (Chapter 1);
- The reasoning for using conceptual and computational modeling as a tool for understanding and resolving conflict (Chapter 2);
- The history of conflict modeling and modeling interventions (Chapter 3); and
- The role of participation by parties involved in conflicts in ensuring successful outcomes of modeling interventions (Chapter 4).

1

Environmental Conflicts in a Complex World

Change means movement. Movement means friction. Only in the frictionless vacuum of a nonexistent abstract world can movement or change occur without that abrasive friction of conflict.

—Saul Alinsky (1971, p. 21), Rules for Radicals, American community organizer

Introduction

In this chapter, we will introduce an argument for reframing how scholars and policy makers approach conflict. At the core of this, we hope to reconsider the types of tools needed to resolve complex, multilateral environmental disputes. We will begin by discussing key conceptual and background issues relating to conflict, cooperation, and resolution, reviewing how relevant disciplines—including geography, urban planning, business and management, dispute resolution, and environmental studies—view and define conflict and cooperation. We will explore differences between interest- and value-based conflicts and examine how conflict histories and surrounding institutional structure are increasingly recognized as key factors determining outcomes and relative likelihood of resolution.

> Throughout this book, we will use the words "stakeholder," "actor," "party," and "agent" interchangeably. While we will rigorously define "stakeholder" in Chapter 4 and "agent" in Chapter 6, these terms are generally meant to mean "those involved in, or affected by, the conflict."

We will then discuss several environmental conflict examples from the literature, including land and water use, energy and climate change, and resource availability conflicts (e.g., fisheries, water quality, forests, biodiversity). Finally, we will review new approaches and strategies for conflict avoidance, prevention, and resolution, including proactive planning processes, agreements, dialogs, participation, public-inclusion policies, and institutional designs. Throughout this chapter, we will focus on *trust* as a conflict determinant, as stakeholder inclusion is a key element in building trust and vesting stakeholders in consensus solutions during mediated or facilitated conflict resolution processes.

What Is an Environmental Conflict?

Since the early 1990s, many scholars and practitioners have investigated how the scarcity of natural resources—such as minerals, water, energy, fish, and land—affects violence and armed struggle. Although we may be tempted to define environmental conflicts as "any dispute with an environmental component," previous investigations lend the topic a more nuanced view. Possibly, the most comprehensive definition of environmental conflict was jointly put forth in the early 1990s by the Environment and Conflict project of the Swiss Peace Foundation and ETH-Zürich Center for Security Studies (Libiszewski 1992, p. 14):

> Environmental conflicts manifest themselves as political, social, economic, ethnic, religious, or territorial conflicts, or conflicts over resources or national interests, or any other type of conflict. They are traditional conflicts induced by *environmental degradation*. Environmental conflicts are characterized by the principal importance of degradation in one or more of the following fields: overuse of renewable resources, overstrain of the environment's sink capacity (pollution), or impoverishment of the "space of living."

For example, studies of environmental conflict frequently note the linkage between environmental disputes and national or international security (Westing 1986), wherein environmental issues become the root causes of violent conflict. In fact, many of the 20th century's greatest armed conflicts were related to natural resources, such as minerals, fossil energy, fisheries, land, or land "products." However, historians or military officials would not normally consider the two World Wars to be "environmental conflicts," regardless of the role that natural resource struggles played in these conflicts (e.g., Yergin 1992 describes oil's role in driving both wars). This is an indication that resource issues are rarely seen as predominant causes of conflict, but instead represent less visible undercurrents that aggravate complex conflict dynamics between groups.

To uncover the role of natural resources and the environment requires us to dig deeper into the specific historical roots of conflicts and the drivers that keep them going. However, some scholars have even challenged the very concept of "environmental conflict," questioning its empirical basis in being different from any other conflict (Urdal 2005; Theisen 2008; Lujala 2009; Buhaug, Gates, and Lujala 2009). Others have argued that the general meaning of "environmental conflict" is unclear (Gleditsch 1997, 1998), with many assertions being virtually untestable, as the causality of the environment–conflict relationship is sometimes reversed (conflicts create resource problems). Scholars such as Jon Barnett (2000) have contended that the "environmental conflict" concept is theoretically rather than empirically driven. Addressing this deficit, Carius, Tänzler, and Winterstein (2006) reviewed 73 empirically recorded environmental conflicts between 1980 and 2005, finding that most disputes had a regional scope and did not present a serious threat to international security. Others have observed that cooperation between parties, particularly regarding joint water use agreements, is increasingly occurring in lieu of conflict (e.g., Medzini and Wolf 2004).

Many social scientists studying dispute resolution have argued that our understanding of environmental conflicts should focus on the chain of causation that links environmental change and degradation to social changes that impact human societies. Put simply, environmental damage does not always lead to, or even directly affect, conflict. There have been major environmental changes that have been offset by technology or social interventions to avoid conflict. For example, an increasing number of environmental agreements demonstrate that under certain circumstances, threatened environmental systems can facilitate interparty cooperation (Mitchell 2018; https://iea.uoregon.edu/).

While a clear correlation between environmental degradation and disputes can be hard to prove for complex conflict constellations, Stephan Libiszewski (1992) argues that conflicts occur when environmental changes affect culturally mediated human behaviors, producing complex, feedback-driven social phenomena (e.g., animosity, revolt; Homer-Dixon 1991) that result in conflict. Here, we can recognize the centuries-old role of environmental policies and decisions as sources of conflict; environmental conditions induce policy changes, which modify behavior and create social cleavage and unrest (Andrews 2006). Libiszewski's (1992) systems-oriented understanding of environmental conflict is the approach that we will draw on throughout this book.

Conflict and Scarcity

Libiszewski (1992) has argued that environmental conflicts revolve around environmental degradation, whereby human actions create a dimension of resource scarcity, including physical scarcity (not enough available), geopolitical scarcity (unequal distribution across space), and socioeconomic scarcity (unequal distribution within societies). That is, environmental conflicts are caused by human-made disturbances of the normal regeneration rate of resources, including overuse of a renewable resource or overstrain of an ecosystem's capacity to process and dilute pollution. Under this framework, we can see that disagreements over the possession of oil or land are not inherently *environmental conflicts*, but rather are economic or social conflicts that have resource components. These disputes become environmental conflicts when environmental *changes* or *degradation* create scarcities (Farber 2000).

There are several ways that resource scarcity and environmental degradation can lead to social and economic disruptions that directly or indirectly induce conflict (Homer-Dixon 1991, 1994; Baechler 1999). Perhaps, the most prominent pathway is through *resource capture*, where powerful groups try to influence resource distributions in their favor or directly capture resources in the face of resistance by other groups. Another phenomenon, *ecological marginalization*, occurs when population growth, resource degradation, and unequal resource distribution contribute to the impoverishment of marginalized social groups. Many additional pathways can lead to the loss of vital resources (e.g., agricultural land), which may also lead to economic decline, weakened institutions, and conflicts. Finally, environmental problems can drive affected people into regions that are already ecologically fragile and conflict prone, where they could further contribute to environmental problems.

In cases where the identities of migrants and local communities fail to align, conflicts may aggravate. Massive environmental damage is often associated with large-scale migrations or human displacements in lesser-developed countries, which itself can result in international conflict. Through a sophisticated exploration of nearly 40 cases of environmentally induced migration, Reuveny (2007) determined that people living in lesser-developed countries may be more likely to leave degraded areas (or areas hit by environmental disaster), which may cause conflict in receiving areas. For example, since the 1950s, mass migration from Bangladesh to India resulting from wide-scale land degradation has created major friction between the two nations. Other historical cases include social tensions arising from dust bowl-induced migration in the 1930s, as American Midwesterners migrated to California, and the massive migration out of southern Louisiana following Hurricane Katrina in 2005 (Reuveny 2008). Similarly, in the late 1960s, migrants from El Salvador entering Honduras led to a war between the countries.

Altogether, environmental change can create complex forms of interference in political, economic, and social conflict processes. For example, population growth, increased demand, and unequal distribution can impair resource availability. Resource destruction can also lead to economic damage and diminish state revenues, which can undermine the capacity to act and the political legitimacy of governments. If basic human needs can no longer be satisfied through diminishing resources, the conflict potential between social groups over resource distribution can increase substantially.

Although environmental destruction and resources scarcity have the potential to contribute to the emergence or aggravation of violent conflicts, as we discussed previously, it may be hard to prove due to complex and multiple causal chains. In many cases, the environment is a catalyzer or multiplier of existing conflicts, rather than an initial conflict cause. For example, local changes in fishery, forests, water, and land typically have a more direct impact on conflicts than the global and long-term phenomena of climate change and ozone depletion (e.g., BenDor et al. 2014). Environmental destruction rarely leads to direct violence. Instead, environmental changes often disturb complex ecological balances and social processes over long time periods and on smaller spatial scales. Thus, the scale of these scarcities—real or perceived—is important.

While we have thus far focused on conflicts when resources become scarce, major conflicts often also occur over resources when they are abundant, leading to a *resource curse* (Collier 2000; De Soysa 2000, 2002; Le Billon 2001). For example, copious natural resources can produce corruption, "rent-seeking" behavior, increasing disparities in resource access, and can promote revenues that finance continuations of conflicts. Studies by Reuveny et al. (2011) and Le Billon (2001) reference numerous conflicts around the world that exemplify this behavior, noting that abundant resources helped drive over a dozen armed conflicts during the 20th century alone.

> "Rent-seeking" behavior occurs when an entity expends resources to obtain returns that do not result in any gain to society (Sandler 2000). For example, a company is rent seeking when it takes actions (e.g., lobbying, rule making) to ensure that it can increase its share of wealth, while not creating additional wealth for society (e.g., limiting competition while decreasing levels of service).

The degree to which environmental risk will actually lead to conflict strongly depends on the societal conditions that are shaped by the conflict history, group identities, the organization and capacity of conflict parties, as well as the specific use of resources for group interests. Whether a latent conflict over water or food manifests, depends on differences in resource access, as well as the power and interest structures between conflict parties. Of particular relevance is whether conflict parties are—justified or not—held responsible for resource and environmental problems. In many cases, responsibility is difficult to prove in the face of long and complicated chains of impacts and in the presence of large uncertainties. Here, perceptions and interpretations of other actors' behavior play an important role where misunderstandings can aggravate conflicts. As a result, actions taken by each party can undermine each other's values and provoke responses that spawn additional losses. Following this, conflicts escalate when actions by the parties in conflict aggravate these social tensions, intensifying the conflict and driving continually unstable interactions.

Whether environmental problems lead to violent or peaceful actions, will be determined by different contextual conditions, including governance structures, institutions, and conflict regulation mechanisms. Weak political institutions, limited carrying capacity, and the dissolution of established living conditions and livelihoods increase conflict potential. Conversely, the development of cooperation and management procedures can diminish conflict potential. Thus, a simple and direct relationship between resource scarcity and violent conflict is difficult to verify. More common are less-visible intrastate conflicts, where the contribution of resource scarcity often remains opaque and indirect.

Why Are Environmental Conflicts Worth Resolving?

Thomas Dietz (2001), a noted scholar of conflict and environmental sociology, has argued that there are two primary reasons we should be interested in environmental disputes. First, as the 21st century unfolds into an epoch of urbanization, cultural integration, and climate change, environmental conflicts will become more complex and severe, and emerging disputes will become increasingly abundant. Reuveny et al. (2011) further note that climate change, in particular, will likely reduce or redistribute the availability of renewable resources, such as arable land, fresh water, fish, timber, and numerous valuable commodities. Furthermore, increasing resource extraction will reduce or redistribute the availability of nonrenewable resources such as oil, and mineral and metal commodities that currently do not have good substitutes. As demand shortfalls increase, it becomes increasingly likely that groups will resort to conflict as a means for obtaining resources. The complexity of these conflicts will be further exacerbated as technology and social media give stronger voices to a larger number of conflicting parties, while making conflicts more visible to a larger number of international observers.

Second, Dietz (2001) argues that environmental conflicts represent a bellwether of numerous other types of clashes that will surface in the coming decades. This is because environmental conflicts often suffer from the same drivers as other types of disputes, including the muddling of facts and values, uncertainty, and conflicting visions of what our future should hold (Sowell 2007). Although they are often lower on the public's radar, many environmental conflicts have important implications for our world. Moreover, many major conflicts have origins as resource disputes, including conflicts between Iran and Iraq over access to oil or water usage (Iran/Iraq 1975), and clashes along the India–Bangladesh border due to years of land degradation in Bangladesh (Myers 1993; BBC 2001).

The emergence of new technologies, geopolitical realities, and the increasing complexity of our everyday interactions, through increasing diversity and newly developed modes of contact, all represent new opportunities for conflict to form. Globalization has created a multipolar world in which decision making and conflict are determined by a variety of actors and factors, which are coupled in a complex manner (Scheffran 2008). Coupled with environmental complexity are rapidly changing security conditions, which can provoke instabilities and the outbreak of violence, as was vividly demonstrated by the destructive, chaotic dynamics in the Balkan region during the 1990s (e.g., in the violent struggles on Bosnia and Kosovo). The sudden and violent breakup of existing political structures is often associated with increased "entropy" in the fragmented societies of a region, where new stable coalitions evolve slowly (we see this today across the African continent and in the Middle East). Security is not only a function of military arsenals but also shaped by technological and environmental factors, and social and economic developments, from local to global levels.

A third impetus to focus on environmental disputes, as argued by Homer-Dixon (2001) and Reuveny et al. (2011), involves the fact that environmental conflict may reduce society's longer-term ability to innovate as environmental degradation undermines economic and societal conditions, decreases population well-being, and compounds stress factors that contribute to conflict (Homer-Dixon 1994; Scheffran 1999). For example, Sandler (2000) discusses how enduring conflict can disincentivize production, since there remains a strong probability that one day, the fruits of individual labor and production may be stolen by conflicting groups.[1] We should note that this argument is debated; for example, more conservative scholars, including Julian Simon (1996) and Ester Boserup (1981), dispute Thomas Malthus's (1798 [1970]) early thinking around the links between population growth and conflict. They instead argue that these same adverse forces can also generate innovation, as "necessity is the mother of invention."

In recent decades, one important development in conflict studies has been the increasingly accepted view of conflicts as *investment strategies*. This idea, elaborated in Reuveny and Maxwell (2001), Findley (2008), and Reuveny et al. (2011), focuses around the reality that entities (agents) involved in conflict have limited resources. As we will explore throughout this book, when we interpret conflict as a form of investment where time, energy, money, and human capital are expended on conflicts that instead of other purposes, we can consider conflict's opportunity costs (Ward and Becker 2015). In doing so, we can determine if the induced effects of conflict on our society involve foregone development or economic growth (i.e., the choice of whether to fight over resources or instead devise policies to manage resource use). We can also determine if conflicting parties that have to spend a substantial portion of their resources forcefully advancing their interests,[2] or guarding themselves and their possessions, are left with fewer resources for productive activities.

The Goals of Environmental Conflict Resolution

Many scholarly efforts to find solutions for complex environmental conflicts point out that in order to examine any dispute intervention, we must explicitly identify goals, both for the intervention process and for the eventual post-conflict system. Reuveny et al. (2011) dig into this idea in a provocative way by discussing two hypothetical groups engaged in an armed conflict over a resource (e.g., forest timber) and asking: what does a resolved conflict look like? What are the goals for the resolution process?

It is reasonable to assume that societies strive to reduce resource conflict and promote cooperation. Reuveny et al. (2011) note that cutting off the supply of weapons to the conflicting groups could reduce the duration of the conflict, thereby reducing the risk that the combatants would adopt conflict as a way of life and become very good at it. A third party could also limit access to the resource (i.e., forested regions) to both groups. If the fighting is very destructive or costly, we might also conclude that the groups would have more resource wealth if they stop fighting, cooperate in their timbering efforts, and share

---

[1] Substantial economics and political science literature has demonstrated how governments emerge from a natural state of anarchy to protect property rights, in return for taxes to support this protection (e.g., Schlager and Ostrom 1992).

[2] In the United States, for example, environmental lawsuits may result in higher legal fees than mediated settlements between plaintiffs and defendants (Florio 2000).

the proceeds. However, if the conflict had diverted efforts from timbering activities to low-intensity fighting, then the economy may have actually received long-term benefit from the conflict, prolonging the life of the forested regions. That is, if the conflict was resolved, and all the time and energy from the conflict were diverted to timbering the area at an even faster rate, the society may deplete timber resources faster than they can grow.

In this example, we can see the need for explicit goals set around post-conflict systems. Kenneth Thomas (1992) explores the variety of goals that we can establish, including our choice of beneficiary and time frame. For example, we can optimize the outcomes of one of the parties ("partisan" outcome), all parties ("joint-welfare" outcome), or the larger system of which conflicting parties are a part ("systemic" outcome).

In thinking about the goals of conflict resolution, we can also consider different modes of handling conflicts. Thomas (1992) approaches this from a management perspective, considering the different modes of conflict engagement along dimensions of parties' assertiveness and cooperativeness. His approach—shown in Figure 1.1—neatly summarizes the range of behavioral modes among conflicting parties. These modes, which can change throughout the course of a conflict, can range from cooperation—whereby parties continue to assert their own values while also attempting to satisfy the values of other agents—to the complete avoidance of conflict and any attempts on the part of an agent to satisfy neither their own values nor those of other parties. As we will explore in several of our applications in Part III of this book, conflicts are often tentatively and insufficiently "resolved" through movements from avoidance or competition modes toward the space of compromise. We will spend significant time exploring the role of cooperation in achieving long-lasting solutions that meet agents' goals and build their value in the face of ongoing agent interactions.

Sustainability as a Conflict Resolution Target

"Sustainability" is a powerful management concept that arose from the 1987 World Commission on Environment and Development or "Brundtland commission," named for the lead author, former Prime Minister of Norway, Gro Harlem Brundtland (1987). The Commission's final report stated that environmental management should generally focus

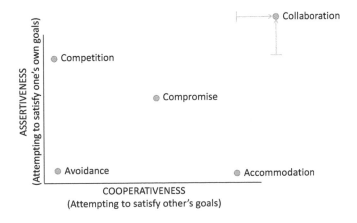

FIGURE 1.1
Conflict management modes as given by dimensions of assertiveness and cooperativeness of conflicting parties. In contrast to "compromise," "collaboration" can be viewed as a way of achieving one's own goals as well as the goals of others. Adapted from Thomas (1992).

on sustainability, which endeavors to "...meet the needs of the present generation without compromising the ability of future generations to meet their own needs." Although sustainability is now an indefinite and contested concept in the environmental sphere (Costanza and Patten 1995; Marcuse 1998; Vucetich and Nelson 2010), the concept of a "sustainable" peace has long been a central organizing principle and goal for conflict resolution processes (Lederach 1997; Buchholz, Volk, and Luzadis 2007; Oswald Spring 2008; Oswald Spring and Brauch 2011; Scheffran 2016). We often hear from those hoping to resolve conflicts that both sides in a dispute must be able to find "sustainable" solutions, where their collective needs are met and conflicted resources are kept in a reasonably stable state.

The pursuit of sustainability goals in conflict environments requires intimate knowledge of three complex, dynamic, interconnected, and overlapping systems: the biosphere, the economy, and human society (Videira et al. 2010). Classically, this knowledge arises from integrated assessments of social, economic, and ecological values, as well as social learning components that collect and synthesize information for decision making from stakeholder participation (discussed in Chapter 4; Buchholz et al. 2007; Reed 2008).

Ecologist C.S. Holling (2001, p. 399) distinctively defines sustainability as "the capacity to create, test, and maintain adaptive capability," which means that systems are sustainable when they possess now—and into future—the necessary infrastructure and capability (e.g., material wealth) to adapt to new situations. By applying an explicitly dynamic definition to sustainability, Holling argues that sustainability in a conflict environment is process-oriented rather than goal-oriented, existing in a social sphere that encompasses a variety of human values, perceptions, and political interests.

Buchholz et al. (2007) argue that sustainability is, by its nature, controversial; it can be thought of as a social value, which requires that we consider a broad set of social and economic criteria for sustainability. For instance, if we seek conflict solutions that balance social, economic, and ecological factors of a conflict system on the same level, we may create a different goal than if we had sought a solution where sustainability goals were nested; i.e., giving ecological constraints—which may limit or direct possible solutions—precedence over social and economic factors (similar to a purely biophysical view of sustainability; Gowdy 1999). While this classic understanding of sustainability has driven substantial work in complexity theory, adaptive systems, ecosystem ecology, urban planning, geography, and may other fields, in this book (Chapter 7) we will offer a more sophisticated and analytical approach to creating and evaluating conflict resolution goals, which involve identifying the specific dimensions of a strategy for post-conflict sustainability, the tools available to implement each strategy, and the set of indicators to measure and monitor results of conflict interventions (Arena et al. 2009).

Linking Sustainability to Conflict Management

The use of sustainability as a conflict resolution goal implies a long-term planning horizon, which can amplify many environmental management challenges. For example, questions around the long-term viability of post-conflict ecological conditions can be easily pushed into the realm of messy or "wicked" managerial problems, where stakeholders can draw into question the definition or even the existence of the problem (Stave 2010; Balint et al. 2011). Luckily, the last several decades have seen major improvements in the theoretical and scientific underpinnings of environmental management programs. New ideas about how to manage the environment have evolved alongside our understanding of the physical world.

Early environmental management approaches typically focused on a single facet of a resource, which often led to isolated and disjoint management and monitoring approaches and further resource degradation (Berkes and Folke 1998; Cockerill et al. 2007). Take two infamous and well-studied examples of this: forest management by fire suppression has led to larger wildfires, and urban flood management through the rapid discharge of stormwater into streams has increased the magnitude of downstream flooding. Realizing the shortcomings of these strategies, decision makers eventually began to employ "systems" approaches, which acknowledge that multi-faceted problems can't be managed in piecemeal and isolated ways. Researchers recognized that managing "ecological systems" was a more effective way of viewing the natural world, and that ecosystem *behavior* is not derived solely from any individual piece of the ecosystem, but is the result of *interactions* between parts of the system (Allen and Hoekstra 1993). While a major improvement, these approaches still possess limited facilities for considering human well-being, including social and economic sustainability (Buchholz et al. 2007; Cockerill et al. 2007).

Dietz (2001) offers a firsthand example of "wicked" problems, whereby an ecologist and economist began to debate the loss of old-growth tropical forest. While they were able to rapidly arrive at an agreement over the scientific status of the resource, they quickly realized that they held completely incongruent views over whether forest loss was even a problem.

A vast literature in dispute resolution and negotiation theory has developed in the last two decades as research begins to bridge the interface between environmental management and sustainable economic and social conflict resolution. As we move into this new age of systems-based environmental and conflict management, it is clear that a major frontier lies where systems approaches to environmental management attempt to interface with the social and economic systems that determine conflict outcomes (Berkes and Folke 1998; Buchholz et al. 2007; Cockerill et al. 2007).

The History and Evolution of Conflict Resolution

Louis Kriesberg (2009) provides a broad and helpful overview of the history and evolution of alternative dispute resolution. His key argument is that all conflict resolution (not just environmental conflict resolution) remains a complicated practice because no general theory of conflict mediation exists. Conflict resolution is important in that it deviates strongly from the traditional means of ending conflict, where the "winner take all" principle imposes the will of the victor on the loser (i.e., winning a lawsuit or a war). Instead, conflict resolution seeks mutual gains for conflicting parties, where agents can gain consensus over future actions that they seek to take (Susskind and Landry 1991; Margerum 2011). The history of conflict resolution began during the 18th-century Enlightenment, blossoming out of philosophical concepts such as respect for persons and conflict mitigation in society. After this period, Kriesberg (2009) identifies four broad, but distinct periods of conflict resolution practice, wherein historical currents influenced major conflict resolution initiatives as well as the academic literature of the time.

- First, a period of "preliminary developments" (1914–1945) in the field prompted efforts to build institutions that reduced the causes of war and to develop the collective security needed to halt wars outright.

- Second, a period he terms, "laying the groundwork" (1946–1969) saw government and nongovernmental efforts to foster additional institutions and reconciliation, as well as expanded academic analysis of conflict (both analytical and ethnographic). These analyses revealed the dangers as well as the desirable social changes that could emerge from large-scale disputes.

- Third, an era of "expansion and institutionalization" (1970–1989) involved the establishment of dispute resolution centers throughout the United States, the creation of centers for research on conflict resolution, new professional associations, and the publication of influential journals (e.g., the *Negotiation Journal*) and texts, such as Fisher and Ury's (1981) famous text, *Getting to YES*. Along these lines, it becomes increasingly clear that research can drive peace efforts in areas like the Middle East, where research organizations can contribute over the long term toward conflict resolution efforts (Greene 2008).

- Finally, Kriesberg defines a period, since 1989, of "diffusion and differentiation," wherein the end of the cold war helped establish a world with increasing economic integration, a growing adherence of human rights norms and protections, proliferating democracies, and more specialized and sophisticated conflict research activities.

Susskind and McKearnan (1999) further break down the evolution of dispute resolution applied to public policy from the early 1970s to 1999, focusing on four particular milestones, including successful efforts to resolve environmental disputes through mediation, dialogs bringing multiple levels of government agencies together to improve public investment strategies, attempts by federal agencies to use consensus-based approaches like "negotiated rule making" to augment conventional approaches, and as Kriesberg (2009) identifies, the development of community-based conflict resolution centers around the United States. Many debates continue in the conflict resolution literature, including the alignment of theory and practice, the way we define conflicts, the importance of resolution process versus outcomes, and the manner by which we choose parties to participate in resolution processes (Wondolleck et al. 1996; National Research Council 2008).

Conflict Resolution Efforts across Many Disciplines

Group decision making and the difficulties in navigating its complexities are well studied in the context of conflict. Additionally, many scholars have studied why certain disputes are able to find specific types of solutions (Fuller 2009). In an interesting example, Tastle and Wierman (2007) take a mathematical approach to measure the level of disagreement and agreement among stakeholders, creating metrics for determining and measuring whether a group of individuals is converging on consensus. They take consensus and dissention to be opposing concepts, whereby a consensus is defined as "an opinion or position reached by a group of individuals acting as a whole; it is also considered general agreement" and dissention is defined as "…a difference of opinion such that strife is caused within the group undertaking to make a decision."

Although achieving conflict resolution usually tends to attract much less attention than the conflicts themselves, numerous fields including dispute resolution, public administration (Bryson 2004), law, economics, water resources planning (Delli Priscoli and Wolf 2009), international relations (Tolba and Rummel-Bulska 1998), ecology (e.g., Baker 2004), environmental studies and justice (e.g., Hurley 1995), and urban planning, have explored the processes by which conflicting parties can resolve their own conflicts for the benefit of all groups involved (Shmueli et al. 2008; Kriesberg 2009). Together, these research efforts—often very disparate—have created an extensive literature in consensus building (Cruikshank and Susskind 1989; Susskind and Landry 1991; Susskind et al. 2000) and negotiation theory (Fisher and Ury 1981; Raiffa 1982; Halpert et al. 2010) that helps us to understand how stakeholders can reach agreement in the face of entrenched, value-based differences.

Urban Planning

Within the field of urban planning, whose practitioners often facilitate multilateral conflicts and mediate citizen disputes to government actions, significant work has tested emerging techniques for improving alternative dispute resolution and consensus building (Susskind et al. 2000; Innes 2011), improved negotiation and public participation approaches (Forester 1987; Godschalk 1992), transformed negotiation frameworks (Godschalk 1992; Brody et al. 2003), expanded capacity for involving citizens in public decision making, and improved "civic infrastructure" (i.e., enabling civic involvement; Sirianni 2007). Perhaps the best discussion of conflict resolution in urban planning is found in Shmueli, Kaufman, and Ozawa (2008), who draw on a series of case studies to offer a robust discussion of how three fundamental aspects of negotiation theory—interests, mutual gains, and information—are injected into planning processes and products. The authors argue that the classic, "rational" model of planning—based on the idea that planners are expert decision makers—is being slowly replaced by communicative and collaborative models, which specifically highlight how joint knowledge production occurs at the core of many planning decisions. That is, planners often inherently frame negotiations as a form of interactive problem solving (Kelman 1996), applying negotiation theory to improve collaborative interactions and enhance the odds of improved decisions being implemented.

Economics

Within economics, a diverse set of traditions in addressing conflict have emerged. Of particular interest are voices that offer thoughts on fundamental philosophies for crafting environmental policies. For example, many conservative leaning or libertarian economists (e.g., Hayek 1944; Friedman 1962) argue that those entities who demand a stop to pollution or environmental damage are really demanding a transfer—without compensation—of a property right from the polluter to some other group, such as a land conservancy or the public in general. These entities, they argue, view many conflicts as a problem of "market failures," whereby actions of certain entities impose externalities (harmful or beneficial effects) that cannot be traced to the originator (e.g., pollution, overfishing) or underprice environmental degradation (e.g., mining, carbon emissions). As a result, vocal critics of past environmental neglect, like Barry Commoner (1971) and Paul Ehrlich (1968), find solutions in top-down legislative action, where legal and administrative rules are enacted to promote better outcomes.

Increasingly, the "neoliberal" economic approach of using markets as a means of reducing conflicts has gained traction around the world. Under this view, economists (e.g., Harlow 1974; Gardener and Simmons 2012) argue that pollution and environmental degradation are the result of too little private property. They point out that when market techniques are used (i.e., a pricing system for a degraded or scarce good), conflicts are diverted from a centralized political sphere to a decentralized market process. Furthermore, they argue that conflicts are settled as (1) rights to pollute or degrade are freely bought and sold by different actors and (2) the market removes what we often see as "extreme" consequences of groups that fail to meet standards set by environmental policies or negotiated agreements (e.g., loss of land-use rights under the U.S. Endangered Species Act [Berkey and BenDor 2012]; harsh penalties under U.S. Corporate Average Fuel Economy standards [BenDor 2012]).

"Neoliberalism" is an economic philosophy that advocates for greater privatization, free trade, deregulation, and open markets in order to enhance the role of the private sector in the economy (Campbell and Pedersen 2001).

Using market structures, it is clear that some conflicts (e.g., determining who gets permits for environmental damage and why) will disappear, as a price mechanism can help individuals make choices as to how to allocate their limited resources. Fewer rules exist and no one politically sets universal rules over what is and what is not allowed. That is, conflicts will be resolved by "market decisions" (pricing) and bargaining between parties (e.g., different owners of the rights being traded), instead of by regulators who may selectively enforce rules. However, in this type of scenario, it is inevitable that other conflicts will begin to appear at more visible points in the political process. Enforcement of rights, overall structure of the market, and a myriad of other issues can complicate market transactions, creating failures and conflicts throughout (BenDor and Riggsbee 2011; BenDor, Riggsbee, and Doyle 2011). In all of these situations, individuals' choices are affected by the dynamics of society's changing values and technological capacities.

As we will discuss extensively in Chapter 9's exploration of carbon trading, market-based solutions also commonly fail to effectively intervene in environmental conflicts where no inherent property rights over a resource or pollution exist (collective goods; e.g., carbon or water quality). In these cases, markets inherently become political entities as rights are created selectively by governments (e.g., Ellerman, Convery, and Perthuis 2010).

Water Resources Management

Beyond economics, other fields have also contributed to modern thinking on conflict management and resolution, particularly transboundary (international) conflict. Seminal work by Delli Priscoli and Wolf (2009) offers a comprehensive overview of water resources management efforts that aim to shift conflicts into cooperative opportunities, discussing the challenges of water-based conflicts and the rhetoric of "water wars" around the world (Islam and Susskind 2012), such as the Ili-Balkhash Basin Conflict between China and Kazakhstan (Sievers 2002; Peyrouse 2007) or conflicts over irrigation water in Bhutan or Thailand (Becu et al. 2003; Gurung, Bousquet, and Trébuil 2006). Their work shows that international treaties, public participation mechanisms, and efforts to build institutional capacity to mediate and enforce agreements are incredibly important in laying the foundation for negotiated agreements.

International Relations

In the context of international relations approaches aimed at understanding how and why parties comply with international regulatory agreements, Chayes and Chayes (1995) argue that international agreements operate better as management systems, rather than as systems of enforcement or coercion. They contend that within a management system, the

| ◄──────────── Increased coercion and chances of one-sided, win-lose outcome ──────────► | | | | |

Efforts by conflicting parties				Private third-party decisions		Legal (public), authoritative third-party involvement		Coercive, extra-legal actions	
Conflict Avoidance	Informal discussion and problem solving	Negotiation	Mediation	Administrative decision	Arbitration	Judicial decision	Legislative decision	Nonviolent direct action	Violence

FIGURE 1.2
Continuum of conflict management and resolution approaches. Adapted from Moore (1996).

prevailing atmosphere is collaborative, where underperformance by a party represents not an offense to be punished, by a problem that can be solved by collaborative consultation and analysis. Chayes and Chayes (1995, p. 27) point out that

> ...for all but a few self-isolated nations, sovereignty no longer consist[s] in the freedom of states to act independently, in their perceived self-interest, but in membership in reasonably good standing in the regimes that make up the substance of international life. ...Isolation from the pervasive and rich international context means that the state's potential for economic growth and political influence will not be realized.

How can effective agreements be reached in situations where competing parties have different objectives? Raiffa (1982) suggests that negotiation is part art and part science; art as in "interpersonal skills" and persuasion tactics, and science as in "systematic analysis" of situations, which together can have synergistic interactions that improve negotiations.

Along these lines, substantial work concentrates on the role of facilitators, mediators, and arbitrators—a continuum in terms of their relative power in the resolution process (Figure 1.2; Moore 1996)—to successfully resolve conflicts quickly and fairly. A good deal of this work is linked to international diplomatic solutions to environmental problems, where negotiations may be particularly intractable due to the sheer volume of conflicting parties and the complexity of their relationships with by government-level actors that purport to represent them (Susskind, Moomaw, and Hill 1997; Tolba and Rummel-Bulska 1998). In particular, practitioners and researchers like Raiffa (1982) and Moore (2014) make strong arguments that society needs better mediators to guide negotiations that reach win–win solutions.

Mutual Gains, Conflict Frames, and Joint Fact-Finding

Drawing on several decades of workshops with stakeholders, role-playing simulations, and a wide variety of real-life case-study applications (e.g., Susskind, Sarah, and Jennifer 1999), dispute resolution pioneer Lawrence Susskind and his colleagues at MIT and Harvard developed a groundbreaking framework for consensus-building processes known as the "mutual-gains approach" (Susskind and Landry 1991; Susskind and Field 1996; Susskind, Levy, and Thomas-Larmer 2000). Contrasting classic approaches to settling disputes where parties "split the difference," the mutual gains approach is a collaborative process that aims to produce benefits for all parties and establishes a formidable justification for participants to maintain collaborative efforts (Selin and Chevez 1995; Shmueli, Kaufman, and Ozawa 2008). This approach is founded on the idea that litigious environmental negotiations often reach unsatisfying settlements after protracted battles, and that a better method involves open negotiation among regulators, businesses, interest groups, and the public (Crowfoot and Wondolleck 1990). The mutual gains approach directs that specific actions be undertaken during each of the four distinct stages of multiparty negotiations, which include preparation, value creation, value distribution, and "follow-through" (i.e., execution of negotiated agreements). The idea is that outcomes reached under their framework should be satisfactory, efficient, and relationship enhancing for all parties.

These ideas, including the emphasis on preserving relationships between conflicting parties and preventing resentment, are echoed in Roger Fisher and William Ury's (1981) seminal work, *Getting to Yes: Negotiating Agreement Without Giving In*. In this book—which is heralded as *the* archetypical example of mainstream professional negotiation and conflict mediation techniques—the authors describe a process with four key elements that include (1) separating actors from the problem; (2) focusing on actors' interests, not their positions; (3) generating a variety of alternatives before deciding what to do; and (4) instituting that the resulting agreement be based on some objective standard.

Interests and Positions

A key facet of Fisher and Ury's approach involves getting disputing parties to think beyond their own positions and consider the perspective that others bring to the table. They highlight what is now a standard negotiation concept (Shmueli, Kaufman, and Ozawa 2008)—the "Best Alternative to a Negotiated Agreement"—which can act as a baseline to which alternatives can be compared to assess the extent of the benefits from collaboration. In fact, any mutual gains analysis requires conflicts to consider *interests* rather than *positions* to develop alternative courses of action that are based around each party's priorities; this is now a paramount factor in modern negotiation theory (Fisher and Ury 1981; Lax and Sebenius 1986; Moore 2014). Shmueli, Kaufman, and Ozawa (2008) define *positions* as either the demands made by parties or a statement of their preferred means for addressing specific problems (i.e., what they *believe*). Contrasting this, *interests* are the actual concerns motivating parties to advocate for specific solutions; that is, interests are the "why" behind stated positions that get to what parties actually want.

For example, while studying a small community in rural Thailand, where an economically successful, upland minority group is often identified as the driving force behind environmental destruction, Becu et al. (2003) describe the difficulty among disputing parties in separating biophysical realities of the watershed from ethnic prejudice and socioeconomic envy. Positions of stakeholders, forged by long and bitter histories, are starkly divided from their interests in resolving a languishing dispute. The history of conflicts and the institutional structures driving them are key factors that determine conflict outcomes and relative likelihood of resolution (National Research Council 2008).

A growing body of negotiation literature has argued that positions offer a poor foundation for resolving disputes. Worrying about what conflicting parties believe—rather than what they actually want—can lead conflict resolution widely astray. Shmueli, Kaufman, and Ozawa (2008) note that while positions "bundle" together an actor's underlying interests, it gives their interests a very low resolution. For example, asking someone if they agree with local urban planning efforts (a broad position or philosophical stance) is very different from asking about their interest in strong local schools or well-maintained trail systems (a specific and measurable concern), which may be the direct result of those planning activities.

Even though two parties may have similar interests, their positions may be diametrically opposed, thus hindering any attempt to find jointly agreed upon courses of action.

Scholars have observed a diverse and complex range of interests that parties bring to the table, including essential concerns about objective and measurable gains or losses experienced—legal and political precedent, property rights—and more intangible issues, including relationships, sense of control, informal or formal authority structures, and respect from other stakeholders (Fisher and Shapiro 2005; Galster 2012).

Framing

Another cornerstone concept in environmental dispute resolution involves conflict "framing" (see, e.g., Gray 2004; Whittemore and BenDor 2018), where frames function as lenses (or filters) through which parties interpret conflict dynamics, and interpret conflicts as tractable or intractable resolution, and these interpretations construct the conflict as more or less tractable. Lewicki, Gray, and Elliott (2003) note that conflict framing affects how we interpret conflict dynamics and alters how we interpret a conflict as tractable (easy to settle) or intractable (resolution-resistant). Along these lines, Shmueli, Kaufman, and Ozawa (2008) and Elliott and Kaufman (2003) also argue that frames lead actors to differentially weight the value of information they receive in a conflict based on the credibility they attribute to the information source.

Joint Fact-Finding: Expanding the Concept

A cornerstone of alternative dispute resolution practice involves developing techniques for conflicting parties to engage in "joint fact-finding," a process for collaboratively constructing an information base that leverages scientific expertise and can be jointly analyzed to achieve a consensus solution (Ehrmann and Stinson 1999). The idea is that joint fact-finding may be less prone to perceptions of bias (e.g., framing; Kaufman 1999) and other judgmental issues that create concerns about the strategic manipulation of information or bias in data presentation (Ehrmann and Stinson 1999; McCreary, Gamman, and Brooks 2001). Shmueli, Kaufman, and Ozawa (2008) argue that this is important because, during negotiations, conflicting parties typically bring information with them derived from diverse perceptions, expertise, experience, and input from other parties and entities that they trust. In doing so, they construct their own information base regarding a given conflict situation. The problem is that parties typically have very limited ability to process diverse, complex, and uncertain information. This is further hampered by differing levels of trust that they place in sources, disparities in data access, and the extensive time required to understand the technical details of conflict issues, their own impacts, conflict models, and future projections (Shmueli, Kaufman, and Ozawa 2008).

While joint fact-finding typically increases the time needed to collect relevant conflict data (e.g., rates of species decline and resource usage, or relative costs of certain interventions), the process specifically aims to reveal the value in specific pieces of information to all parties. This helps to develop a shared vocabulary, something that does not occur when experts deliver a prepackaged, "black-box" data set. Joint fact-finding also often includes efforts to translate detailed scientific information for the general public (e.g., see discussion in Rogers 2006). Therefore, the extra time spent on this effort is gained back when less time is spent in disputing the "facts" at a later date (Shmueli, Kaufman, and Ozawa 2008). The logic of joint fact-finding is embodied by the mantra, "going slow to go fast."

As we will discuss in Chapters 4 and 5, the process of joint fact-finding described in negotiation theory—and now commonly used in environmental dispute resolution—shares many of the consensus-seeking elements of participatory modeling. The techniques that we present in this book are very much an extension of the concepts of joint fact-finding. They are a way of representing the actions, behaviors, and understanding of conflict by all parties involved in a very explicit, transparent manner.

We should note that a major criticism of conflict resolution efforts, generally, and joint fact-finding processes, in particular, centers on how unequal power relationships can exacerbate intractable conflicts between parties. Recognizing and managing these directional relationships is often exceedingly challenging, so much that Kritek (2002) argues that many conflict resolution efforts (and joint fact-finding exercises) can actually exacerbate conflict, supporting the same socially created and sanctioned inequities that arise from power discrepancies between conflicting parties (e.g., Crowfoot and Wondolleck 1990; McCreary, Gamman, and Brooks 2001). While we discuss this issue in detail in Chapter 4, we can read this as part of an emerging and broader critique of "rational" public policy and planning processes more generally (e.g., local government decision making; Davidoff 1965; Berke et al. 2006), which are based on the idea that good outcomes in public decision making are reached through centrally implemented consensus-driven processes.

The Convergence of Social Science and Modeling Approaches to Conflict

Consensus Processes

In considering consensus-building processes within the framework of complexity science, we can build on the efforts of Cioni (2008), who frames consensus-building processes as efforts to bring disputing interests (or perspectives) toward an equilibrium point. This equilibrium point exists where conflicting parties see their expectations best—or at least partly—satisfied. Under this view, consensus among parties can be viewed as a search process, which composes a variety of actions that may satisfy parties. We can build on this by framing the results of the consensus process as separate from the process itself; the results may be a solution that satisfies some set of requirements, but the process represents a way of searching the conflict space for many possible solutions.

Given this, Voinov and Brown Gaddis (2008) point out that scientific input into consensus processes must be reinforced with local knowledge and participatory interactions to derive solutions that are well understood, politically feasible, and scientifically sound. In Chapter 4, we will detail the role of participatory decision making in conflict situations, particularly in the use of technical information and education as a means for reinforcing consensus efforts. An emphasis on *participatory* decision making is reinforced throughout this book, particularly in Part III, where we apply agent-based modeling to several environmental conflicts.

Agent-Based Analysis for Dispute Resolution

Throughout this book, we will argue that mathematical methods and models can help us gain a deeper understanding of the processes and interactions in conflict and cooperation, as well as act as new tools for improve conflict resolution, cooperation, and peace building. Alongside the rapid growth in computer technology, our capacity for creating sophisticated representation of conflicts can now extend beyond the elegant, but simplistic models of armed conflict, arms races (Lanchester 1916; Richardson 1960), and game theoretic interactions of hypothetical disputes (Axelrod and Hamilton 1981). We can use models—formally or informally—to become clearer thinkers, to use and comprehend data, and to understand why conflicts exist and how they develop and change over time. Models informing conflict resolution can also help to stabilize conflict interactions by helping actors to learn and adjust their actions and interactions until agreement is reached (Scheffran and Hannon 2007). Under this theoretical and analytical framework, we view cooperation as a process by which actors adjust their goals and actions to achieve mutual benefits.

In the context of the literature we have discussed throughout this chapter, we can view modeling as a quantitative translation of the insights of negotiation theory and other social science examinations of conflict histories and motivations. The transition from conflict to cooperation involves adaptation toward common positions and mutually beneficial actions that stabilize interactions. Whether this transition is successful depends on the capacity of conflicting parties to establish institutions and governance structures to prevent or manage conflicts. In the absence of these institutions, prevailing patterns of conflict or cooperation depend on the responses of each actor, which may be destabilizing or chaotic.

Chaos theory is the study of the dynamic behavior of systems that are sensitive to initial conditions (Lewin 1992). The most popular conceptual example is the "butterfly effect," in which a butterfly's wing movements in South America are said to trigger hurricane patterns thousands of miles away due to nonlinear and unpredictable responses in the complex atmosphere (Hilborn 2004).

In the conflict models that we introduce throughout this book, actors' responses can be mathematically represented by their decision rules and action priorities, as well as through their potential for learning and adaptation. Complex systems analysis and nonlinear dynamics provide a particularly useful framework for conflict analysis. For example, elements of *chaos theory* have been used as the basis for modeling an arms race and war outbreak, demonstrating that even simple arms race models—which are usually nonlinear and deterministic (i.e., there is no randomness involved in developing future states of the system)—can lead to completely unpredictable behavior (Saperstein 1984, 1986; Grossmann and Mayer-Kress 1989). Other examples of nonenvironmental conflicts that are readily modeled include legal maneuvers over property, labor conflicts (e.g., strikes, lockouts), and terrorism (Enders and Sandler 1995). In each of these examples, outcomes can be determined by contest "success functions," which define an appropriative outcome based on the "inputs" of fighting effort, and are the foundation for a variety of economic analyses of conflict (Sandler 2000). Building on this, we have seen a variety of other sophisticated analytical techniques with conflict resolution applications, including game theory (Fudenberg and Tirole 1983; Myerson 1997) and network analysis (Flint et al. 2009).

Viability Analysis and System Resilience

In this book, we present an initial effort to unify elements of cooperative and dynamic game theory with agent-based modeling—a sophisticated simulation technique we will discuss at length in Chapter 6—as a means for further developing conflict decision-support tools.

In employing a game-theoretic computational approach to agent-based conflict analysis, we introduce the use of *viability theory* as a way of reinterpreting the goals of environmental conflict resolution.

Viability theory, developed in the early 1990s by mathematician Jean-Pierre Aubin (1991), describes a decision framework for keeping a dynamic system within constraints, defined by either objective limits or value-based judgments (Aubin and Saint-Pierre 2007; Aubin, Bayen, and Saint-Pierre 2011). If the evolution of the system stays within the boundaries defined by these constraints, then a system can be said to be "viable" over a set period of time. In this book, we use viability theory to reinterpret "sustainability" as a mathematically defined and evolving complex system. Our use of viability as a systems-focused paradigm for sustainability assessment allows us to formalize conflict analyses that bridge the chasm separating sustainable environmental management concepts and the social and economic realities of conflict systems. For example, if we want to model a fishery conflict (see Chapter 8), we can assess how the fishers' behaviors and decision rules affect not only the long-term viability of fish stocks but also the economic productivity of fishers with heterogeneous abilities, information, and resources.

In this context, we can also assess the *resilience* of systems to exogenous shocks (e.g., an oil spill in a depleted fishery) and the role of these shocks in prolonging and complicating conflict. Along these lines, Reuveny (2007) argues that in the context of conflict avoidance, a major issue is the relative size of the disaster as well as the infrastructure of parties to withstand it. For example, all else equal, a region with relatively small populations and large reservoirs will likely suffer less from drought conditions than densely populated areas with few water reserves.

"Game theory" is an area of mathematics and economics that studies and models conflict and cooperation, usually between parties that are intelligent and make rational decisions (von Neumann 1928; Von Neumann and Morgenstern 1953). Dynamic games occur when parties have to make decisions over multiple time periods.

Resilience is a broad concept referring to the extent of change that a system can endure while still maintaining its underlying configuration (Atwell, Schulte, and Westphal 2011). Resilience can also refer to a system's ability to self-organize, learn, and adapt to changing conditions (Brand and Jax 2007). Resilience analyses often attempt to determine how systems can bolster desirable configurations (e.g., Alberti and Marzluff 2004; Beatley 2009) while avoiding movement toward undesirable states (e.g., Walker et al. 2002; Anderies, Walker, and Kinzig 2006; Oh, Deshmukh, and Hastak 2010). As part of this book's conflict viability analysis, we will analyze how conflicting actors *adapt* to experience or maintain systems in their present state or *transform* systems into alternative states that they desire.

Summary

In this chapter, we began to explore the types of tools needed to resolve complex, multilateral environmental disagreements. We have reviewed key conceptual, definitional, and background issues relating to conflict, cooperation, and conflict resolution, presenting an outline of how relevant disciplines—including geography, urban planning, business and management, conflict resolution, and environmental studies—approach conflict and cooperation. We have also explored different categories of conflicts (e.g., interest- and value-based conflicts) as well as the major approaches and strategies for conflict avoidance, prevention, and conflict resolution, including proactive planning processes, agreements, dialogs, participation, public-inclusion polices, and institutional designs. Finally, we began to argue that the approaches of many disparate fields concerned with environmental conflict can be unified using concepts from complexity sciences, enhancing conflict understanding and resolution from technical and anthropological standpoints.

In the next chapter, we discuss the basic philosophy around models, their representation of systems, and the concept of "mental modeling." In particular, we address an important question: how can modeling help to resolve conflict?

Questions for Consideration

1. In this chapter, we argue that environmental conflict complexity can be exacerbated by some technologies, such as social media, giving stronger voices to a larger number of conflicting parties, while making conflicts more easily visible to a large number of international observers. *Do you agree or disagree with this? Can you think of other ways that technologies impact environmental conflicts?*

2. Jerome Delli Priscoli and Aaron Wolf's seminal book, *Managing and Transforming Water Conflicts* (Cambridge University Press, 2009), overviews numerous water resources management efforts that have successfully shifted conflicts into cooperative opportunities. Their work shows that international treaties, public participation mechanisms, and efforts to build institutional capacity to mediate and enforce agreements are highly important in laying the foundation for negotiated agreements. When considering Delli Priscoli and Wolf's work, along with the growing International Environmental Agreements Database (https://iea.uoregon.edu/), it appears that there are more environmental agreements than ever. Look into both of these resources and consider: *What types of environmental conflicts tend to result in agreements? What types do not? What are the reasons for this? What are "institutions" and how do they play a role in addressing conflict?*

3. Pick an ongoing or recent conflict discussed in one of the Additional Resources section of this chapter. Considering that disputes become environmental conflicts in the face of environmental changes or degradation, *what environmental changes took place to create or extend these conflicts? Why did these changes occur?*

4. The joint fact-finding process attempts to do many things, including development of shared vocabulary among all parties. We note that this does not happen when experts deliver a prepackaged, "black-box" data set. Consider the same conflict as you did in the last question (3) in the context of the rapidly growing "big data" revolution (see, e.g., Mayer-Schönberger and Cukier 2013). *How could this conflict be exacerbated with more accurate and higher resolution data? How could better data availability aid conflict resolution?*

Additional Resources

Interested in learning more about ongoing conflicts? For more information, see

- The GDELT project (www.gdeltproject.org/) is a near real-time database that monitors news media to categorize human activity, including conflict events such as riots, protests, and diplomatic exchanges.
- The UCDP/PRIO Armed Conflict Dataset (www.prio.org/Data/Armed-Conflict/ UCDP-PRIO) was created by the Upsala Conflict Data Program's (UCDP) and the Peace Research Institute Oslo (PRIO).
- The Center for Documentation of Environmental Conflicts (www.cdca.it).
- For more data on fisheries and fisheries conflicts, see the State of World Fisheries and Aquaculture (SOFIA; www.fao.org/fishery/sofia/en) created by the Food and Agriculture Organization of the United Nations (FAO-FAD 2010).
- Rafael Reuveny, Maxwell, and Davis (2011) discuss numerous environmental disputes driven by resource scarcity around the world, including transboundary disputes between India and Bangladesh, Ethiopia and Somalia, Brazil and Paraguay, South Africa and Lesotho, Senegal and Mauritania, and many others, as well as numerous internal conflicts in areas like Darfur, Yemen, China, and Ethiopia.

Interested in learning more about conflict research? Many researchers have investigated natural resources and violent or armed conflict since the 1990s, including the Toronto Project on Environment, Population and Security (Homer-Dixon 1991, 1994); the Environment and Conflict Project (ENCOP) of the Swiss Peace Foundation and the ETH Zürich Center for Security Studies (Baechler 1999); the International Peace Research Institute in Oslo (Gleditsch 1997); the Woodrow Wilson Center's Environmental Change and Security Project (Dabelko and Dabelko 1995); and Adelphi, formerly the Institute for International and European Environmental Policy (Alexander Carius and Lietzmann 1999; Alexander Carius, Tänzler, and Maas 2008). For an overview, see Breitmeier (2009) and the studies prepared for WBGU (2008).

Interested in conflicts that are specifically related to water? Delli Priscoli and Wolf (2009) offer a comprehensive overview of water resources management efforts that aim to shift conflicts into cooperative opportunities, discussing the unique challenges of water-based conflicts and the rhetoric of "water wars" seen around the world (Islam and Susskind 2012), such as the Ili-Balkhash Basin Conflict between China and Kazakhstan (Sievers 2002; Peyrouse 2007) or conflicts over irrigation water in Bhutan or Thailand (Gurung, Bousquet, and Trébuil 2006; Becu et al. 2008). For more on water conflict, check out the Water Diplomacy Network (http://waterdiplomacy.org/).

Interested in the mechanics of environmental conflict resolution?

- McCreary, Gamman, and Brooks (2001) provide a great guide to setting up a joint fact-finding process and compare joint fact-finding with other means of establishing scientific facts.
- For those interested in alternative dispute resolution, generally, we also recommend the seminal work of Lawrence Susskind (e.g., Susskind and Field 1996;

Susskind, McKearnen, and Thomas-Lamar 1999) and excellent resources by Kritek (2002) and Furlong (2009). Readers should also check out Fisher and Ury's (1981) best seller, *Getting to Yes: Negotiating Agreements without Giving In*.

- If you want to practice your performance in negotiations, check out Noah Eisenkraft's (2017) negotiation simulator: www.customnegotiations.org/.

Interested in *viability theory*? We explain this theory in detail in Chapter 7. However, the wider theoretical underpinnings of viability theory are much more complex and mathematically intensive than the method we present in this book. We recommend going directly to the original sources by Aubin (1991) and Aubin, Bayen, and Saint-Pierre (2011) for this material. For a more extensive treatment of viability theory in fishery and natural resource management, see Eisenack, Scheffran, and Kropp (2006) and Eisenack et al. (2007).

Interested in "wicked" problems? Wicked problems occur when not everyone can consistently define a problem or even agree about its existence. Peter Balint et al.'s (2011) book, *Wicked Environmental Problems: Managing Uncertainty and Conflict* (Island Press/Center for Resource Economics), provides a good overview of these types of problems.

References

Alberti, M., and J. Marzluff. 2004. "Resilience in Urban Ecosystems: Linking Urban Patterns to Human and Ecological Functions." *Urban Ecosystems* 7: 241–65.

Alinsky, Saul. 1971. *Rules for Radicals: A Practical Primer for Realistic Radicals*. New York: Random House.

Allen, Timothy, and Thomas W. Hoekstra. 1993. *Toward a Unified Ecology*. New York: Columbia University Press.

Anderies, J. M., B. Walker, and A. Kinzig. 2006. "Fifteen Weddings and a Funeral: Case Studies and Resilience-Based Management." *Ecology and Society* 11 (1): 21.

Andrews, Richard N. L. 2006. *Managing the Environment, Managing Ourselves: A History of American Environmental Policy* (2nd Edition). New Haven, CT: Yale University Press.

Arena, Marika, Natalia Duque Ciceri, Sergio Terzi, Irene Bengo, Giovanni Azzone, and Marco Garetti. 2009. "A State-of-the-Art of Industrial Sustainability: Definitions, Tools and Metrics." *International Journal of Product Lifecycle Management* 4 (1–3): 207–51.

Atwell, Ryan C., Lisa A. Schulte, and Lynne M. Westphal. 2011. "Tweak, Adapt, or Transform: Policy Scenarios in Response to Emerging Bioenergy Markets in the U.S. Corn Belt." *Ecology and Society* 16 (1): 10.

Aubin, Jean-Pierre. 1991. *Viability Theory*. Berlin: Birkhäuser.

Aubin, Jean-Pierre, Alexandre M. Bayen, and Patrick Saint-Pierre. 2011. *Viability Theory: New Horizons* (2nd Edition). Berlin: Springer-Verlag.

Aubin, Jean-Pierre, and Patrick Saint-Pierre. 2007. "An Introduction to Viability Theory and Management of Renewable Resources." In *Advanced Methods for Decision Making and Risk Management in Sustainability Science*, edited by Jürgen P. Kropp and Jürgen Scheffran, 56–95. New York: Nova Science Publishers.

Axelrod, Robert, and William D. Hamilton. 1981. "The Evolution of Cooperation." *Science* 211: 1390–6.

Baechler, A. 1999. "Environmental Degradation in the South as a Cause of Armed Conflict." In *Environmental Change and Security: A European Perspective*, edited by A. Carius and K. Lietzmann, 107–30. Berlin: Springer.

Baker, M. 2004. *Socioeconomic Characteristics of the Natural Resources Restoration System in Humboldt County, California*. Taylorsville, CA: Forest Community Research.

Balint, Peter J., Ronald E. Stewart, Anand Desai, and Lawrence C. Walters. 2011. *Wicked Environmental Problems: Managing Uncertainty and Conflict*. Washington, DC: Island Press/Center for Resource Economics.

Barnett, J. 2000. "Destabilizing the Environment-Conflict Thesis." *Review of International Studies* 26 (2): 271–88.

BBC. 2001. "India-Bangladesh Border Battle." http://news.bbc.co.uk/2/hi/south_asia/1283068.stm

Beatley, Timothy. 2009. *Planning for Coastal Resilience: Best Practices for Calamitous Times*. Washington, DC: Island Press.

Becu, Nicolas, Andreas Neef, Pepijn Schreinemachers, and Chapika Sangkapitux. 2008. "Participatory Computer Simulation to Support Collective Decision-Making: Potential and Limits of Stakeholder Involvement." *Land Use Policy* 25 (4): 498–509.

Becu, Nicolas, P. Perez, A. Walker, O. Barreteau, and C. Le Page. 2003. "Agent Based Simulation of a Small Catchment Water Management in Northern Thailand: Description of the CATCHSCAPE Model." *Ecological Modelling* 170: 319–31.

BenDor, Todd. 2012. "The System Dynamics of U.S. Automobile Fuel Economy." *Sustainability* 4: 1013–42.

BenDor, Todd, and J. Adam Riggsbee. 2011. "A Survey of Entrepreneurial Risk in U.S. Wetland and Stream Compensatory Mitigation Markets." *Environmental Science and Policy* 14: 301–14.

BenDor, Todd, J. Adam Riggsbee, and M.W. Doyle. 2011. "Risk and Ecosystem Service Markets." *Environmental Science and Technology* 45 (24): 10322–30.

BenDor, Todd, Douglas A. Shoemaker, Jean-Claude Thill, Monica A. Dorning, and Ross K. Meentemeyer. 2014. "A Mixed-Methods Analysis of Social-Ecological Feedbacks between Urbanization and Forest Persistence." *Ecology and Society* 19 (3): 3.

Berke, P., D. Godschalk, E. Kaiser, and D. A. Rodriguez. 2006. *Urban Land Use Planning* (5th Edition). Urbana, IL: University of Illinois Press.

Berkes, F., and C. Folke. 1998. "Linking Social and Ecological Systems for Resilience and Sustainability." In *Linking Social and Ecological Systems: Management Practices and Social Mechanisms for Building Resilience*, edited by F. Berkes, C. Folke, and J. Colding, 1–25. Cambridge, England: Cambridge University Press.

Berkey, Kathleen Oppenheimer, and Todd BenDor. 2012. "A Comprehensive Solution to the Biofouling Problem for the Endangered Florida Manatee and Other Species." *Environmental Law* 42 (2): 415–67.

Boserup, E. 1981. *Population Growth and Technological Change*. Chicago, IL: Chicago University Press.

Brand, F. S., and K. Jax. 2007. "Focusing the Meaning(s) of Resilience: Resilience as a Descriptive Concept and a Boundary Object." *Ecology and Society* 12 (1): 23.

Breitmeier, Helmut. 2009. *Klimawandel Und Gewaltkonflikte*. Osnabrück: Deutsche Stiftung Friedensforschung.

Brody, Samuel D., David R. Godschalk, and Raymond J. Burby. 2003. "Mandating Citizen Participation in Plan Making: Six Strategic Planning Choices." *Journal of the American Planning Association* 69 (3): 245–64.

Brundtland, G., ed. 1987. *Our Common Future: The World Commission on Environment and Development*. Oxford, England: Oxford University Press.

Bryson, John M. 2004. "What to Do When Stakeholders Matter: Stakeholder Identification and Analysis Techniques." *Public Management Review* 6 (1): 21–53.

Buchholz, Thomas S., Timothy A. Volk, and Valerie A. Luzadis. 2007. "A Participatory Systems Approach to Modeling Social, Economic, and Ecological Components of Bioenergy." *Energy Policy* 35 (12): 6084–94.

Buhaug, H., S. Gates, and P. Lujala. 2009. "Geography Rebel Capability, and the Duration of Civil Conflict." *Journal of Conflict Resolution* 53 (4): 544–69.

Campbell, John L., and Ove K. Pedersen, eds. 2001. *The Rise of Neoliberalism and Institutional Analysis*. Princeton, NJ: Princeton University Press.

Carius, Alexander, and Kurt M. Lietzmann, eds. 1999. *Environmental Change and Security - A European Perspective*. Berlin: Springer.

Carius, Alexander, Dennis Tänzler, and Achim Maas. 2008. *Climate Change and Security - Challenges for German Development Cooperation*. Eschborn: Gesellschaft für technische Zusammenarbeit.

Carius, Alexander, Dennis Tänzler, and Judith Winterstein. 2006. *Weltkarte von Umweltkonflikten*. Berlin: Adelphi.

Chayes, Abram, and Antonia Chayes. 1995. *The New Sovereignty: Compliance with International Regulatory Agreements*. Cambridge, MA: Harvard University Press.

Cioni, Lorenzo. 2008. *Participative Methods and Consensus Theory (Technical Report: TR-08-23)*. Pisa, Italy: Dipartimento di Informatica, Universit`a di Pisa.

Cockerill, Kristan, Vincent C. Tidwell, Howard D. Passell, and Leonard A. Malczynski. 2007. "Cooperative Modeling Lessons for Environmental Management." *Environmental Practice* 9 (1): 28–41.

Collier, Paul. 2000. *Economic Causes of Civil Conflict and Their Implications for Policy*. Washington, DC: World Bank. https://siteresources.worldbank.org/DEC/Resources/econonmic_causes_of_civilwar.pdf [August 3, 2018].

Commoner, Barry. 1971. *The Closing Circle*. New York: Random House.

Costanza, R., and B.C. Patten. 1995. "Defining and Predicting Sustainability." *Ecological Economics* 15: 193–96.

Crowfoot, J. E., and J. M. Wondolleck. 1990. *Environmental Disputes: Community Involvement in Conflict Resolution*. Washington, DC: Island Press.

Cruikshank, Jeffrey, and Lawrence Susskind. 1989. *Breaking the Impasse: Consensual Approaches to Resolving Public Disputes*. New York: Basic Books.

Dabelko, Geoffrey D., and David D. Dabelko. 1995. "Environmental Security: Issues of Conflict and Redefinition." In *Environmental Change and Security Project Report 1*, edited by P.J. Simmons, 3–13. Washington, DC: Woodrow Wilson Center.

Davidoff, Paul. 1965. "Advocacy and Pluralism in Planning." *Journal of the American Institute of Planners* 31 (4): 331–8.

De Soysa, Indra. 2000. "The Resource Curse: Are Civil Wars Driven by Rapacity or Paucity?" In *Greed and Grievance: Economic Agendas in Civil Wars*, edited by Mats Berdal and David M. Malone, 113–36. Boulder, CO: Lynne Rienner.

De Soysa, Indra. 2002. "Paradise Is a Bazaar? Greed, Creed, and Governance in Civil War, 1989–99." *Journal of Peace Research* 39 (4): 395–416.

Delli Priscoli, J., and A. T. Wolf. 2009. *Managing and Transforming Water Conflicts*. International Hydrology Series. Cambridge, England: Cambridge University Press.

Dietz, Thomas. 2001. "Thinking about Environmental Conflicts." In *Celebrating Scholarship*, edited by Lisa M. Kadous, 31–54. Fairfax, VA: College of Arts and Sciences, George Mason University.

Ehrlich, Paul. 1968. *The Population Bomb*. New York: Ballatine Books.

Ehrmann, J. R., and B. L. Stinson. 1999. "Joint Fact-Finding and the Use of Technical Experts." In *The Consensus Building Handbook: A Comprehensive Guide to Reaching Agreement*, edited by Lawrence Susskind, McKearnen Sarah, and Thomas-Lamar Jennifer, 375–500. Thousand Oaks, CA: SAGE Publications.

Eisenack, Klaus, Matthias KB Lüdeke, Gerhard Petschel-Held, Jürgen Scheffran, and Jürgen P. Kropp. 2007. "Qualitative Modelling Techniques to Assess Patterns of Global Change." In *Advanced Methods for Decision Making and Risk Management in Sustainability Science*, edited by Jürgen P. Kropp and Jürgen Scheffran, 99–146. New York: Nova Science Publishers.

Eisenack, Klaus, Jürgen Scheffran, and J. Kropp. 2006. "Viability Analysis of Management Frameworks for Fisheries." *Environmental Modeling and Assessment* 11: 69–79.

Eisenkraft, Noah. 2017. "CustomNegotiations.Org: A Free Resource for Creating Custom Negotiation Simulations." *Negotiation Journal* 33 (3): 239–53.

Ellerman, A. Denny, Frank J. Convery, and Christian de Perthuis. 2010. *Pricing Carbon: The European Emissions Trading Scheme*. Cambridge, England: Cambridge University Press.

Elliott, M., and S. Kaufman. 2003. "Building Civic Capacity to Manage Environmental Quality." *Environmental Practice* 5 (3): 265–72.

Enders, Walter, and Todd Sandler. 1995. "Terrorism: Theory and Applications." In *Handbook of Defense Economics*, Vol. 1, edited by Keith Hartley and Todd Sandler, 213–49. Amsterdam: North-Holland.

FAO-FAD. 2010. *State of World Fisheries and Aquaculture (SOFIA)*. Rome, Italy: Fisheries and Aquaculture Department, Food and Agriculture Organization of the United Nations.

Farber, Stephen. 2000. "*Welfare-Based Ecosystem Management: An Investigation of Trade-Offs.*" *Environmental Science and Policy* 3: S491–S98.

Findley, Michael G. 2008. "Agents and Conflict: Adaptation and the Dynamics of War." *Complexity* 14 (1): 22–35.

Fisher, R., and D. Shapiro. 2005. *Beyond Reason: Using Emotions as You Negotiate*. New York: Penguin.

Fisher, R., and William Ury. 1981. *Getting to Yes: Negotiating Agreements without Giving In*. New York: Penguin Books.

Flint, Colin, Paul Diehl, Jürgen Scheffran, John Vasquez, and Sang-hyun Chi. 2009. "Conceptualizing ConflictSpace: Toward a Geography of Relational Power and Embeddedness in the Analysis of Interstate Conflict." *Annals of the Association of American Geographers* 99 (5): 827–35.

Florio, Kerry D. 2000. "Attorney's Fees in Environmental Citizen Suits: Should Prevailing Defendants Recover?" *Boston College Environmental Affairs Law Review* 27 (4): 707–40.

Forester, John. 1987. "Planning in the Face of Conflict: Negotiation and Mediation Strategies in Local Land Use Regulation." *Journal of the American Planning Association* 53 (3): 303–14.

Friedman, M. 1962. *Capitalism and Freedom*. Chicago, IL: University of Chicago Press.

Fudenberg, D., and J. Tirole. 1983. *Game Theory*. Cambridge, MA: MIT Press.

Fuller, Boyd. 2009. *Moving Through Value Conflict: Consensus Building and Trading Zones for Resolving Water Disputes*. Weisbaden: VDM Verlag.

Furlong, Gary T. 2009. *The Conflict Resolution Toolbox: Models and Maps for Analyzing, Diagnosing, and Resolving Conflict*. New York: John Wiley & Sons.

Galster, George. 2012. *Driving Detroit: The Quest for Respect in the Motor City*. Philadelphia, PA: University of Pennsylvania Press.

Gardener, D., and R. Simmons, eds. 2012. *Aquanomics: Water Markets and the Environment*. San Francisco, CA: Independent Institute Press.

Gleditsch, Nils Petter. 1997. "Environmental Conflict and the Democratic Peace." In *Conflict and the Environment*, edited by Nils Petter Gleditsch, 91–106. Dordrecht: Kluwer.

Gleditsch, Nils Petter. 1998. "Armed Conflict and the Environment: A Critique of the Literature." *Journal of Peace Research* 35 (3): 381–400.

Godschalk, David R. 1992. "Negotiating Intergovernmental Development Policy Conflicts: Practice-Based Guidelines." *Journal of the American Planning Association* 58 (3): 368–78.

Gowdy, J. 1999. "Hierarchies in Human Affairs: Microfoundation and Environmental Sustainability." In *Sustainability in Question: The Search for a Conceptual Framework*, edited by Jörg Köhn, 67–84. Northampton, MA: Edward Elgar.

Gray, B. 2004. "Strong Opposition: Frame-Based Resistance to Collaboration." *Journal of Community and Applied Social Psychology* 14: 166–76.

Greene, Michael. 2008. "A Force for Peace in the Middle East." *Science* 322 (5905): 1192.

Grossmann, S., and G. Mayer-Kress. 1989. "Chaos in the International Arms-Race." *Nature* 337: 701–4.

Gurung, Tayan Raj, Francois Bousquet, and Guy Trébuil. 2006. "Companion Modeling, Conflict Resolution, and Institution Building: Sharing Irrigation Water in the Lingmuteychu Watershed, Bhutan." *Ecology and Society* 11 (2): 36.

Halpert, Jane A., Alice F. Stuhlmacher, Jeffrey L. Crenshaw, Christopher D. Litcher, and Ryan Bortel. 2010. "Paths to Negotiation Success." *Negotiation and Conflict Management Research* 3 (2): 91–116.

Harlow, Robert L. 1974. "Conflict Reduction in Environmental Policy." *Journal of Conflict Resolution* 18 (3): 536–52.

Hayek, F. A. 1944. *The Road to Serfdom*. Chicago, IL: University of Chicago Press.

Hilborn, Robert C. 2004. "Sea Gulls, Butterflies, and Grasshoppers: A Brief History of the Butterfly Effect in Nonlinear Dynamics." *American Journal of Physics* 72 (4): 425–27.

Holling, C. S. 2001. "Understanding the Complexity of Economic, Ecological, and Social Systems." *Ecosystems* 4 (5): 390–405.

Homer-Dixon, Thomas. 1991. "On the Threshold: Environmental Changes as Causes of Acute Conflict." *International Security* 16 (2): 76–116.

Homer-Dixon, Thomas. 1994. "Environmental Scarcities and Violent Conflict: Evidence from Cases." *International Security* 19 (1): 5–40.

Homer-Dixon, Thomas. 2001. *Environment, Scarcity, and Violence.* Princeton, NJ: Princeton University Press.

Hurley, A. 1995. *Environmental Inequalities: Class, Race, and Industrial Pollution in Gary, Indiana, 1945–1980.* Chapel Hill, NC: University of North Carolina Press.

Ide, Tobias. 2015. "Why Do Conflicts Over Scarce Renewable Resources Turn Violent? A Qualitative Comparative Analysis." *Global Environmental Change* 33: 61–70.

Innes, Judith E. 2011. "Coordinating Growth and Environmental Management through Consensus Building, Vol. 1," December. https://escholarship.org/uc/item/308983c0

Iran/Iraq. 1975. "Agreement between Iran and Iraq Concerning the Use of Frontier Watercourses," February. http://ocid.nacse.org/tfdd/tfdddocs/380ENG.pdf

Islam, Shafiqul, and Lawrence Susskind. 2012. *Water Diplomacy: A Negotiated Approach to Managing Complex Water Networks.* New York: Routledge.

Kaufman, Sanda. 1999. "Framing and Reframing in Land Use Change Conflicts." *Journal of Architectural and Planning Research* 16 (2): 165.

Kelman, Herbert C. 1996. "Negotiation as Interactive Problem Solving." *International Negotiation* 1 (1): 99–123.

Koubi, Vally, Gabriele Spilker, Tobias Böhmelt, and Thomas Bernauer. 2014. "Do Natural Resources Matter for Interstate and Intrastate Armed Conflict?" *Journal of Peace Research* 51 (2): 227–43.

Kriesberg, Louis. 2009. "The Evolution of Conflict Resolution." In *The Sage Handbook of Conflict Resolution,* edited by Jacob Bercovitch, Victor Kremenyuk, and I. William Zartman. Thousand Oaks, CA: SAGE.

Kritek, P. B. 2002. *Negotiating at an Uneven Table: Developing Moral Courage in Resolving Our Conflicts.* Jossey-Bass Health Series. Hoboken, NJ: Jossey-Bass.

Lanchester, Frederick W. 1916. *Aircraft in Warfare: The Dawn of the Fourth Arm.* London, England: Constable and Company, Ltd.

Lax, D. A., and J. K. Sebenius. 1986. *The Manager as Negotiator: Bargaining for Cooperation and Competitive Gain.* New York: Free Press.

Le Billon, P. 2001. "The Political Ecology of War: Natural Resources and Armed Conflicts." *Political Geography* 20: 561–84.

Lederach, John Paul. 1997. *Building Peace: Sustainable Reconciliation in Divided Societies.* Washington, DC: Institute of Peace Press.

Lewicki, R. J., B. Gray, and M. Elliott, eds. 2003. *Making Sense of Intractable Environmental Conflicts: Concepts and Cases.* Washington, DC: Island Press.

Lewin, R. 1992. *Complexity: Life at the Edge of Chaos.* New York: Macmillan.

Libiszewski, Stephan. 1992. *What Is an Environmental Conflict? (Occasional Paper No. 1).* Zurich, Switzerland, Berne: Environment and Conflicts Project, Swiss Peace Foundation, Center for Security Studies and Conflict Research.

Lujala, P. 2009. "Deadly Combat over Natural Resources: Gems, Petroleum, Drugs, and the Severity of Armed Civil Conflict." *Journal of Conflict Resolution* 53 (1): 50–71.

Malthus, Thomas R. 1798. *An Essay on the Principle of Population.* New York: Penguin.

Marcuse, Peter. 1998. "Sustainability Is Not Enough." *Environment and Urbanization* 10 (2): 103–11.

Margerum, Richard D. 2011. *Beyond Consensus: Improving Collaborative Planning and Management.* Cambridge, MA: MIT Press.

Mayer-Schönberger, Viktor, and Kenneth Cukier. 2013. *Big Data: A Revolution That Will Transform How We Live, Work, and Think*. Reprint edition. London, England: Eamon Dolan/Houghton Mifflin Harcourt.

McCreary, Scott T., John K. Gamman, and Bennett Brooks. 2001. "Refining and Testing Joint Fact-finding for Environmental Dispute Resolution: Ten Years of Success." *Mediation Quarterly* 18 (4): 329–48.

Medzini, Arnon, and Aaron T. Wolf. 2004. "Towards a Middle East at Peace: Hidden Issues in Arab-Israeli Hydropolitics." *International Journal of Water Resources Development* 20 (2): 193–204.

Mitchell, Ronald B. 2018. "Environmental Agreements Database Project (Version 2018.1)." http://iea.uoregon.edu

Moore, Christopher W. 1996. *The Mediation Process* (2nd Edition). San Francisco, CA: Jossey-Bass.

Moore, Christopher W. 2014. *The Mediation Process: Practical Strategies for Resolving Conflict* (4th Edition). New York: John Wiley & Sons.

Myers, Norman. 1993. *Ultimate Security: The Environmental Basis of Political Stability* (1st Edition). New York: W.W. Norton.

Myerson, Roger B. 1997. *Game Theory: Analysis of Conflict*. Cambridge, MA: Harvard University Press.

National Research Council. 2008. *Public Participation in Environmental Assessment and Decision Making*. Washington, DC: The National Academies Press.

Neumann, J. von. 1928. "Zur Theorie Der Gesellschaftsspiele." *Mathematische Annalen* 100 (1): 295–320.

Nillesen, Eleonora, and Erwin Bulte. 2014. "Natural Resources and Violent Conflict." *Annual Review of Resource Economics* 6 (1): 69–83.

Oh, Eun Ho, Abhijeet Deshmukh, and Makarand Hastak. 2010. "Disaster Impact Analysis Based on Inter-Relationship of Critical Infrastructure and Associated Industries: A Winter Flood Disaster Event." *International Journal of Disaster Resilience in the Built Environment* 1 (1): 25–49.

Peyrouse, Sebastien. 2007. "Flowing Downstream: The Sino-Kazakh Water Dispute." *China Brief* 7 (10): 132.

Raiffa, H. 1982. *The Art and Science of Negotiation*. Cambridge, MA: Harvard University Press.

Reed, Mark S. 2008. "Stakeholder Participation for Environmental Management: A Literature Review." *Biological Conservation* 141 (10): 2417–31.

Reuveny, Rafael. 2007. "Climate Change-Induced Migration and Violent Conflict." *Political Geography* 26 (6): 656–73.

Reuveny, Rafael. 2008. "Ecomigration and Violent Conflict: Case Studies and Public Policy Implications." *Human Ecology* 36 (1): 1–13.

Reuveny, Rafael, and John W. Maxwell. 2001. "Conflict and Renewable Resources." *Journal of Conflict Resolution* 45 (6): 719–42.

Reuveny, Rafael, John W. Maxwell, and Jefferson Davis. 2011. "On Conflict over Natural Resources." *Ecological Economics* 70 (4): 698–712.

Richardson, Lewis Fry. 1960. *Arms and Insecurity: A Mathematical Study of the Causes and Origins of War*. Ann Arbor, MI: University of Michigan Press.

Rogers, Kevin H. 2006. "The Real River Management Challenge: Integrating Scientists, Stakeholders and Service Agencies." *River Research and Applications* 22 (2): 269–80.

Sandler, Todd. 2000. "Economic Analysis of Conflict." *Journal of Conflict Resolution* 44 (6): 723–29.

Saperstein, Alvin M. 1984. "Chaos - a Model for the Outbreak of War." *Nature* 309 (5966): 303–5.

Saperstein, Alvin M. 1986. "Predictability, Chaos, and the Transition to War." *Security Dialogue* 17 (1): 87–93.

Scheffran, Jürgen. 1999. "Environmental Conflicts and Sustainable Development: A Conflict Model and Its Application in Climate and Energy Policy." In *Environmental Change and Security*, edited by A. Carius and K. M. Lietzmann, 195–218. Berlin: Springer.

Scheffran, Jürgen. 2008. "The Complexity of Security". *Complexity* 14 (1): 13–21.

Scheffran, Jürgen. 2016. "From a Climate of Complexity to Sustainable Peace: Viability Transformations and Adaptive Governance in the Anthropocene." In *Handbook on Sustainability Transition and Sustainable Peace*, edited by H.G. Brauch, Ú. Oswald Spring, J. Grin, and J. Scheffran, 305–46. Hexagon Series on Human and Environmental Security and Peace. Berlin: Springer.

Scheffran, Jürgen, and Bruce Hannon. 2007. "From Complex Conflicts to Stable Cooperation: Cases in Environment and Security." *Complexity* 13 (1): 78–91.

Schlager, Edella, and Elinor Ostrom. 1992. "Property-Rights Regimes and Natural Resources: A Conceptual Analysis." *Land Economics* 68 (3): 249–62.

Selin, Steve, and Deborah Chevez. 1995. "Developing a Collaborative Model for Environmental Planning and Management." *Environmental Management* 19 (2): 189–95.

Shmueli, Deborah F., Sanda Kaufman, and Connie Ozawa. 2008. "Mining Negotiation Theory for Planning Insights." *Journal of Planning Education and Research* 27 (3): 359–64.

Sievers, Eric W. 2002. "Transboundary Jurisdiction and Watercourse Law: China, Kazakhstan and the Irtysh." *Texas International Law Journal* 37: 1–42.

Simon, J. L. 1996. *The Ultimate Resource 2*. Princeton, NJ: Princeton University Press.

Sirianni, Carmen. 2007. "Neighborhood Planning as Collaborative Democratic Design: The Case of Seattle." *Journal of the American Planning Association* 73 (4): 373–87.

Sowell, T. 2007. *A Conflict of Visions: Ideological Origins of Political Struggles*. New York: Basic Books.

Spring, Úrsula Oswald. 2008. "Peace and Environment: Towards a Sustainable Peace as Seen From the South." In *Globalization and Environmental Challenges: Reconceptualizing Security in the 21st Century*, edited by Hans Günter Brauch, Úrsula Oswald Spring, Czeslaw Mesjasz, John Grin, Pál Dunay, Navnita Chadha Behera, Béchir Chourou, Patricia Kameri-Mbote, and P. H. Liotta, 113–26. Berlin: Springer-Verlag.

Spring, Úrsula Oswald, and Hans Günter Brauch. 2011. "Coping with Global Environmental Change Sustainability Revolution and Sustainable Peace." In *Coping with Global Environmental Change, Disasters and Security*, edited by Hans Günter Brauch, Úrsula Oswald Spring, Czeslaw Mesjasz, John Grin, Patricia Kameri-Mbote, Béchir Chourou, Pál Dunay, and Joern Birkmann, 1487–503. Berlin: Springer-Verlag.

Stave, Krystyna. 2010. "Participatory System Dynamics Modeling for Sustainable Environmental Management: Observations from Four Cases." *Sustainability* 2 (9): 2762–84.

Susskind, Lawrence, and P. Field. 1996. *Dealing with an Angry Public: The Mutual Gains Approach to Resolving Disputes*. New York: Free Press.

Susskind, Lawrence, and Elaine M. Landry. 1991. "Implementing a Mutual Gains Approach to Collective Bargaining." *Negotiation Journal* 7 (1): 5–10.

Susskind, Lawrence, Paul Fidanque Levy, and Jennifer Thomas-Larmer. 2000. *Negotiating Environmental Agreements: How to Avoid Escalating Confrontation, Needless Costs, and Unnecessary Litigation*. Washington, DC: Island Press.

Susskind, Lawrence, and Sarah McKearnan. 1999. "The Evolution of Public Policy Dispute Resolution." *Journal of Architectural and Planning Research* 16 (2): 96–115.

Susskind, Lawrence, Sarah McKearnen, and Jennifer Thomas-Lamar, eds. 1999. *The Consensus Building Handbook: A Comprehensive Guide to Reaching Agreement*. Thousand Oaks, CA: SAGE Publications.

Susskind, Lawrence, W. Moomaw, and T. L. Hill. 1997. *Global Environment: Negotiating the Future. Papers on International Environmental Negotiation*. Cambridge, MA: Program on Negotiation (PON Books), Harvard Law School.

Susskind, Lawrence, Mieke Van der Wansem, and Armand Ciccarelli. 2000. *Mediating Land Use Disputes: Pros and Cons*. Cambridge, MA: Lincoln Institute of Land Policy.

Tastle, William J., and Mark J. Wierman. 2007. "Consensus and Dissention: A Measure of Ordinal Dispersion." *International Journal of Approximate Reasoning* 45 (3): 531–45.

Theisen, Ole Magnus. 2008. "Blood and Soil? Resource Scarcity and Internal Armed Conflict Revisited." *Journal of Peace Research* 45 (6): 801–18.

Thomas, Kenneth W. 1992. "Conflict and Conflict Management: Reflections and Update." *Journal of Organizational Behavior* 13 (3): 265–74.

Tolba, M. K., and I. Rummel-Bulska. 1998. *Global Environmental Diplomacy: Negotiating Environment Agreements for the World, 1973–1992*. Cambridge, MA: MIT Press.

Urdal, Henrik. 2005. "People vs. Malthus: Population Pressure, Environmental Degradation, and Armed Conflict Revisited." *Journal of Peace Research* 42 (4): 417–34.

Videira, Nuno, Paula Antunes, Rui Santos, and Rita Lopes. 2010. "A Participatory Modelling Approach to Support Integrated Sustainability Assessment Processes." *Systems Research and Behavioral Science* 27 (4): 446–60.

Voinov, Alexey, and Erica J. Brown Gaddis. 2008. "Lessons for Successful Participatory Watershed Modeling: A Perspective from Modeling Practitioners." *Ecological Modelling* 216 (2): 197–207.

Von Neumann, J., and O. Morgenstern. 1953. *Theory of Games and Economic Behavior.* Princeton, NJ: Princeton University Press.

Vucetich, John A., and Michael P. Nelson. 2010. "Sustainability: Virtuous or Vulgar?" *BioScience* 60: 539–44.

Walker, B., S. Carpenter, J. Anderies, N. Abel, G. S. Cumming, M. Janssen, L. Lebel, J. Norberg, G. D. Peterson, and R. Pritchard. 2002. "Resilience Management in Social-Ecological Systems: A Working Hypothesis for a Participatory Approach." *Conservation Ecology* 6 (1): 14.

Ward, Frank A., and Nir Becker. 2015. "Cost of Water for Peace and the Environment in Israel: An Integrated Approach." *Water Resources Research* 51 (7): 5806–26.

WBGU. 2008. *"World in Transition – Climate Change as a Security Risk."* www.wbgu.de/wbgu_jg2007_engl.html

Westing, A. P. 1986. *Global Resources and International Conflict: Environmental Factors in Strategic Policy and Action.* New York: Oxford University Press.

Whittemore, Andrew H., and Todd K. BenDor. 2018. "Talking about Density: An Empirical Investigation of Framing." *Land Use Policy* 72 (March): 181–91.

Wondolleck, Julia M., Nancy J. Manring, and James E. Crowfoot. 1996. "Teetering at the Top of the Ladder: The Experience of Citizen Group Participants in Alternative Dispute Resolution Processes." *Sociological Perspectives* 39 (2): 249–62.

Yergin, D. 1992. *The Prize: The Epic Quest for Oil, Money and Power.* New York: Simon and Schuster.

2

Why Model? How Can Modeling Help Resolve Conflict?

Remember that all models are wrong; the practical question is how wrong do they have to be to not be useful.
—*George Box and Norman Draper (1987), Empirical Model-Building and Response Surfaces*

Even in the modern age of science and industrialization, social policy decisions are based on incompletely communicated mental models. The assumptions and reasoning behind a decision are not really examinable, even to the decider. The logic, if there is any, leading to a social policy is unclear to most people affected by the policy.
—*Dana Meadows and Jennifer Robinson (1985), The Electronic Oracle*

Introduction

Today's scholars commonly point to the prominent work of Thomas Malthus (1798) as the origin of modern environmental and natural resource conflict studies. Malthus, an English demographer and intellectual, argued more than 200 years ago that the embattled relationship between nature and population growth would create resource scarcities, exacerbate environmental degradation, and eventually lead to conflict.

Within the 21st century, it is expected that substantial conflict will involve Earth's natural assets, including species and habitats, the atmosphere and ozone layer, forests and fisheries, rivers and streams, lakes, oceans, and mineral resources. As growing populations (nearly 10 billion worldwide by 2050; United Nations 2018) strain available resources, particularly in less developed nations, more effort will be put into obtaining additional resources. This is particularly troubling since territorial and resource disputes are already among the major causes of interstate conflicts and civil war (Sandler 2000). Additional resource demands, coupled with increased technology and institutional intricacy, will only serve to increase the complexity of environmental conflicts in the future.

In his landmark book *Arms and Influence*, Nobel Laureate Thomas Schelling (1966) argued that if the costs of fighting were large enough and they had complete information, nations would never go to war. If the triggers that start wars were known, he contended, nations would always find diplomatic solutions. If not, war may be thought of as a result of misunderstanding, misperceptions, incomplete information, or irrationality (Sandler 2000). Although we will discuss cases in the next chapter where this may not be true, it is nevertheless clear that current and expected future resource-based conflicts will have profound impacts on resource allocation, poverty, and income distribution. This emerging reality suggests an urgent need for new and improved methods of understanding conflicts, managing information, and addressing disputes (CEMARE 2002; Reuveny, Maxwell, and Davis 2011).

Although there was a time when virtually all of the work in conflict studies in this area was done by political scientists (primarily focused on violent conflict such as war, civil war, insurrections, externally led interventions, coups, and terrorism), efforts to apply rigorous, and often quantitative, thinking to conflict problems and conflict resolution have commenced in many arenas. Researchers and professionals in a variety of fields, including physicists, mathematicians and complexity scientists (Cho 2009), urban planners (Shmueli, Kaufman, and Ozawa 2008), geographers (Flint et al. 2009), economists, and many others, are now involved in creating and improving methods—including case studies, stakeholder intervention techniques, and modeling—for thinking about individual and aggregate conflict behaviors and patterns, and the systemic structures that create conflict situations.

In this chapter, we will begin by directly addressing the questions, *"what are models?"* and *"why model conflict?"* In doing so, we will discuss "mental modeling," as well as techniques for formalizing models through mathematical approaches and simulations. We will also examine the implications of modeling, including some of the potential drawbacks of using models. Finally, we will provide a brief overview of the general process of building models.

What Are Models?

Some scholars argue that a primary goal of the scientific method is to help explain cause and effect by encoding observational data into forms that are easier for people to understand and use (Cornwell 2004). In light of this, philosopher Ludwig Wittgenstein's (1922) early definition of models as "pictures of reality," suggests a solid theoretical rooting of modeling in the philosophy and practice of scientific inquiry as a means of data reduction or simplification. In this context, models are key tools for reducing the complexity of a system or problem to a degree that is understandable and manageable for the human brain (Buchholz, Volk, and Luzadis 2007). Batty (2001) notes that it was not until the 1950s that modeling became an explicit and widely used concept in social sciences, wherein the definition of a model explicitly transitioned from a "picture" to a "simplification" of reality (e.g., see Lowry 1965).

In a general sense, models form abstract representations of what we understand about reality, which, in open and complex systems, is never fully defined (Winz, Brierley, and Trowsdale 2009). In this way, models are simplified representations of extraordinarily complex realities (i.e., mediators to those realities; Morgan and Morrison 1999), making them theoretical tools of how the world operates when viewed in its entirety. Although models represent theories themselves, they are also used to test and explore more formalized theories about the way the world works. As we will discuss later, we are constantly forming "mental models" that help us make sense of the world that we perceive around us (Doyle and Ford 1998; Lynam and Brown 2012).

Why Model Conflict?

"Why model?" is a foundationally important question that should always be at the root of any effort to solve problems (or intervene in conflicts). One simple motivation is that modeling has historically been, and continues to be, a promising method of understanding and resolving environmental problems and disputes. We point to overwhelming evidence that modeling, in many forms, is a useful tool for understanding and solving urban, social and ecological problems (Hannon and Ruth 1997; Sterman 2000; Agarwal et al. 2002; Barredo et al. 2003; Costanza and Voinov 2004). Historically, some form of modeling—conceptual or otherwise—has always been useful for informing policy makers or stakeholders about the predicted outcomes of any type of action or policy. Almost invariably, however, these models are conceptual and often unconsciously designed. In short, we already subconsciously use "mental models" all the time, for better or for worse.

Mental Modeling

The concept of "mental modeling" was first introduced by psychologist Kenneth Craik (1943), who proposed that people carry in their minds a small-scale model of how the world works. In this sense, mental modeling is everywhere (Costanza and Ruth 1998). In fact, language itself is an expression of mental modeling, which may be a precondition for rational thought. We can comprehensively consider *mental modeling* as defined by Doyle and Ford (1998):

Mental models are abstractions of our perceptions and experiences in the reality that we experience. They are intuitive generalizations from our real world observations that form a vast matrix that dynamically connects familiar facts and concepts. Mental models form deeply ingrained assumptions, conscious and unconscious, which create the fabric of our perceived world that guides our decisions.

The mental model concept progressed significantly through work by Jay Forrester (1961), who was interested in why humans continue to have difficulty with thinking in terms of cause and effect, and in terms of *feedback loops* and nonlinearities, even after extensive training (Figure 2.1; Vennix 1999). Research over the last 60 years has discovered profound limitations in human capacity to process and synthesize information (see the influential work of Herbert Simon 1948, 1985). In thinking about why these cognitive limitations exist, it is important to consider that individuals and groups often employ complex biases and heuristics to reduce mental stress ("cognitive load"; Kahneman, Slovic, and Tversky 1982; Hogarth 1987; Vennix 1999).

Forrester thought of mental models in a more system-oriented sense as the mental image of the world that we carry around in our heads, which contains selected concepts and relationships that we use to represent the real system. Our mental models are grounded on implicit, causal maps that we use to understand systems. These maps consist of the networks of cause and effect that govern

Causal relationships among variables commonly form "feedback loops," whereby changes in a variable are often caused, at least in part, by the state of the variable itself. In *positive* feedback loops ("vicious" or "virtuous" cycles), variables are positively linked to themselves through chains of causation, meaning that if an increase in variable *a* increases variable *b*, then *b* feeds back and increases *a* further ("more begets more"; the same is true with *decreases* to both variables, as "less begets less"). Positive feedback loops lead to run-away behavior, as seen in bank accounts earning interest over time, arms races as nations compete to accrue weapons, and increasingly rapid decline in areas plagued by out-migration and job loss. In negative feedback loops ("balancing loops"), inverse relationships between variables *a* and *b* mean that an increase in *a* causes a decrease in *b*, which then causes a decrease in *a*. This type of feedback prompts systems to seek goals or self-adjust, preventing run-away behavior, even when it is perceived to be good.

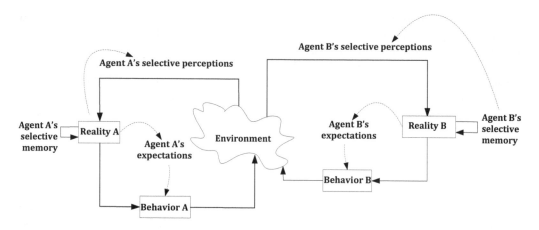

FIGURE 2.1

The social reality of intertwined relationships, where Agent A's behavior affects Agent B's environment, and vice versa. Both Agents A and B select and interpret information in their environments, construct their mental models, and behave according to those models. Adapted from Vennix (1999).

system operation, system boundaries, and time horizons across which we discount actions and frame our articulation of problems (Doyle and Ford 1998).

Much of Forrester's career was devoted to the idea that using modeling and diagramming in simple ways can help uncover a system's organizational structure and reveal complex, and often confusing or counterintuitive, relationships within. Leveraging simple representations of this system structure can lead to profound understanding of a system's otherwise intractable behavior over time (Sweeney and Sterman 2000). For example, in his landmark exploration of industrial dynamics, Forrester (1961) discovered that managers deal continuously with mental and verbal models of corporations, not the real corporations themselves. However, in some cases, these models failed to be useful (Sweeney and Sterman 2000; Saperstein 2008). The models used by managers were not necessarily correct and were sometimes inaccurate substitutes for thinking about the real system that the mental models represent.

Substantial evidence now suggests that people can only mentally simulate extremely simple mental models without making substantial errors (Sweeney and Sterman 2000). In fact, mental models are "fuzzy," imprecise, impermanent, and cognitively limited, as the number of variables that humans can track and relate to each other is very limited (Forrester 1994). When attempting to learn from the failures of mental models, we can focus on their lack of specificity—what are their assumptions and what types of deductions were made using these assumptions? While all models are inherently simplified versions of reality, inevitably omitting details that could be important in complex conflicts means that "unconscious" models (i.e., models we did not even know that we were using) suffer from the same problems, but in a less transparent way.

Formalizing Mental Models through Mathematics and Simulation

Making mental models explicit in some form, whether mathematical or not, starts to shed the vagueness of purely "unconscious" or verbal modeling. This is the reasoning that drives the creation of the dynamic computer simulations that we will discuss in Part II of this book.

In this context, formal mathematical models aim to deliberately and scientifically represent problems or conflicts that concern us. The usefulness of mathematical models is captured in how they allow us to test our theories about the world in artificial settings, without needing to actually change or experiment in the real world, which can often be impossible, unethical (e.g., administering high drug doses to test their effects), or extraordinarily expensive (as in the case of repeatable, controlled scientific experiments).

The rise in mathematical modeling paralleled the computer revolution during the last three decades of the 20th century, and up until today; ever-growing computing power has facilitated the use of increasingly extensive and complex data sets, as well as more sophisticated models that better represent real-world phenomena. Simply put, models today can be more realistic and more representative of real-world conflicts, allowing us to learn more, act faster, and make more informed decisions than ever before.

What is the difference between a *"simulation"* and a *"model"*? We draw on a stylized version of Endy and Brent's (2001, p. 391) definition of *simulation* as a representation of the information contained and created by the model which provides access to the model by allowing computation of system behavior. That is, while the *model* represents the system itself (system structure), a *simulation* is an examination of the output (compressed over space and time) of that modeled system for specific initial conditions of variables and parameter sets. Therefore, *simulation modeling* implies a study of both the system structure and its specific output.

The Implications of Modeling

Robinson et al. (2007) argue that modeling and simulation help us to understand environmental conflicts and coupled human-natural systems, generally, in several ways. First, modeling does this by facilitating the integration of multidisciplinary perspectives. Techniques such as agent-based modeling, system dynamics, evolutionary game theory, and network analysis, all facilitate enormous cross-pollination of ideas and approaches between relevant fields, such as economics, computer science, physics, and sociology. Modeling also helps us understand why observed dynamics occur, while allowing us to test different causes and explanations of observed behavior. This also allows us to explore different scenarios in the model that produce varying future states of the system or conflict. Furthermore, modeling helps us to improve, over time, the way that we collect empirical data about conflicts and problems.

Finally, after understanding the array of alternative futures, modeling is a tool to explore the impacts of different management or policy interventions on the conflict. Along these lines, Starfield (1997) argues that in a society where resource conservation managers are held accountable for their actions and decisions and where public scrutiny plays a role in the implementation of those decisions, the question is not whether to model, but rather how to model in a useful and efficient way.

Models Can Formalize Studies of Conflict

Adding to these advantages of modeling, Detlef Sprinz (1998) presents a strong argument for moving beyond case study explorations of environmental conflicts and their causes. For several decades, perhaps the most compelling work aimed at understanding environmental conflicts has taken the form of in-depth explorations of individual conflicts by researchers like Thomas Homer-Dixon (1991, 1994, 1996). However, to begin looking at the larger issues surrounding how and why conflict occurs, Sprinz (1998) proposes using empirically based, quantitative models to understand the link between environmental scarcity and armed conflict.

Sprinz argues that case study-based research of environmental conflict is incomplete because it is limited to situations that actually *experience* environmental conflict as a result of environmental problems. That is, self-selected case studies of conflict limit the external validity and inferences of any findings since there are no counterfactuals; why are there environmental problems that *do not* create conflict among involved parties? Instead of focusing on individual conflicts, are there *ex ante* research hypotheses that could guide a causal study of more fundamental conflict behavior and causes? Could we determine why conflict does *not* occur in some cases?

This type of criticism is not unique but is rather illustrative of a broader divide between qualitative and quantitative research within the social sciences (Schwartz, Deligiannis, and Homer-Dixon 2000; Gleditsch 1998). Are case studies unique, or are they part of a pattern? How can we connect environmental problems and armed conflicts?

One way of addressing these criticisms would be to introduce quantitative methods, such as statistical (e.g., regression analysis) or simulation modeling, which can assess when environmental impairments (e.g., reduction of environmental quality below some threshold) or other conditions are ripe for actually producing conflict. For example, several studies have analyzed conflicts in search of statistically relevant patterns that explain important aspects of conflict or cooperation (Weidmann and Ward 2010). In this sense, the goal of conflict resolution researchers and practitioners shifts from understanding individual conflicts toward identifying general patterns of behavior that can be treated as universal tendencies to increase or decrease conflict.

This brings us back to the question, "why model conflict?" What if we determined that certain instances of environmental deterioration are a sufficient condition for the outbreak of violent conflict? For example, we may find, through modeling work, that certain nonlinearities (breaking points or thresholds) associated with sudden decreases in environmental quality below certain thresholds can make a huge difference in the likelihood of conflict. This is the type of investigation that may lead to better understanding of how and why conflicts occur, leading to improved policy formation and new forms of conflict resolution.

This is a powerful argument for environmental modeling of any kind. By calling for research that extends beyond conflict case studies, we can anticipate further, wide-ranging questions that build on those of Sprinz: Why does conflict, including violent conflict, occur? How can we stop it from happening? These are important questions, which must be informed by data, experience, and rigorous ways of understanding causal and associative relationships.

Modeling with Caution

It is important to point out a couple of major problems with perceptions that form around modeling, particularly modeling of conflict. The first problem concerns perceptions that modeling produces concrete "truths" about the systems being modeled. These truths, like scientific laws, can then be mistakenly held as being infallible, faithful representations of reality. In many cases, this infallibility may come at the expense of the usefulness of the model for solving problems (Ruth and Hannon 2012); that is, perhaps we can represent conflicts and conflicted systems perfectly, but is this useful and does it lend us any new understanding or way of solving the conflict itself? Therefore, many environmental and conflict modelers instead argue that a better definition of a "model" can sometimes be "a purposeful representation" of a conflicted system (Starfield, Smith, and Bleloch 1994), where it is more useful to think of models as experimental hypotheses and as tools for solving problems, coping with uncertainty, and improving data collection and monitoring (Sterman 2000).

The second problem is that while models of all types have become nearly ubiquitous in policy evaluation and environmental management, models can be misused as a means of legitimizing status quo policy decisions, rather than informing new strategies for conflict resolution. Models, at their core, continue to be crude, yet often indispensable, representations of the subjective and objective realities that we experience. If taken as such, their usefulness exists primarily in their ability to help explain systems, generate new ideas, facilitate more informed decisions, and expand our capabilities for handling complex conflicts (Hannon and Ruth 1997; Ruth and Hannon 2012; Costanza and Ruth 1998).

The Three-Step Modeling Process

In light of those modeling limitations, perhaps the best modeling process is the one that is seen as being logical and defensible in the eyes of those making decisions and scrutinizing their implementation. Goodwin and Wright (1991) outline three broad and important steps that help to structure this process. First, we must know what we want to achieve. To do this, we must *outline* a lucid objective statement. As we will show in the participatory modeling chapter later in this book (Chapter 4), this type of "scoping process" is often more difficult than it seems.

As a second step, we must be able to *measure* how well our strategy for solving a problem performs, particularly with respect to our stated objective. That is, ongoing monitoring and continual evaluation is essential. To do this, we will need a set of indicators or metrics that help us to determine if we are meeting our goals. Negotiation and alternative dispute resolution literature typically evaluate conflicts and their resolution strategies based on whether agreement occurs or not (Halpert et al. 2010). However, researchers involved in modeling efforts for conflict resolution are now moving toward more robust strategies for evaluating the outcomes of consensus processes, including postprocess interviews, surveys, and additional measurements of social learning and pre-/posttests of joint problem understanding (Rouwette, Vennix, and van Mullekom 2002; Rouwette and Vennix 2006; Rouwette et al. 2011; Gray et al. 2018).

Finally, in step three, we need to *establish a procedure* for ranking the alternative options of strategies that emerge from the modeling effort, in terms of the metrics created in step two. We can think about this final step in two ways. First, seeing it through the lens of different actions that stakeholders can take after a modeling intervention, we can evaluate the possible actions of agents (e.g., people, nations, rulers, firms, governments, or institutions) involved in conflict and determine how those actions *might* perform in the real world. Participatory modeling has established many innovative ways of exploring systems and understanding how conflicting agents can affect system performance. On the other hand, if we think about modeling efforts as elements within larger planning processes and conflict resolution strategies, then step three very much becomes a locus for adaptive management. That is, modeling for conflict resolution needs to be iterative; measurements of the outcomes of conflict modeling and conflict resolution processes should be taken into account during future modeling efforts.

Modelers often think about this third step in terms of "validation," which is defined in a stakeholder-centric manner as the process by which stakeholders suggest changes to aspects of the model that they consider did not match reality (Barreteau, Bousquet, and Attonaty 2001; Castella,Trung, and Boissau 2005). Models therefore become "valid" when consensus forms and a model is accepted as a "good" representation of stakeholders' and researchers' reality. Unlike formal methods of model validation that seek to provide statistical or scientific evidence that a model is "correct" (e.g., Barlas and Carpenter 1990), many validation efforts seek to simply legitimize a model with its end users through a collective learning process where the model becomes well understood and is considered to be legitimate by decision makers.

It is important to note that the structure and complexity of any given model is a fundamental issue to consider. Grimm et al. (2005) point out that if a model is too simple, it ignores essential parts of the real world, limiting our ability to understand and test predictions of the problem, system, or conflict that the model hopes to represent. Conversely, when a model is too complex, our ability to analyze it becomes very difficult, and we will likely be bogged down in detail.

Summary

In this chapter, we have explored a number of important issues around the fundamental roles of modeling in addressing environmental conflicts. We have explored the motivations for creating models and modeling processes, arguing that modeling is useful in describing and resolving conflicts for which middle ground is difficult to find in other conflict resolution settings. We have also argued that computer modeling helps us enhance our own limited cognitive abilities and expand our capacity for understanding complex situations. Moreover, modeling allows us to perceive the consequences of our assumptions, to understand the inherent complexity in the system we are trying to understand, and to mediate between opposing points of view. Simply put, modeling provides a path for us to see through environmental problems and presents a platform for understanding and learning from complexity.

We have also argued for conflict modeling as a tool for understanding and exploring conflict resolution potentials, noting that explicit modeling, particularly rigorous mathematical modeling, increases transparency and provides a forum for participatory input, for better or for worse. In the next chapter, we will explore more of the history of modeling and the types of computer simulation and mathematical modeling that have been used to understand and resolve conflict.

Questions for Consideration

1. Douglass Lee (1973) penned a very well-known criticism of computer modeling, particularly as applied to urban systems. *Find this paper and summarize his criticisms. Do you think they are valid? Why or why not?*

2. Almost twenty-one years later, Douglass Lee (1994) followed up his 1973 critique, offering reflections on his original arguments. *Do you think his arguments are still valid? Do models today suffer from the same problems they did in 1973? What about in 1994?*

3. *Reflecting on Douglass Lee's arguments in his 1973 and 1994 papers, what weaknesses do you think models will have in the future?*

Additional Resources

We refer readers interested in learning more about modeling, generally, to Alexey Voinov's (2010) excellent text, *Systems Science and Modeling for Ecological Economics* (Academic Press).

Interested in mental modeling? A 2012 special issue of the journal *Ecology and Society* (Volume 17, Issue 3) titled *Mental Models in Human–Environment Interactions: Theory, Policy Implications, and Methodological Explorations*, discussed mental models in detail and is an excellent, open-access resource for those interested in the concept. There was also a great overview of mental modeling in the same journal the previous year, *Mental Models: An Interdisciplinary Synthesis of Theory and Methods*, written by Natalie Jones et al. (2011).

We are constantly forming "mental models" that help us make sense of the world that we perceive around us. However, our ability to generate information far exceeds our ability to understand it. Determining patterns and meaningful connections within complex data is emerging as one of the primary challenges of the 21st century. Manuel Lima's (2011) book, *Visual Complexity: Mapping Patterns of Information* (Princeton Architectural Press), provides a great overview of ways in which researchers and designers are using technology, color, graphics, and participatory techniques to help make sense of large data sets ("Big Data").

Interested in causal thinking? Even after extensive training, humans have incredible difficulty with thinking causally and in terms of feedback loops and nonlinearities. Jac Vennix (1999) provides an in-depth discussion of why this is and what can be done to more easily understand systems in causal terms. He also provides an excellent literature review in this area.

Much of the success of the system dynamics field (Chapter 5) has been due to its ability to leverage simple representations of system structure into profound insights into otherwise-intractable dynamic behavior. See Sweeney and Sterman (2000) for an innovative survey of this phenomenon.

What fields have applied models to conflict? Many new fields have been involved in creating new methods for understanding conflict, including physicists and complexity scientists (Cho 2009), urban planners (Shmueli, Kaufman, and Ozawa 2008), geographers (Flint et al. 2009), economists, and many others. Sandler (2000) aptly cites the groundbreaking work that influential researchers in numerous fields—whose names number too many to print here—have completed in studying the causes and progression of conflict and its causes.

Thomas Malthus remains a highly controversial, yet influential scholar. Check out Henrik Urdal's (2005) analysis of Malthus's writing in the context of modern resource conflicts such as those in the Democratic Republic of the Congo, in his paper, *People vs. Malthus: Population Pressure, Environmental Degradation, and Armed Conflict Revisited*.

References

Agarwal, Chetan, Glen M. Green, J. Morgan Grove, Tom P. Evans, and Charles M. Schweik. 2002. *A Review and Assessment of Land-Use Change Models: Dynamics of Space, Time and Human Choice*

(Technical Report NE-297) Northeastern Research Station: US Forest Service. www.fs.fed.us/ne/newtown_square/publications/technical_reports/pdfs/2002/gtrne297.pdf

Barlas, Y., and S. Carpenter. 1990. "Philosophical Roots of Model Validation." *System Dynamics Review* 6 (2): 148–66.

Barredo, Jose I., Marjo Kasanko, Niall McCormick, and Carlo Lavalle. 2003. "Modelling Dynamic Spatial Processes: Simulation of Urban Future Scenarios through Cellular Automata." *Landscape and Urban Planning* 64: 145–60.

Barreteau, O., F. Bousquet, and J. M. Attonaty. 2001. "Role-Playing Games for Opening the Black Box of Multi-Agent Systems: Method and Lessons of Its Application to Senegal River Valley Irrigated Systems." *Journal of Artificial Societies and Social Simulation* 4 (2): 12.

Batty, Michael. 2001. "Models in Planning: Technological Imperatives and Changing Roles." *International Journal of Applied Earth Observation and Geoinformation* 3 (3): 252–66.

Box, George E. P., and Norman Richard Draper. 1987. *Empirical Model-Building and Response Surfaces*. Wiley Series in Probability and Mathematical Statistics Applied Probability and Statistics. New York: Wiley.

Buchholz, Thomas S., Timothy A. Volk, and Valerie A. Luzadis. 2007. "A Participatory Systems Approach to Modeling Social, Economic, and Ecological Components of Bioenergy." *Energy Policy* 35 (12): 6084–94.

Castella, Jean-Christophe, Tran Ngoc Trung, and Stanislas Boissau. 2005. "Participatory Simulation of Land-Use Changes in the Northern Mountains of Vietnam: The Combined Use of an Agent-Based Model, a Role-Playing Game, and a Geographic Information System." *Ecology and Society* 10 (1): 27.

CEMARE. 2002. "The Management of Conflict in Tropical Fisheries, Final Technical Report." www.fmsp.org.uk/Documents/r7334/R7334_FTR.pdf

Cho, Adrian. 2009. "Ourselves and Our Interactions: The Ultimate Physics Problem?" *Science* 325 (5939): 406–8.

Cornwell, J. 2004. *Explanations: Styles of Explanation in Science*. Oxford, England: Oxford University Press.

Costanza, Robert, and Alexey Voinov. 2004. "Introduction: Spatially Explicit Landscape Simulation Models." In *Landscape Simulation Modeling: A Spatially Explicit, Dynamic Approach*. New York: Springer-Verlag.

Costanza, Robert, and Matthias Ruth. 1998. "Using Dynamic Modeling to Scope Environmental Problems and Build Consensus." *Environmental Management* 22 (2): 183–95.

Craik, Kenneth J.W. 1943. *The Nature of Explanation*. Cambridge, England: Cambridge University Press.

Doyle, James K., and David N. Ford. 1998. "Mental Models Concepts for System Dynamics Research." *System Dynamics Review* 14 (1): 3–29.

Endy, D., and R. Brent. 2001. "Modelling Cellular Behaviour." *Nature* 409 (6818): 391–95.

Flint, Colin, Paul Diehl, Jürgen Scheffran, John Vasquez, and Sang-hyun Chi. 2009. "Conceptualizing ConflictSpace: Toward a Geography of Relational Power and Embeddedness in the Analysis of Interstate Conflict." *Annals of the Association of American Geographers* 99(5): 827–35.

Forrester, Jay W. 1961. *Industrial Dynamics*. Cambridge, MA: MIT Press.

Forrester, Jay W. 1994. "Policies, Decisions, and Information Sources for Modeling." In *Modeling for Learning Organizations*, edited by J. D. W. Morecroft, and John D. Sterman, 51–84. Portland, OR: Productivity Press.

Gleditsch, Nils Petter. 1998. "Armed Conflict and the Environment: A Critique of the Literature." *Journal of Peace Research* 35 (3): 381–400.

Goodwin, Paul, and George Wright. 1991. *Decision Analysis for Management Judgment*. Chichester, England; New York: Wiley.

Gray, Steven, Alexey Voinov, Michael Paolisso, Rebecca Jordan, Todd BenDor, Pierre Bommel, and Pierre Glynn, et al. 2018. "Purpose, Processes, Partnerships, and Products: Four Ps to Advance Participatory Socio-Environmental Modeling." *Ecological Applications* 28 (1): 46–61.

Grimm, Volker, Eloy Revilla, Uta Berger, Florian Jeltsch, Wolf M. Mooij, Steven F. Railsback, Hans-Hermann Thulke, et al. 2005. "Pattern-Oriented Modeling of Agent-Based Complex Systems: Lessons from Ecology." *Science* 310(5750): 987–91.

Halpert, Jane A., Alice F. Stuhlmacher, Jeffrey L. Crenshaw, Christopher D. Litcher, and Ryan Bortel. 2010. "Paths to Negotiation Success." *Negotiation and Conflict Management Research* 3 (2): 91–116.

Hannon, B., and M. Ruth. 1997. *Dynamic Modeling of Economic Systems*. New York: Springer-Verlag.

Hogarth, R. 1987. *Judgment and Choice* (2nd Edition). Chichester, England: Wiley.

Homer-Dixon, Thomas. 1991. "On the Threshold: Environmental Changes as Causes of Acute Conflict." *International Security* 16 (2): 76–116.

Homer-Dixon, Thomas. 1994. "Environmental Scarcities and Violent Conflict: Evidence from Cases." *International Security* 19 (1): 5–40.

Homer-Dixon, Thomas. 1996. "Strategies for Studying Causation in Complex Ecological-Political Systems." *Journal of Environment and Development* 5 (2): 132–48.

Jones, Natalie, Helen Ross, Timothy Lynam, Pascal Perez, and Anne Leitch. 2011. "Mental Models: An Interdisciplinary Synthesis of Theory and Methods." *Ecology and Society* 16 (1): 46.

Kahneman, D., P. Slovic, and A. Tversky. 1982. *Judgment under Uncertainty: Heuristics and Biases*. Cambridge, England: Cambridge University Press.

Lee, Douglass B. 1973. "Requiem for Large-Scale Models." *Journal of the American Institute of Planners* 39 (3): 163–78.

Lee, Douglass B. 1994. "Retrospective on Large-Scale Urban Models." *Journal of the American Planning Association* 60 (1): 35–40.

Lima, Manuel. 2011. *Visual Complexity: Mapping Patterns of Information*. Princeton, NJ: Princeton Architectural Press.

Lowry, I. S. 1965. "A Short Course in Model Design." *Journal of the American Institute of Planners* 31: 158–66.

Lynam, Timothy, and Katrina Brown. 2012. "Mental Models in Human-Environment Interactions: Theory, Policy Implications, and Methodological Explorations." *Ecology and Society* 17 (3): 24.

Malthus, Thomas R. 1798. *An Essay on the Principle of Population*. New York: Penguin.

Meadows, Donella H., and J. M. Robinson. 1985. *The Electronic Oracle: Computer Models and Social Decisions*. Chichester, England: Wiley.

Morgan, Mary S., and Margaret Morrison. 1999. *Models as Mediators: Perspectives on Natural and Social Science*. Vol. 52. Cambridge, England: Cambridge University Press.

Reuveny, Rafael, John W. Maxwell, and Jefferson Davis. 2011. "On Conflict over Natural Resources." *Ecological Economics* 70 (4): 698–712.

Robinson, Derek T., Daniel G. Brown, Dawn C. Parker, Pepijn Schreinemachers, Marco A. Janssen, Marco Huigen, Heidi Wittmer, et al. 2007. "Comparison of Empirical Methods for Building Agent-Based Models in Land Use Science." *Journal of Land Use Science* 2 (1): 31–55.

Rouwette, Etiënne A. J. A., Hubert Korzilius, Jac A. M. Vennix, and Eric Jacobs. 2011. "Modeling as Persuasion: The Impact of Group Model Building on Attitudes and Behavior." *System Dynamics Review* 27 (1): 1–21.

Rouwette, Etiënne A. J. A., and Jac A. M. Vennix. 2006. "System Dynamics and Organizational Interventions." *Systems Research and Behavioral Science* 23 (4): 451–66.

Rouwette, Etiënne A. J. A., Jac A. M. Vennix, and Theo van Mullekom. 2002. "Group Model Building Effectiveness: A Review of Assessment Studies." *System Dynamics Review* 18 (1): 5–45.

Ruth, Matthias, and Bruce Hannon. 2012. *Modeling Dynamic Economic Systems* (2nd Edition). New York: Springer-Verlag.

Sandler, Todd. 2000. "Economic Analysis of Conflict." *Journal of Conflict Resolution* 44 (6): 723–29.

Saperstein, Alvin M. 2008. "Mathematical Modeling of the Interaction between Terrorism and Counter-Terrorism and Its Policy Implications." *Complexity* 14 (1): 45–49.

Schelling, Thomas C. 1966. *Arms and Influence*. New Haven, CT: Yale University Press.

Schwartz, Daniel M., Tom Deligiannis, and Thomas Homer-Dixon. 2000. "The Environment and Violent Conflict: A Response to Gleditsch's Critique and Some Suggestions for Future

Research." In *Environmental Change & Security Project Report*, 77–93. Washington, DC: Woodrow Wilson Center.

Shmueli, Deborah F., Sanda Kaufman, and Connie Ozawa. 2008. "Mining Negotiation Theory for Planning Insights." *Journal of Planning Education and Research* 27 (3): 359–64.

Simon, H. A. 1948. *Administrative Behavior: A Study of Decision-Making Processes in Administrative Organizations*. New York: Macmillan.

Simon, H. A. 1985. "Human Nature in Politics: The Dialogue of Psychology with Political Science." *The American Political Science Review* 79: 293–204.

Sprinz, Detlef F. 1998. "Modeling Environmental Conflict." In *Environment and Security: Challenges for International Policy*, edited by A. Carius and K. M. Lietzmann, 195–208. Berlin, Germany: Springer.

Starfield, Anthony M. 1997. "A Pragmatic Approach to Modeling for Wildlife Management." *The Journal of Wildlife Management* 61 (2): 261–70.

Starfield, Anthony M., Karl A. Smith, and A. L. Bleloch. 1994. *How to Model It: Problem Solving for the Computer Age*. Edina, MN: Burgess International Group.

Sterman, John D. 2000. *Business Dynamics: Systems Thinking and Modeling for a Complex World*. New York: Irwin/McGraw-Hill.

Sweeney, Linda Booth, and John D. Sterman. 2000. "Bathtub Dynamics: Initial Results of a Systems Thinking Inventory." *System Dynamics Review* 16 (4): 249–86.

United Nations. 2018. *World Population Prospects: The 2017 Revision*. Geneva, Switzerland: United Nations. www.un.org/development/desa/publications/world-population-prospects-the-2017-revision.html

Urdal, Henrik. 2005. "People vs. Malthus: Population Pressure, Environmental Degradation, and Armed Conflict Revisited." *Journal of Peace Research* 42 (4): 417–34.

Vennix, Jac A. M. 1999. "Group Model-Building: Tackling Messy Problems." *System Dynamics Review* 15 (4): 379–401.

Voinov, Alexey A. 2010. *Systems Science and Modeling for Ecological Economics*. San Diego, CA: Academic Press.

Weidmann, Nils B., and Michael D. Ward. 2010. "Predicting Conflict in Space and Time." *Journal of Conflict Resolution* 54 (6): 883–901.

Winz, Ines, Gary Brierley, and Sam Trowsdale. 2009. "The Use of System Dynamics Simulation in Water Resources Management." *Water Resources Management* 23 (7): 1301–23.

Wittgenstein, Ludwig. 1922. *Tractatus Logico-Philosophicus*. International Library of Psychology, Philosophy and Scientific Method. New York: Harcourt, Brace & Company.

3

The History and Types of Conflict Modeling

[I]t is often more important to clarify the deeper causes behind a given problem and its consequences than to describe the symptoms of the problem and how frequently they occur.
—*Bent Flyvbjerg (2006, p. 229), Five Misunderstandings about Case-Study Research*

Introduction

In this chapter, we will explore the history of environmental conflict modeling far beyond its analytical origins in the writings of Thomas Malthus (1798). While this review is in no way exhaustive, it is meant to illustrate several of the many types of methods and applications of conflict modeling. We will start this chapter by discussing the military origins of conflict modeling and their focus on modeling the resources that drive military clashes, including weapons and troops. We will chart how this evolved into more sophisticated models that are specifically focused on environmental conflict. While much of this chapter focuses on modeling conflict and its causes, we will discuss the use of modeling for conflict *resolution* purposes more explicitly in the next chapter.

Models of War and Arms Races

The application of mathematical modeling in the study of parties involved in competitive, international conflict has a very long history. Historically, some sort of theoretical modeling has always been instrumental for informing policy makers or military commanders about the predicted outcomes of security policies or military actions. In many cases throughout history, these models were verbal and often unconsciously held. Some of these analyses turned into "laws of war," such as the treatise, "The Art of War," by the ancient Chinese strategist Sun Tzu, or the comprehensive study, "On War," by the German general Carl von Clausewitz (1832). In modern times, numerous guidelines for warfare have evolved. In an example of informal military modeling, Marshall (1947) argued that 5,000 "fresh" troops would defeat 15,000 "worn out" ones.

Moll and Luebbert (1980) have characterized the early literature on military conflict/combat

Historically, models used during wars were verbal and often subconsciously created. "[I]t has often been said that Generals prepare to fight the next war exactly as they fought the previous war. We know that such models often fail to be useful. In attempting to learn from the failures of such verbal models, we are confronted with their lack of specificity: what exactly was assumed; what deductions were made from these assumed premises, and were the deductions correctly made?" (Saperstein 2008, p. 46).

models into two groups: *Lanchester-style* "arms-using" models, which describe how troop "attrition" (i.e., fatality) occurs in battle, and "arms-building" models, which include *Richardson-style* studies describing how nations build or disarm their military forces. In both cases, these mathematical models use systems of ordinary differential equations (i.e., equations describing rates of change with respect to a single independent variable) to understand how many variables can characterize underlying conflict systems, whether they are focused on troop attrition and longevity (Lanchester) or arms construction (Richardson).

Perhaps one of the first mathematical modeling efforts for arms-using purposes was Frederick Lanchester's (1916) model of mass combat, which depicted two opposing forces, engaged in different types of deterministic military "duels," where the outcomes (troop attrition on both sides) are based on the numbers and types of forces, the ability to concentrate firepower, and other influences (Moll and Luebbert 1980). Consider a conflict between two sides, which we will call red r and blue b. Lanchester's model—a system of two ordinary differential equations—tracks the rate of change of the total number of troops on both sides, consisting of n_r and n_b units, respectively, each of which can fire at any enemy unit with a single shot probability of hitting their enemy, p_r and p_b, and rate of fire, γ_r and γ_b (Weiss 1983).

$$\frac{dn_b}{dt} = -\gamma_r p_r n_r$$

$$\frac{dn_r}{dt} = -\gamma_b p_b n_b$$

These equations calculate the rates of change of the number of troops on each side. They state that if you increase either the probability p of hitting your enemy or your rate of fire γ, you will increase the rate that your enemy's army force n dies off. Moreover, this increase will occur in proportion to the size of your own troop force n.

What is an *"arms race?"* From its original military use, we can define an arms race as a situation in which multiple antagonistic parties compete in terms of military effectiveness, leading to rapid increases in arms quantity or quality (Gray 1971). Over the decades since the 1960s, however, the term has expanded to include a broad range of situations in which parties are competing with no absolute goal and are driven only by the relative need to stay ahead of the other competitors in rank, knowledge, or resource access. This is seen when the accumulation of a resource by one party forces competitors to stockpile resources in order to defend themselves. Examples of "arms races" include aspects of automobile safety (i.e., your vehicle size relative to other vehicles; Bradsher 2002) and races between authors of computer viruses and antivirus software.

In contrast to Lanchester's work, Lewis Fry Richardson (1960) developed a seminal model of arms races, which aims to describe the rates of change of two countries' (again, given as red r and blue b) weapon stockpile sizes, given as x_r and x_b. To model these rates, Richardson used "fear" (or "reaction") factors, a_r and a_b, which represent the desire of one country to increase arms proportionally to the size of the arms stockpile of their opponent. He also used "fatigue" (or "restraint") factors, m_r and m_b, to represent each nation's desire to reduce arms stockpiles proportionally to what they already possess. Finally, Richardson employed "grievance" factors (g_r and g_b), which represent many factors, including each country's ambition to obtain arms, revenge motives, international pressure, and other factors not directly linked to the stockpile of each nation. Like Lanchester, Richardson's model consists of two ordinary differential equations, this time representing weapon stockpile size as it changes over time:

$$\frac{dx_r}{dt} = a_r x_b - m_r x_r + g_r$$

$$\frac{dx_b}{dt} = a_b x_r - m_b x_b + g_b$$

These two modeling types differ significantly, as *arms-using* (Lanchester) models mathematically describe how losses can occur in a military conflict (or series of battles). Arms-use models start where *arms-building* models (Richardson) leave off—at the start of a conflict that signals the end of arms stockpiling. Using information about arms stockpiling, the arms-using model projects the military effectiveness and resource vulnerabilities of two opposing powers. In the early days of this type of modeling, these basic theoretical models were very useful for assessing the balance of military engagement and the security of both nations.

The early theoretical studies by Lanchester and Richardson were important contributions that have supported the technology of making and managing security policy, while theories of arms races prior to the 1940s were based on abstract and unproven mathematical models, work in the 1950s was driven by "rational" methods of extending the statistical findings from World War II operations research. In turn, cultural and subjective factors were added to models in the 1960s, domestic political processes in the 1970s, and explicit social–psychological models in the 1980s. These more recent models move beyond Richardson's simplistic assumptions about the influence of various parameters on national arms-building effort and toward modeling domestic political and bureaucratic processes, modern terrorism, the social and psychological factors creating and prolonging conflicts, and a range of newer techniques for more fully explaining armament and disarmament behavior.

Modeling Conflict vs. Modeling the Causes of Conflict

Perhaps the most important lessons that we can learn from these early conflict models relate to the style and structure of the models themselves. Each of the models boils down to a statement of causal influences between resources, factors that promote or prevent conflict, and the outcome of the conflict itself. Lanchester and Richardson are not modeling the underlying causes of the resource limitations or territorial intrusions that may lead to interstate conflict. In their models, however, the underlying causes of the conflict were not of interest; the central questions that they studied surrounded the dynamics of troop attrition and arms stockpiling. In this sense, these models are very much focused at simulating the dynamics of the *conflict itself*, rather than modeling the underlying *causes of conflict*.

More recent applications of conflict modeling have turned toward this latter goal. For example, Hirshleifer (2000) introduced a conceptual conflict model in which two contenders decide how to allocate resources between some sort of productive activity (e.g., government economic investments) and combat with each other. In this model, each opponent's relative contribution toward the fighting helps to determine his or her likelihood of winning. This model is an example of a shift toward an underlying cause of conflict as Hirshleifer focuses extensively on the role of technology governing the conflict and its effects on the connection between participants' fighting effort and their probability of success.

A General Typology of Environmental Modeling

As we will show, models built for military conflicts have strong implications for environmental conflict theory. Since the early 1990s, scientists have studied and debated whether damage to renewable resources, including deforestation, land use change and degradation, fisheries depletion, and food and water mismanagement, are increasingly contributing to violent conflict (Myers 1993; Baechler 1998; Homer-Dixon 2001). Climate–conflict linkages have also attracted interest in the research community (see Scheffran, Brzoska, Kominek, et al. 2012; Scheffran, Brzoska, Brauch, et al. 2012). Several scholars have pointed out that the environment is one among several conflict factors and suggest that conflict more likely occurs during economic recessions, periods of large population migration (Reuveny 2007) or socio-economic and political schisms, and in areas with weakened political institutions.

To consider the capabilities of environmental conflict models, we can reflect on the categorization scheme for natural resource models created by Bots and van Daalen (2008). The authors consider whether a model can reasonably represent the physical, social, and physical–social interaction aspects of resource usage and interaction. For example, a model of river dynamics could be limited to representations of riverbed dynamics or hydrogeomorphology. Instead, the model could be broadened to consider social, cultural, and economic aspects of river irrigation, navigation, water quality, river governance, and democratic regulatory processes, i.e., the social processes through which river governance is developed (Bots and van Daalen 2008).

Bots and van Daalen (2008) allocate resource modeling efforts into five major categories (Figure 3.1), delineated by models' representation of single or multiple parties and their

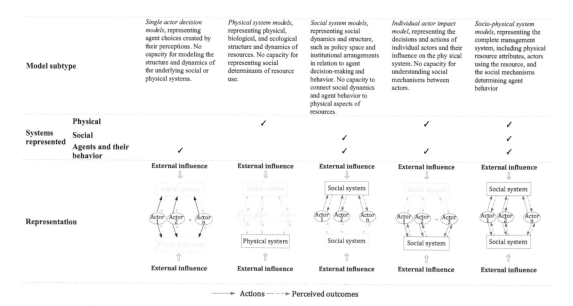

FIGURE 3.1
Typology of natural resource models. Adapted from Bots and van Daalen (2008).

capabilities in linking parties to the social and/or physical systems in which they are embedded. The extent to which each of these types of models can be reasonably applied to environmental *conflict* depends on the structure of the conflict and the specific relationships between parties and the environmental systems at the center of the conflict.

Game Theory and Conflict Simulation

When we think about conflict modeling generally, there are several fields that have historically approached the topic in very different ways. Over the last several decades, *game theory* has become the premier theoretical tool for modeling conflict in economics and political science (e.g., see Sandler 2000; Reuveny and Maxwell 2001). Game theory is particularly useful for understanding the strategic behavior of agents engaged in a conflict when their choices are interdependent with those of other agents. The interactivity of these choices leads to sometimes-complex strategies for maximizing the impacts of agent choice on well-being (called "payoff" or "value" in economics terms).

Game theory is the branch of mathematical analysis dealing with conflict and cooperation between intelligent and rational decision makers (Myerson 1997). Originally, game theory concerned "zero-sum games" between two participants, in which one competitor's gain was the result of another's loss (e.g., von Neumann 1928). However, today, game theory concerns many aspects of behavioral relations, and is treated generally as an umbrella concept for the science and logic of decision making.

For example, one player may think that an opponent (or competitor) will behave in a certain way, and therefore, their choices will reflect this belief. However, when an opponent anticipates the other player's belief-based actions and chooses a new strategy based on this belief, the strategies of all agents become reflective of this (and the cycle continues). For the better part of 60 years, game theory has grown into a prominent tool for analyzing conflict, particularly those in which a few opponents are engaged in intense struggles with one another over resources or property (Myerson 1997).

Dynamic Models of Conflict

Static models only simulate one time period. As a result, the equations of static models do not describe or analyze how a system changes or evolves over time, which is sometimes referred to as system "trajectories" or "equations of motion." *Dynamic* models, on the other hand, explicitly simulate change over time. Here, decisions or strategic actions by conflicting agents may occur repeatedly and can even change from one period to the next.

Historically, most resource conflict models have generally neglected the dynamic nature of population, natural resources, and conflict itself. Many of these models have considered *time* in conflicts in very basic ways, usually modeling disputes around static, game-theoretic scenarios. For example, early work by Jack Hirshleifer (1988, 1989, 1991) established a class of *static*, game-theoretic models for which conflict is understood as a rational activity, and the conflicted resource does not fall under established property rights (Harlow 1974). Here, actors are limited to single interactions or "one-shot games," which cannot address instances of prolonged conflict where actors take many actions through time.

This emphasis on static modeling has meant that until recently, many unresolved questions remained in the conflict modeling literature. For example, what are the feedback effects between the conflict, agent wealth, and conflict effort (investment of agents' resources in the conflict) over time? How do resources and agent groupings (e.g., coalitions) change these dynamics? Once a conflict occurs, does wealth reduce or exacerbate resource conflicts? Only by modeling conflicts as they change over multiple time periods can we gain any insight into these types of important questions.

Relatively early on, Reuveny and Maxwell (2001) began to move beyond Hirshleifer's static, game-theoretic framework by focusing specifically on the *dynamic* interplay between conflict and the contested resources. Reuveny and Maxwell relied jointly on ecological models of competition and economic models of conflict to create their modeling framework, which they applied to simulating the emergence, maturation, and conflict-driven societal collapse on Easter Island. Their modeling approach was based on two important ideas: (1) environmental conflict commonly arises over scarce renewable resources (e.g., forest timber, water) and (2) conflicted resources usually lack well-defined or enforceable property rights.

Ecological competition derives from the seminal predator–prey modeling work of Alfred Lotka (1925) and Vito Volterra (Volterra and Brelot 1931), who specify a system of ordinary differential equations describing the population states of predator- and prey-species, as well as resources being consumed, over time. In their model, Reuveny and Maxwell (2001) situate humans as the predator and the renewable resource as the prey. However, unlike traditional Lotka–Volterra models, the population sizes do not respond automatically to each other, and instead are controlled by the rival actors' decisions on allocation choices of efforts toward conflict. More importantly, groups fight over resources and wealth partly for the ability to invest their revenues to increase their own resources (i.e., capital) in the future, which can be used for further productive activities, or for more conflict.

For our purposes, it is important to briefly touch on two major limitations of Reuveny and Maxwell's approach. First, they focus exclusively on the limited case where conflicts involve only two parties. This is partly the consequence of modeling using a series of complex, closed-form equations, rather than a numerical simulation—a more flexible approach that we will introduce in detail in Chapters 5 and 6.

Second, while Reuveny and Maxwell extend the capabilities of individual agents to optimize their decisions beyond Hirshleifer-style models, agents are still not able to create strategies that account for future consequences of their chosen actions. Although they do not address this latter shortcoming directly, Reuveny and Maxwell suggest that improvements should be made to the model to simulate agents' foresight in predicting the actions of other agents, and subsequently allowing agents to optimize their own decisions over a future time path (i.e., forecasting). In the following sections, we will discuss modeling dynamic strategies in conflict. The conceptual framework that Reuveny and Maxwell (2001) introduce goes along with much of the early initial basis for the VIABLE modeling technique that we present in Chapter 7.

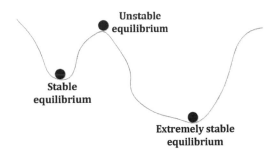

FIGURE 3.2
A typical metaphor for stability in a dynamic system is a ball on a hill, table, or in a depression. Depending on the system, perturbation—e.g., nudging the ball—may result in system collapse, a very different system state, or no change in the system at all. The subtle slope to the left of the "extremely stable" equilibrium point indicates more time required to return to the equilibrium depending on the direction that the system is pushed.

Simulating Strategy in Conflicts

Perhaps the most well-known (but certainly not the earliest) modeling exploration of conflict strategies is Robert Axelrod's application of Darwinian natural selection concepts to dynamic, game-theoretic simulations of conflicting agents. In his groundbreaking book, *The Evolution of Cooperation*, Axelrod (1984) sought out the evolutionary origins of conflict and cooperation, searching for strategies that achieve greater cooperative outcomes and evolve to be stable over time. In general, a system (or strategy in this case) is dynamically "stable" if, after being perturbed slightly, it returns to its original state (Figure 3.2). In Axelrod's case, the decision of each prisoner in the Prisoner's Dilemma, a classic game-theoretic scenario, returns to its original state after the behavior of their opponent has been slightly modified (they choose a new strategy).

The Prisoner's Dilemma is a classic scenario that demonstrates why two individuals may fail to cooperate, even if it first appears that it is in their interest to do so (Rapoport and Chammah 1965). In the classic *Prisoner's Dilemma*, two people are arrested, separated, and offered the same deal.

1. If one prisoner testifies against his/her partner (defects or betrays the partner), and the other remains silent (cooperates or assists the partner), the betrayer is set free and the cooperator receives the full (long) jail sentence.
2. If both prisoners remain silent, both are sentenced to very little jail time (or are set free).
3. If the prisoners betray each other, each prisoner receives a midrange sentence.

Each prisoner must independently choose to cooperate or not. What should they do to avoid mutual noncooperation and the penalties that result?

To explore this topic, Axelrod created a repeated Prisoner's Dilemma "tournament," inviting scholars from all over the world to submit strategies, which would be programmed to play repeated Prisoner's Dilemma games against other submitted strategies. In experiments (perhaps some of the earliest crowd-sourced research), a simple rule called "Tit-for-Tat" became the tournament winner. Players following this strategy would start

by cooperating with each other, and in subsequent
rounds, would do whatever the other player did
in the preceding move. If both players continue
to cooperate in any one round, they would continue to cooperate in the next, unless one defected,
which would lead to a defection by the other. This
is essentially a strategy of cooperation based on
reciprocity; the stability of the solution arises as
the decision of each prisoner returns to its original
state after the behavior of the opponent has been
slightly modified (they choose a new strategy).

When agents' strategies and population composition change over time,
this becomes known as an *evolutionary game* framework (Huberman and
Glance 1993; Vega-Redondo 1996).
The central concept in evolutionary
game theory is that the better a strategy performs in the present, more
often it will be used in the future. As
a result, the elements of an evolutionary game strategy—stability, fitness,
and payoffs against other strategies—
are fundamentally interconnected
(BenDor and Swistak 1997).

Following the tournament, Axelrod used an
"agent-based" simulation (described in Chapter 6)
to demonstrate why Tit-for-Tat strategies can evolve
as the dominant strategy starting from various distributions of initial strategy populations (Janssen
and Ostrom 2006). As Axelrod's findings were widely publicized, they attracted widespread
criticism from pure game theorists. For example, Binmore (1998) questions the need for the
computer simulations in the first place:

> While Axelrod did us a service of focusing attention on the importance of evolution
> selecting strategies for solving conflicts, what is the point of running a complicated
> simulation to find some of the equilibria of a game when they can often be easily com-
> puted directly?… This is a common mistake among social scientists that use simulation
> techniques. We need to pull from theory underlying conflict studies.

More recent work (e.g., Danielson 2002) has demonstrated that while reciprocity is
"necessary to engender cooperative behavior, it is not sufficient to ensure sustained cooperation, because cooperators can be exploited by those with whom they have struck up
apparently reciprocal relationships" (Bryant 2004, pp. 87–88; Bryant 2015). Subsequent work
also determined that *The Evolution of Cooperation* (Axelrod and Hamilton 1981; Axelrod
1984) did not *solve* the strategic conflict questions that Axelrod had posed. In fact, later
results appear to contradict some of Axelrod's main findings (BenDor and Swistak 1997).
Perhaps this is best demonstrated in Boyd and Lorberbaum's (1987) very specifically titled
paper, *No Pure Strategy Is Evolutionarily Stable in the Repeated Prisoner's Dilemma*.

Notwithstanding the controversial nature of Axelrod's work, his simulation experiments
publicly demonstrated the flexibility of simulation modeling for understanding competitive and cooperative behaviors. Like Lanchester and Richardson, much of the impact of
Axelrod's seminal and well-publicized work was in the subsequent explosion of conflict
research they prompted. Axelrod instigated enormous interest in the evolution and stability
of cooperation, making them issues of foremost importance for the science of politics and
conflict. His work also led to subsequent research demonstrating the importance of cooperation in systems with no central authority (e.g., Axelrod 1981; Milgrom, North, and Weingast
1990), and that informal cooperation is often the glue holding together many formal institutions where not all actions are specified by formal roles or contracts (Chisholm 1989).

Conflict as an Investment Strategy

Over the past four decades, one of the important contributions of the growth of conflict
modeling has been the general reconception of conflicts as investment strategies. This idea

centers on the reality that conflicting parties have limited resources. If parties then have to spend a substantial portion of their resources forcefully advancing their interests, or guarding themselves and their possessions, there are fewer resources for productive activities, such as economic growth or other investments.

In these situations, actors may devote effort to protecting resources through conservation (e.g., the production of technology and efficiency), harvesting resources, or through the appropriation of resources from rival actors through conflict (i.e., aggression or war) to maximize their income or reach their goals. These individual-level decisions lead to complex dynamic interactions between the agent populations and the resource.

Reuveny and Maxwell (2001) make the key observation that decisions to shift investments between resource harvesting and conflict may require substantial alterations in inputs, particularly inputs that are difficult (either cost- or time-intensive) to redirect between investment options (e.g., diverting heavy machinery towards military uses, thereby slowing economic growth or resource harvesting). This suggests that conflict can have non-intuitive destructive effects, which could potentially reduce income and result in system de-stabilization. Although the idea of conflict, particularly violent conflict, as a rational strategy for gaining or protecting resources is not new, it is a sophisticated and useful framework for organizing conflict resolution strategies that we will use throughout this book.

Optimal Strategies and Bounded Rationality

Historically, many conflict modeling efforts (particularly game-theoretic models) have argued that conflicting individuals "optimize" their behavior. For example, Reuveny, Maxwell, and Davis (2011, p. 700) surveyed the vast literature on natural resource conflict modeling, summarizing, "...we can say that states or nonstate actors often fight repeatedly over resource extraction. The outcome of encounters is not known in advance, though an overall victory may arise, ending the conflict. One may model such conflict in several ways, though it is reasonable to assume that the actors seek to maximize their gain." But, what does "maximize" mean exactly?

The *rational actor paradigm* suggests that humans are rational decision makers, who take into account all information and make decisions optimally to achieve a specific goal. However, as Janssen and Ostrom (2006, p. 3) aptly state, "...[i]t is now relatively well established, as a result of experimental research on social dilemmas, that the narrow model of 'economic man' focused primarily on monetary returns—which has been the primary model of human behavior adopted by many social scientists—is not a good foundation for explaining behavior outside of open competitive situations. Scholars should no longer presume that individuals seek only short-term material benefits for themselves in either experimental or field settings outside of competitive situations (including markets as well as elections and other competitive political situations)." Seminal work on "bounded rationality" by economist/computer scientist Herbert Simon (1957) instead suggested that when individuals make decisions, their rationality is limited by their cognitive limitations, their ability to understand the decision problem (its tractability), and the time they have to make the decision. In defining an alternative mode to optimization, Simon (1956, p. 136) coined the term "satisficing," where an agent finds "...a path that will permit satisfaction at some specified level of all its needs."

Many modeling efforts, have begun to represent individual behavior in a "boundedly" rational fashion, as it is affected by biased past experiences, personal expectations, and limited information processing capacity. As a result, conflicting parties highly weigh successes from past behavior, straying from it only slightly if it produces satisfying results. As a result, actors are said to "satisfice" (Simon 1956), rather than "optimize" their behavior.

In light of this, Reuveny, Maxwell, and Davis's (2011) work presents a major advance over previous efforts by representing the bounded decision making of two conflicting agents as they occur in a dynamic, multigame situation in which resource levels fluctuate and the parties must compete over multiple time periods. The authors introduce modeling innovations such as (1) agents with asymmetric power, (2) stochastically (i.e., calculated with a random component) represented victory probabilities, (3) fighting and resource extraction capabilities that are specific to each agent, (4) resource access that is a function of accumulated power, and (5) different types of ending scenarios to the conflict.

In Reuveny's model, the victor in the conflict receives all of the resources extracted during that time period ("winner takes all") and converts these spoils into additional power for the ensuing conflict period. It is important to point out that victories are modeled stochastically, with the probability of victory increasing with fighting effort. During each subsequent conflict, contesting groups allocate their capabilities to either resource extraction or conflict over extraction levels. It is worth noting that the authors explicitly contextualize this conflict in an anarchic geopolitical environment, which lacks property rights, universally accepted authorities, and social norms. Here, agents reject legal frameworks and social norms establishing peace.

Out of all this, Reuveny, Maxwell, and Davis (2011) found that in situations where resources are more plentiful—implying increased resource carrying capacity and growth rates—resource extraction (e.g., mining, deforestation, or harvests) rates can rise, thereby yielding additional power and nonintuitively leading to more intense conflicts. Moreover, given the lack of long-term strategy "optimization," groups that fight more efficiently tend to allocate fewer resources (effort) to conflict, are less likely to surrender, and appear to have bigger marginal gains to continuing conflict. Conversely, groups that are more efficient at extracting resources tend to fight more and surrender more readily. These findings imply that aid interventions aiming to strengthen resource-dependent economies may inadvertently exacerbate conflicts and should instead force negotiations by threatening to create further power imbalances between conflicting agents.

In Chapter 6, we will discuss how we can transform Reuveny's style of modeling into a modified form using agent-based modeling theory, where many additional actors can be involved, thus mirroring the multilateral conflicts more commonly seen in the real world. Following this, the method that we introduce in Chapter 7 will continue to build on the logic driving Reuveny, Maxwell, and Davis's (2011) approach, as well as incorporate improved methods of modeling agent strategies and allocation of resources (exploitation of resources vs. conflict).

Simulating Complex, Multiparty Conflicts

Today, simulation modeling is often seen as a more effective method for analyzing complex situations than game theory alone. For example, Galtier et al. (2012) argue that simulations are more effective for situations such as decentralized markets, where transactions occur between many parties and repeat over long periods of time. They also rightly point out that game theory and simulation modeling are useful, *complimentary* tools for exploring conflicts and the dynamics that arise in conflict systems.

For example, Axelrod's (1997) later work branched out from the Prisoner's Dilemma tournament, toward more sophisticated, agent-based adaptive models of "convergent social influence." Unlike previous models of social influence or cultural change that consider the effects of variables one at a time, Axelrod attempted to take into account the interactions of many different factors to understand how "local" convergence of ideas or beliefs among neighboring or closely connected agents can generate "global" polarization within the population as a whole. In particular, he wanted to know, if increased interactions tend to make people more alike in their behavior, beliefs, and attitudes, why don't all such differences eventually disappear?

While this question is just as fundamental to the nature of conflict as in his previous work, Axelrod's later studies reflect the substantial advances in conflict research that occurred during the 1980s and 1990s, including the use of much more sophisticated agent representations that specify the mechanics of change for many local actors. Consistent with the classic agent-based approach (discussed in Chapter 6), is Axelrod's lack of any central coordinating agent in his cultural dissemination model, such as powerful authorities like church leaders or political rulers. Moreover, agents are not necessarily rational actors, whose decisions are based on costs and benefits or typical, game-theoretic strategic analysis (e.g., Howard 1971; Fang, Hipel, and Kilgour 1993), but are rather assumed to follow simple, adaptive rules about influencing other agents around them. Computer simulation aids this approach and is focused on discovering the emergent properties of the system when many actors with simple behaviors interact.

Although this type of modeling has existed for decades, it is particularly demonstrative to trace the evolution of conflict modeling in Axelrod's work, from the almost purely game-theoretic style of simulation modeling, toward a "multi-agent" (i.e., more than two parties) simulation where individual agent behavior is explored to form theories about the structure of the larger conflict system. We will explore the relationship between conflict system structure and individual agent behavior in detail in Chapter 6.

Geography of Conflict

So far, we have discussed some of the many conflict modeling efforts over the years that attempted to extend conflict modeling to deal with multiple conflicting parties and dynamic conflicts. However, we would be remiss if we did not touch on how this modeling has approached conflict geography.

A number of studies have approached geography from a spatial statistical perspective. For example, Weidmann and Ward (2010) explore the geographic components of forecasting political violence using a spatial and temporal (time-series) regression model of conflict events in Bosnia (1992–1995). The use of models to predict future behavior represents a best practice for model validation across many statistically oriented sciences, but is often avoided in conflict analysis. As a predictive approach, Weidmann and Ward (2010) note that inclusion of both spatial and temporal conflict characteristics vastly improves the accuracy of violent conflict forecasts. The authors demonstrate strong spatial and temporal dimensions to violent outbreaks in Bosnia, which rightly leads them to make a strong case for predictive modeling of conflict; they note that "[s]tatistical models of health risks and carcinoma, for example, are intended not as a theory of disease but as a guide to individual action (p. 884)." Indeed, they ask, "[w]hy would it be a good idea to assume that [our] models only apply to the data we already have in hand?"

However, while many scholars have become interested in the role of geography in conflict, few studies have focused on developing sophisticated mechanisms for addressing geography. Instead, traditional conflict analysis has, unfortunately, considered space simply as a fixed, unchanging attribute between pairs of conflicting parties. These studies have also approached geography using simplified metrics like contiguity (e.g., do nations share a border?) and physical distance, which may largely fail to capture meaningful spatial relationships (O'Loughlin 2000).

This gives little nuance to the manner by which power relationships are established (and change) spatially, and leaves us largely unable to address questions, like: how do environmental conflicts spread? What roles do geography and network relationships—including power and political relationships—play in the spread of resource-based conflicts? After all, Sheppard (2002) argues that an agent's decision to engage in conflict, or support another agent in conflict, can be considered an outcome of a power relationship that is a function of spatial, network, and social attributes. Power can also be viewed as a relational concept, rather than a set of absolute attributes (e.g., roles, titles, resource volumes) that are possessed by each agent. Therefore, the complexity of conflict movement across space might be better handled through network analysis of the multiple types of relationships that occur between conflicting (or nonconflicting) agents.

Network Analysis and Conflict Modeling

Network representations of geography can capture conflict dynamics across noncontiguous spatial relationships, such as those that might occur between a mainland and distant islands (BenDor and Kaza 2012). For a strong example of this, we refer to Flint et al.'s (2009) development of a novel, network-based conflict approach, called "ConflictSpace," which facilitates the analysis of interstate conflict data and starts to get at the question of why certain actors get involved in conflicts.

Network analysis (or network theory) is an area of mathematical graph theory that helps us to understand the relationships (i.e., connectivity; represented as "links") between entities (i.e., agents; represented as "nodes"). *Social network analysis* involves the use of networks for structural analysis of social interaction. Network techniques have been increasingly applied to the study of conflict, generally (Flint et al. 2009; Maoz 2010), though not frequently for environmental conflicts. For example, models of the diffusion of material or information across uneven networks have been developed to analyze the spread of diseases and also have been used to study the spread of social behavior patterns (BenDor and Kaza 2012), describe the proliferation of violence and conflict (Flint et al. 2009), and explore the diffusion of technical innovations and social practices for climate mitigation and adaptation (Scheffran, Link, and Schilling 2012).

The work by Flint et al. (2009) is very much representative of a new wave of creativity in conflict modeling techniques, which breaks from the traditional international relations view of conflict and power that we have discussed so far: conflict as a zero-sum game where two competing agents gain and lose power, and as a result, one agent can force the other agent to involuntarily accede to their wishes.

Instead, Flint et al. (2009) argue that the value of the traditional approach is confined to an idealized scenario where two isolated countries compete over limited resources (environmental or otherwise) until one conquers the other. However, in a globalized world, conflicts have multidimensional contexts, multiple issues at stake, usually involve more than two agents, and almost always have impacts on other actors not involved in the dispute. Instead, they argue that we need to start thinking about resource-based conflicts by integrating representations of conflict geography with our understanding of the politics of networked agent relationships.

Building on newer viewpoints that scholars and professionals use to understand power relationships, the authors started thinking about the spatial context of conflict as a combination of where countries are located (i.e., "territorial embeddedness" is modeled using spatial analysis; Anselin 2002), as well as the way that they are connected to each other ("network embeddedness," which is modeled using social network analysis; Scott 2000). Leitner, Sheppard, and Sziarto (2008) and Flint et al. (2009) refer to this as the "multiple spatialities" of conflict. Flint et al.'s (2009) approach breaks from traditional methods of understanding spatial conflict (purely through agent proximity) by considering the relationships between individual pairs of conflicting parties (as measured by the presence of interactions like trade, conflict, migration, or alliances) and populating these relationships into a network of social relations across which conflict spreads.

Flint et al. (2009) note that Switzerland remained neutral in World War I, which might be unexpected during that conflict, considering the spatial context of the country. In fact, relative geographic position alone does not accurately determine the spread of World War I; in their analysis, they show that only 14 of the 37 pairs of warring countries shared borders.

For example, the ability of a country to avoid becoming embroiled in a war is only partially a function of its shared border with a country that is already at war. Rather, contiguity is just one expression of the geography of conflict and is inherently coupled with the network relationships among conflicting agents. By placing these relationship pairs within this network, they are able to simultaneously analyze how conflict spreads in terms of political power relationships ("network space") as well as across physical, geographic space.

To address classic criticisms of social network theory's, "static and uni-dimensional concentration" on network structure (e.g., Sheppard 2002), Flint et al. (2009) designed their network model to be a dynamic "surface" over which the complex feedbacks between space and power relationships play out, thus defining the spread of conflicts. Flint et al. (2009) is a good example of the emerging style of cross-disciplinary work that is transforming modern conflict modeling.

Summary

Over the last several decades, the parallel developments in computing resources and modeling paradigms—including advances in game theory, system dynamics, and agent-based modeling—have led to a proliferation of environmental conflict modeling techniques and applications. In this chapter, we have surveyed a sample of the vast amount of work done in this area, focusing on the evolution of conflict modeling from simplified models of war and arms races into sophisticated representations of conflict dynamics and agent behaviors.

Many studies have observed that the traditional analytical methods used for addressing complex societal and policy problems have been hindered by a stark absence of stakeholder participation (Stoorvogel, Bouma, and Orlich 2004; Videira et al. 2010). Costanza and Ruth (1998) argue that it is absolutely essential for any new modeling approaches to make consensus building and participation an integral part of the conflict modeling process. In fact, many models of natural resource management are instigated by questions posed directly by individuals or groups who hold a stake in resources or conflicts over resources. In recent years, participatory modeling, particularly using system dynamics models, has made a significant mark in the conflict resolution profession. In the following chapter, we will explore the role of stakeholder participation in forming models of conflicts as well as start to examine models of conflict resolution, wherein modeling efforts are taken to be an element of larger dispute resolution processes.

Questions for Consideration

1. The *rational actor paradigm* suggests that humans are rational decision makers who take into account all information and make decisions optimally to achieve a specific goal. Given the weight of the literature over the last half century arguing against this model of human decision making (e.g., see Barnes and Sheppard's [1992] critique in the context of neoclassical economics), *why do you think we still so frequently represent humans as rational decision makers in computer models?*

2. Herbert Simon's (1957) concept of "bounded rationality" suggests that when individuals make decisions, their rationality is limited by their cognitive limitations. *Give some examples of this in your life. How do you think your decision making is affected by other factors (e.g., complete access to information)?*

3. In defining an alternative mode to optimization, Herbert Simon (1956, p. 136) coined the term "satisficing," which involves finding, "...a path that will permit satisfaction at some specified level of all its needs." *What are some examples of satisficing that you have experienced? Why do you believe that they were satisficing instead of a different (maybe more "global") type of optimization?*

Additional Resources

Interested in learning more about differential equations? We recommend consulting a textbook such as Tenenbaum and Pollard (1985).

Interested in learning more about game theory? You can find John von Neumann's original (in English) discussion as he developed the ideas in his seminal 1953 book, *Theory* *of Games and Economic Behavior* (Princeton University Press). Or, you can read it in its original German form in his 1928 paper, *Zur Theorie der Gesellschaftsspiele* (in the journal, *Mathematische Annalen*). If you are really interested in game theory, there are research organizations that specialize in this area, such as the *International Society of Dynamic Games* (e.g., see Bernhard, Gaitsgory, and Pourtallier 2009).

Reuveny, Maxwell, and Davis (2011) discuss extensions to build on Hirshleifer's original techniques, including additional resource stocks, conservation norms and property rights (Suzuki and Iwasa 2009), and integration of more sophisticated labor allocation models (dividing resources between resource extraction and fighting; e.g., Welsch 2008). Unlike their previous work (e.g., Reuveny and Maxwell 2001), Reuveny, Maxwell, and Davis's (2011) model the dynamic behavior of two agents numerically, as the mathematical complexity makes analytical (closed-form) solutions to the model equations impossible.

Interested in specialized techniques for modeling conflict? For example, check out Brams, Kilgour, and Klamler's (2014) mathematical technique for fairly dividing indivisible goods between two people (e.g., in a divorce) in their paper, *Two-Person Fair Division of Indivisible Items: An Efficient, Envy-Free Algorithm*. You can also find more examples in Robert Trappl's (2005) book, *Programming for Peace: Computer-Aided Methods for International Conflict Resolution and Prevention* (Springer).

References

Anselin, Luc. 2002. "Under the Hood: Issues in the Specification and Interpretation of Spatial Regression Models." *Agricultural Economics* 27 (3): 247–67.

Axelrod, Robert. 1981. "The Emergence of Cooperation among Egoists." *American Political Science Review* 75: 306–18.

Axelrod, Robert. 1984. *The Evolution of Cooperation*. New York: Basic Books.

Axelrod, Robert. 1997. "The Dissemination of Culture." *Journal of Conflict Resolution* 41 (2): 203–26.

Axelrod, Robert, and William D. Hamilton. 1981. "The Evolution of Cooperation." *Science* 211: 1390–96.

Baechler, G. 1998. "Why Environmental Transformation Causes Violence: A Synthesis." In *Environmental Change and Security Project Report (Issue 4)*, 24–44. Washington, DC: Woodrow Wilson Center.

Barnes, Trevor J., and Eric Sheppard. 1992. "Is There a Place for the Rational Actor? A Geographical Critique of the Rational Choice Paradigm." *Economic Geography*, 68(1): 1–21.

Belt, Marjan van den, Oscar A. Bianciotto, Robert Costanza, Serge Demers, Susana Diaz, Gustavo A. Ferreyra, Evamaria W. Koch, et al. 2006. "Mediated Modeling of the Impacts of Enhanced UV-B Radiation on Ecosystem Services." *Photochemistry and Photobiology* 82 (4): 865–77.

BenDor, Jonathan, and Piotr Swistak. 1997. "The Evolutionary Stability of Cooperation." *American Political Science Review* 91 (2): 290–307.

BenDor, Todd, and Nikhil Kaza. 2012. "A Theory of Spatial System Archetypes." *System Dynamics Review* 28 (2): 109–30.

Bernhard, Pierre, Vladimir Gaitsgory, and Odile Pourtallier. 2009. *Advances in Dynamic Games and Their Applications: Analytical and Numerical Developments*. New York: Springer Science & Business Media.

Binmore, Ken. 1998. "'The Complexity of Cooperation by Robert Axelrod,' a Long Review." *Journal of Artificial Societies and Social Simulation* 1 (1): Review 1.

Bots, Pieter, and C. van Daalen. 2008. "Participatory Model Construction and Model Use in Natural Resource Management: A Framework for Reflection." *Systemic Practice and Action Research* 21 (6): 389–407.

Boyd, Robert, and Jeffrey P. Lorberbaum. 1987. "No Pure Strategy Is Evolutionarily Stable in the Repeated Prisoner's Dilemma Game." *Nature* 327 (6117): 58–59.

Bradsher, Keith. 2002. *High and Mighty: SUVs-The World's Most Dangerous Vehicles and How They Got That Way*. New York: PublicAffairs.

Brams, Steven J., M. Kilgour, and Christian Klamler. 2014. "Two-Person Fair Division of Indivisible Items: An Efficient, Envy-Free Algorithm." *Notices of the AMS* 61 (2): 130–41.

Bryant, J. W. 2004. "Drama Theory as the Behavioural Rationale in Agent-Based Models." In *Analysing Conflict and its Resolution*, 87–90. Oxford, England: The Institute of Mathematics and its Applications.

Bryant, J. W. 2015. *Acting Strategically Using Drama Theory*. New York: Taylor & Francis.

Chisholm, Donald. 1989. *Coordination without Hierarchy*. Berkeley, CA: University of California Press.

Clausewitz, Carl von. 1832. *On War*. Berlin, Germany: Ferdinand Dummler.

Costanza, Robert, and Matthias Ruth. 1998. "Using Dynamic Modeling to Scope Environmental Problems and Build Consensus." *Environmental Management* 22 (2): 183–95.

Danielson, P. 2002. "Competition among Co-Operators: Altruism and Reciprocity." *Proceedings of the National Academy of Sciences* 99 (Suppl. 3): 7237–42.

Fang, Liping, Keith Hipel, and D. Kilgour. 1993. *Interactive Decision Making: The Graph Model for Conflict Resolution*. New York: John Wiley & Sons.

Flint, Colin, Paul Diehl, Jürgen Scheffran, John Vasquez, and Sang-hyun Chi. 2009. "Conceptualizing ConflictSpace: Toward a Geography of Relational Power and Embeddedness in the Analysis of Interstate Conflict." *Annals of the Association of American Geographers* 99 (5): 827–35.

Flyvbjerg, Bent. 2006. "Five Misunderstandings about Case-Study Research." *Qualitative Inquiry* 12 (2): 219–45.

Galtier, Franck, François Bousquet, Martine Antona, and Pierre Bommel. 2012. "Markets as Communication Systems." *Journal of Evolutionary Economics* 22 (1): 161–201.

Gray, C. 1971. "The Arms Race Phenomenon." *World Politics* 24: 39–79.

Harlow, Robert L. 1974. "Conflict Reduction in Environmental Policy." *Journal of Conflict Resolution* 18 (3): 536.

Hirshleifer, Jack. 1988. "The Analytics of Continuing Conflict." *Synthese* 76: 201–33.

Hirshleifer, Jack. 1989. *The Dimensions of Power as Illustrated in a Steady-State Model of Conflict (RAND Note N-2889-PCT)*. Santa Monica, CA: RAND Corporation.

Hirshleifer, Jack. 1991. "The Paradox of Power." *Economics and Politics* 3: 177–200.

Hirshleifer, Jack. 2000. "The Macrotechnology of Conflict." *Journal of Conflict Resolution* 44: 772–91.

Homer-Dixon, Thomas. 2001. *Environment, Scarcity, and Violence*. Princeton, NJ: Princeton University Press.

Howard, N. 1971. *Paradoxes of Rationality*. Cambridge, MA: MIT Press.

Huberman, B. A., and N. S. Glance. 1993. "Evolutionary Games and Computer Simulations." *Proceedings of the National Academy of Sciences* 90 (16): 7716–18.

Janssen, Marco A., and Elinor Ostrom. 2006. "Empirically Based, Agent-Based Models." *Ecology and Society* 11 (2): 37.

Lanchester, Frederick W. 1916. *Aircraft in Warfare: The Dawn of the Fourth Arm*. London, England: Constable and Company, Ltd.

Leitner, H., E. Sheppard, and K. M. Sziarto. 2008. "The Spatialities of Contentious Politics." *Transactions of the Institute of British Geographers* 33 (2): 157–72.

Lotka, Alfred J. 1925. *Elements of Physical Biology*. Baltimore, MD: Williams and Wilkins Company.

Malthus, Thomas R. 1798. *An Essay on the Principle of Population*. New York: Penguin.

Maoz, Z. 2010. *Networks of Nations: The Evolution, Structure, and Impact of International Networks, 1816–2001*. New York: Cambridge University Press.

Marshall, S. L. A. 1947. *Men Against Fire*. New York: William Morrow Publishing Co.

Milgrom, Paul, Douglass North, and Barry Weingast. 1990. "The Role of Institutions in the Revival of Trade." *Economics and Politics* 2: 1–23.

Moll, Kendall D., and Gregory M. Luebbert. 1980. "Arms Race and Military Expenditure Models: A Review." *Journal of Conflict Resolution* 24 (1): 153–85.

Myers, Norman. 1993. *Ultimate Security: The Environmental Basis of Political Stability* (1st Edition). New York: W.W. Norton.

Myerson, Roger B. 1997. *Game Theory: Analysis of Conflict*. Cambridge, MA: Harvard University Press.

Neumann, J. von. 1928. "Zur Theorie Der Gesellschaftsspiele." *Mathematische Annalen* 100 (1): 295–320.

O'Loughlin, J. 2000. "Geography as Space and Geography as Place: The Divide between Political Science and Political Geography Continues." *Geopolitics* 5: 126–37.

Rapoport, Anatol, and Albert M. Chammah. 1965. *Prisoner's Dilemma*. Ann Arbor, MI: University of Michigan Press.

Reuveny, Rafael. 2007. "Climate Change-Induced Migration and Violent Conflict." *Political Geography* 26 (6): 656–73.

Reuveny, Rafael, and John W. Maxwell. 2001. "Conflict and Renewable Resources." *Journal of Conflict Resolution* 45 (6): 719–42.

Reuveny, Rafael, John W. Maxwell, and Jefferson Davis. 2011. "On Conflict over Natural Resources." *Ecological Economics* 70 (4): 698–712.

Richardson, Lewis Fry. 1960. *Arms and Insecurity: A Mathematical Study of the Causes and Origins of War*. Ann Arbor, MI: University of Michigan Press.

Sandler, Todd. 2000. "Economic Analysis of Conflict." *Journal of Conflict Resolution* 44 (6): 723–29.

Saperstein, Alvin M. 2008. "Mathematical Modeling of the Interaction between Terrorism and Counter-Terrorism and Its Policy Implications." *Complexity* 14 (1): 45–49.

Scheffran, Jürgen, Michael Brzoska, Hans Günter Brauch, P. Michael Link, and Janpeter Schilling, eds. 2012. *Climate Change, Human Security and Violent Conflict: Challenges for Societal Stability*. Berlin, Germany: Springer.

Scheffran, Jürgen, Michael Brzoska, Jasmin Kominek, P. Michael Link, and Janpeter Schilling. 2012. "Climate Change and Violent Conflict." *Science* 336 (6083): 869–71.

Scheffran, Jürgen, P. Michael Link, and Janpeter Schilling. 2012. "Theories and Models of Climate-Security Interaction: Framework and Application to a Climate Hot Spot in North Africa." In *Climate Change, Human Security and Violent Conflict: Challenges for Societal Stability*, edited by Jürgen Scheffran, Michael Brzoska, Hans Günter Brauch, P. Michael Link, and Janpeter Schilling, 91–132. Berlin, Germany: Springer Verlag.

Schlüter, Maja, and Claudia Pahl-Wostl. 2007. "Mechanisms of Resilience in Common-Pool Resource Management Systems: An Agent-Based Model of Water Use in a River Basin." *Ecology and Society* 12 (2): 4.

Scott, John P. 2000. *Social Network Analysis: A Handbook* (2nd Edition). Thousand Oaks, CA: Sage Publications.

Sheppard, E. 2002. "The Spaces and Times of Globalization: Place, Scale, Networks and Positionality." *Economic Geography* 78: 307–30.

Simon, H. A. 1956. "Rational Choice and the Structure of the Environment." *Psychological Review* 63 (2): 129–38.

Simon, H. A. 1957. *Models of Man: Social and Rational*. New York: Wiley.

Stoorvogel, J. J., J. Bouma, and R. A. Orlich. 2004. "Participatory Research for Systems Analysis: Prototyping for a Costa Rican Banana Plantation." *Agronomy Journal* 96: 323–36.

Suzuki, Y., and Y. Iwasa. 2009. "Conflict between Groups of Players in Coupled Socioeconomic and Ecological Dynamics." *Ecological Economics* 68: 1106–15.

Tenenbaum, Morris, and Harry Pollard. 1985. *Ordinary Differential Equations: An Elementary Textbook for Students of Mathematics, Engineering, and the Sciences*. Mineola: Courier Dover Publications.

Trappl, Robert. 2005. *Programming for Peace: Computer-Aided Methods for International Conflict Resolution and Prevention*. New York: Springer Science & Business Media.

Vega-Redondo, Fernando. 1996. *Evolution, Games and Economic Behavior*. Oxford, England: Oxford University Press.

Videira, Nuno, Paula Antunes, Rui Santos, and Rita Lopes. 2010. "A Participatory Modelling Approach to Support Integrated Sustainability Assessment Processes." *Systems Research and Behavioral Science* 27 (4): 446–60.

Volterra, Vito, and Marcel Brelot. 1931. *Leçons Sur La Théorie Mathématique de La Lutte Pour La Vie*. Paris, France: Gauthier-Villars et cie.

Von Neumann, J., and O. Morgenstern. 1953. *Theory of Games and Economic Behavior*. Science: Economics. Princeton, NJ: Princeton University Press.

Weidmann, Nils B., and Michael D. Ward. 2010. "Predicting Conflict in Space and Time." *Journal of Conflict Resolution* 54 (6): 883–901.

Weiss, H. K. 1983. "Requirements for the Theory of Combat." In *Mathematics of Conflict*, edited by Martin Shubik, 73–88. Amsterdam, Netherlands: Elsevier Science Publishers.

Welsch, H. 2008. "Resource Abundance and Internal Armed Conflict: Types of Natural Resources and the Incidence of 'New Wars.'" *Ecological Economics* 67: 503–13.

4

Participatory Modeling and Conflict Resolution

I have never met a man so ignorant that I couldn't learn something from him.
>—*Galileo Galilei (from Durant and Durant 1961, p. 605)*

Being aware of trade-offs is the essence of informed decisions.
>—*Thomas Buchholz, Volk, and Luzadis (2007, p. 6091)*

Introduction

What is a *"stakeholder?"* Stakeholders—a term that we use interchangeably with "actors," "parties," and "agents" throughout this book—are typically defined as those entities who are affected by a decision or event. The exact relationship between stakeholder and a decision or event can range widely and can include those who are directly affected positively or negatively (primary stakeholders), those who are indirectly affected (secondary stakeholders; "intermediaries"), and those that have influence within entities considered to be primary or secondary stakeholders (Mitchell et al. 1997). *Stakeholder analysis* is the process of assessing the extent of the impact of a decision or event and determining, ultimately, who is a stakeholder.

Perhaps the first step in using computer-based modeling for conflict resolution purposes is ensuring that stakeholders, including conflicting agents, government agencies, and subject matter experts, are comfortable with the modeling approach; this is what Anthony Starfield (1997) calls a "management-oriented" modeling environment. Unfortunately, many modeling efforts have been directly responsible for breakdowns of consensus processes among stakeholders.

The classic situation involves outside consultants developing large, "black-box" models, which stakeholders have had no hand in building or informing. Models created in this situation commonly make inappropriate assumptions or articulate relationships that may not even come close to realistically representing the system under conflict. These black-box models, typically created by "top-down," expert-led efforts, can also produce outputs that have little relevance to the concerns of stakeholders undergoing conflicts (Funtowicz and Ravetz 1994; Funtowicz, Ravetz, and Connor 2004). Moreover, studies are increasingly demonstrating that existing, "monodisciplinary" models can be inadequate for predictive or educational use in complex, dynamic, and "ill-structured" conflict systems (Giordano et al. 2007; Prell et al. 2007).

Voinov and Brown Gaddis (2008, p. 198) refer to this as science occurring "outside of the policy process." Marjan van den Belt et al. (2006) offer a rich discussion about model credibility, noting that models, similar to other scientific techniques, derive legitimacy based on their ability to reproduce measured observations (technical legitimacy) as well as the degree to which the community accepts them. Contrary to building legitimacy, the "black-box" style of modeling intervention instead breeds suspicion among stakeholders, creates distance between the modelers and the stakeholders (or governments) creating the model, and engenders fear that models may be abused and used against the interests of certain stakeholders. As a result, models are frequently either rejected or simply ignored, particularly when their results suggest unpopular decisions.

In this chapter, we will explore concepts around *"participatory modeling,"* which Krystyna Stave (2010) defines broadly as "...an approach for including a broad group of stakeholders in the process of formal decision analysis." Voinov and Brown Gaddis (2008, p. 197) define the practice similarly:

> Participatory modeling is the process of incorporating stakeholders, often including the public, and decision makers into an otherwise purely analytic modeling process to support decisions involving complex natural resources questions.

"Participatory modeling" is perhaps the most generic term used to describe stakeholder-involved or led modeling; it is widely used throughout the literature and is therefore a term that we will use throughout this text. Participatory modeling is also variously called "group model building" (GMB; Vennix 1999), "mediated" modeling (van den Belt 2004), "computer-aided dispute resolution" (Stephenson et al. 2007), "shared vision planning" (Palmer 1999; Lorie 2009), "cooperative modeling" (Cockerill, Passell, and Tidwell 2006; Cockerill et al. 2007), or "companion" modeling (Gurung, Bousquet, and Trébuil 2006).

The word *"collaborative"* can be a loaded term, in that it connotes a cooperative stance. Therefore, we use the term *collaborative* to mean any conflict resolution effort that "... involves multiple actors intentionally working together toward a collectively determined end (Shmueli, Kaufman, and Ozawa 2008, p. 360)."

Collectively, these techniques have a relatively long history; the use of models in collaborative decision-making processes dates back to the early 1960s (see Rouwette, Vennix, and Mullekom 2002; Tidwell and Van Den Brink 2008). While many of these techniques vary substantially in how they approach modeling, stakeholder involvement, and knowledge solicitation, they are united by their primary focus on developing models through the eyes of the very stakeholders affected by conflict. These stakeholders jointly work toward conflict solutions that are grounded in shared languages, knowledge, and conceptual understanding of both the conflict system and potential solutions (Scheffran 2006a).

In this chapter, we begin by discussing participation and decision making, generally, which we follow with explorations of participatory modeling, social learning, the practical mechanics of participatory modeling, and descriptions of modeling techniques. We end this chapter with a discussion of complexity and participatory modeling. We should note that, while we do not directly incorporate the input or involvement of stakeholders in the models we present in Part III of this book, in those chapters we instead aim to demonstrate how stakeholders could adapt each model for use in real environmental disputes. We hope that, drawing on our detailed reviews of best practices and advice from the participatory modeling literature, readers will use and build on these techniques for your own concrete dispute resolution efforts.

Participation and Decision Making

What's the difference between *participation* and *inclusion*? Quick and Feldman (2011, p. 272) argue that they are independent dimensions of engaging the public, whereby *inclusion* means, "efforts to increase public input oriented primarily to the content of programs", while *participation* references efforts to "continuously [create] a community involved in co-producing processes, policies, and programs for defining and addressing public issues".

The literature on conflict resolution and modeling has reached a fairly decisive consensus around the idea that stakeholder-led decisions are commonly implemented with less conflict and more success than top-down decisions made by experts or authorities (Voinov and Bousquet 2010). Participatory approaches to gathering data and making decisions allow us to meet multiple goals, including obtaining better data, and effectively engaging people in the decision process (Buchholz, Volk, and Luzadis 2007).

Today, participatory decision making is a hallmark of environmental assessment (National Research Council 2008) and local planning processes (Shmueli, Kaufman, and Ozawa 2008). After the advent of American open records and meeting laws in the 1970s, even regulatory decision making has been made more transparent and publicly accessible (e.g., Foerstel 1999). We should note that public involvement rarely automatically lends support and legitimacy to policies.

However, how much participation should we strive to have during dispute modeling or resolution efforts? We can consider a simple continuum concerning the relative level of participation at one end lie "fully" participatory processes, in which stakeholders participate in structuring the conflict problem, describing the underlying system leading to problems, creating operational models of the system, and using those models to strategize future interventions that help to alleviate conflicts (Stave 2010). At the other end of the continuum are "less" participatory approaches, perhaps the near extreme of which (before "no participation") occurs when models are simply used to aid stakeholders' understanding of previously selected management decisions (Stave 2010). Alternatively, Bots and van Daalen (2008) categorize four types of approaches that create different outcomes in terms of social learning (discussed later on in this chapter), including interventions where (1) there is no participation, (2) individual stakeholders are involved separately and independently, (3) stakeholders are involved as a homogenous group that holds similar interests and conflict perceptions (e.g., a group of environmental activists), and (4) stakeholders are involved and hold conflicting interests and divergent perceptions of the conflict.

Whatever the precise extent of participation, participatory approaches to conflict resolution are grounded on the idea that those who live and work in a conflict system are well informed about its processes, and have observed aspects of the system that would not necessarily be captured through expert analysis only (Brown Gaddis, Vladich, and Voinov 2007). The idea that stakeholders are "system experts" means they are uniquely able to provide local knowledge detailing the history and inner workings both of the natural and socio-economic systems. This insider knowledge includes not only partially obscured facts about the system but also information about local values, including ideas about what should be developed, protected, honored, and sustained (Forester 1999; Antunes, Santos, and Videira 2006).

Since the 1960s, participatory modeling has emerged as a powerful tool for empowering stakeholders with improved knowledge about conflict systems and how they change over time. Better conflict understanding allows stakeholders to improve their own decision-making abilities (Korfmacher 2001), democratically engage in collaborative learning with other stakeholders (Laird 1993; Daniels and Walker 2001; Voinov and Brown Gaddis 2008), and build coalitions that improve trust and drive stakeholder buy-in to complex decisions (Prell et al. 2007). This mode of decision making can be thought of as a "bottom-up" approach (Prell et al. 2007), whereby stakeholders play a role in the decision-making process. This can work well in democratic societies, where unpopular decisions that come from the top-down direction (i.e., government institutions, elites) are difficult to implement.

The Goals of Participatory Modeling

Participatory modeling has several distinct aims. First, it aims to enrich stakeholder understanding of environmental problems and conflicts, while exploring difficult trade-offs and ways to overcome the obstacles that challenge the management of the conflict into the future (Cockerill, Passell, and Tidwell 2006; Andersson et al. 2008). A number of studies have shown that participatory modeling can increase understanding of conflict systems and lead to shared knowledge about the dynamics that result from different decisions (Rouwette, Vennix, and Mullekom 2002). For example, Stave's (2003) analysis of water management issues in Las Vegas, Nevada, lent model users enormous insight into water management practices and the complex effects of simple management decisions.

In this sense, participatory modeling is a useful tool for bridging the gap between science and policy, while supporting collaborative processes that engage nonscientists, the public, and conflict stakeholders (van den Belt et al. 2006; Stave 2010). Furthermore, Costanza and Ruth (1998, p. 185) argue that "…[w]e need to see the modeling process as one that involves not only the technical aspects, but also the sociological aspects involved with… build[ing] consensus about the way the system works and which management options are most effective." It is now evident that this consensus needs to cross the gulf separating individual academic disciplines, and the canyon that cuts off the public from science and policy communities. A hope of participatory modeling, generally, is that appropriately designed and implementing modeling interventions can help to bridge these divides.

Data collected through extensive participatory modeling case studies have demonstrated increased consensus about how conflict problems are understood as well as about the opportunities that are available for stakeholders to jointly resolve conflicts and develop innovative policy solutions (Rouwette, Vennix, and Mullekom 2002; Cockerill, Passell, and Tidwell 2006). However, it is important to note that Rouwette, Vennix, and Mullekom (2002) point to research design problems in identifying unsuccessful modeling efforts, making it difficult to make rigorous assertions about the widespread success and usage of participatory modeling. This is still a problem today.

A second aim of participatory modeling is to enable stakeholders to use models to more effectively contribute to the political processes that select future courses of action and resolve conflicts. Today, the value of collaboration between stakeholders and policy makers, as they address important conflict issues, has been well documented in a variety of case studies (e.g., Selin, Schuett, and Carr 2000; Wondolleck and Yaffee 2000; Connick and Innes 2003). Table 4.1 depicts the wide range of modes of public participation seen in

TABLE 4.1

Spectrum of Public Participation and Involvement Levels

Level of Public Interaction	Decision Makers (DM) and Stakeholder Agreements
Inform	DMs keep stakeholders informed.
Consult	DMs inform, listen to, and acknowledge concerns of stakeholders. DMs provide feedback about the manner by which public and stakeholder input influenced their decisions.
Involve	DMs work with stakeholders to ensure that concerns and issues are directly reflected in alternatives developed. DMs provide stakeholders feedback about how public input influenced their decisions.
Collaborate	DMs look to stakeholders for advice and innovation in creating solutions. DMs incorporate advice and recommendations into decisions to the maximum extent possible.
Empowerment and self-determination	DMs implement stakeholder-led decisions.

Adapted from Videira et al. (2003).

participatory management and modeling and chronicled by the International Association of Public Participation.

Third, by involving affected groups, the participatory modeling process aims to increase stakeholder confidence in the model as an accepted tool for conflict resolution as well as stakeholder ownership of the model results and conclusions. As part of this, the modeling process removes exclusive responsibility for conflict resolution from government hands, creating shared ownership with the business community, nongovernmental organizations, and ordinary citizens (Tidwell and Van Den Brink 2008).

Fourth, many scholars, such as van Eeten, Loucks, and Roe (2002), view participatory modeling as a method for contributing useful information to the political debates surrounding conflict, rather than searching for "optimal" environmental policies, which is often the goal of black-box, top-down modeling efforts. In cases where models are fully built with stakeholder involvement, participatory modeling processes can require modelers and stakeholders with varying backgrounds to learn how to work together, thereby ensuring early on that models can be meaningfully and usefully integrated (Prell et al. 2007). This process of shared construction can become the organizing framework of an entire subsequent decision-making process.

In the 1980s, analysts began to realize that much of the understanding of conflict problems occurs during the actual process of constructing conflict models (Cabrera, Breuer, and Hildebrand 2008). In fact, many participatory modeling practitioners do not view *models* as the primary outcome of their work, but rather as the side effect of participatory modeling *processes* (Castella, Trung, and Boissau 2005). This is because participation enriches models by including subjective sources of knowledge in addition to the objective knowledge derived from theories and empirical studies (Geurtz and Joldersma 2001). For example, Gurung, Bousquet, and Trébuil's (2006) companion modeling intervention (discussed later in this chapter) involved stakeholder role-playing games, which led to the successful creation of the Watershed Management Committee in Bhutan's Lingmuteychu watershed.

Voinov and Bousquet (2010) argue that two additional factors are important goals of participatory modeling. First, participatory modeling is increasingly understood as a way of facilitating social learning among policy makers, experts, and stakeholders (Andersson et al. 2008; Muro and Jeffrey 2012). For example, some scholars have contended that group-led modeling efforts are optimally suited as methods for structuring problems, while understanding and learning about systems (as opposed to actually providing solutions; Buchholz, Volk, and Luzadis 2007; Andersson et al. 2008; Reed et al. 2010). Second, participatory modeling is useful for understanding the impacts of suggested solutions to problems, including the results of policies, regulations, management regimes, and other decisions. We will start by defining and discussing social learning.

Social Learning and Participatory Modeling

Scholars of the sociology of science have argued that because social processes are essential in the production of scientific information, citizens must be involved (Tesh 1999; Korfmacher 2001). For example, Korfmacher (2001) has argued that watershed management is fundamentally social in nature (see Rhoads et al. 1999), and as a result, the field has been one of the pioneering areas for stakeholder participation in environmental modeling efforts.

Voinov and Bousquet (2010) link participatory modeling to a broader educational approach known as *collaborative learning* (also interchangeably called *social learning*), which developed in the 1960s as a technique for enhancing group-based problem solving (Mason 1970; Slavin 1983). Collaborative learning is rooted in the understanding that learning is a naturally occurring social activity, which necessitates communication, aids in problems solving (Cardenas, Stranlund, and Willis 2000), and can intensify when stakeholders with different perceptions are brought together in engaging situations (Mostert et al. 2007).

Reed et al. (2010, p. 1) define collaborative learning processes as those, where (1) individuals involved demonstrate changes in their understanding of a problem or topic, (2) those changes extend beyond the individual and become "situated within wider social units or communities of practice," and (3) those changes occur "through social interactions and processes between actors within a social network." Significant work has demonstrated the important role of collaborative learning in organizational settings, where knowledge is circulated among, and within, groups (Nonaka 1994), enabling members to create a shared understanding of environmental problems, while developing and jointly pursuing new solutions.

Efforts of stakeholders to jointly pursue conflict resolution is enabled by strong *"social capital,"* which Stave (2010) describes as "the social bonds and norms that enable and regulate the interactions of people in communities." Trust and social connections, particularly those enabled through participatory modeling, help to construct social capital, building foundations for long-term engagement in conflict resolution processes.

Social psychology theories around collaborative learning were first developed by noted psychologist Albert Bandura (1977, 1986), who viewed collaborative learning as the effect on individuals that come from observing others and their social interactions within a group. This view emerged from Bandura's (1977) demonstrations that many human behaviors can be explained by observing and modeling the behaviors, attitudes, and emotional reactions between individuals. These reactions are continuously and reciprocally influenced by both their environment and the behavioral and cognitive styles of other individuals in interpersonal settings (Muro and Jeffrey 2012); that is, the learner changes both the learning environment and others in it, which then feedback to change the learner (Pahl-Wostl 2006).

Furthermore, developing new relationships (and stronger existing relationships) creates relational changes that transform interpersonal perceptions and perceived placements within groups, particularly groups engaged in conflict. The hope of collaborative learning is that individuals transform the way that they see problems into collectively held views, shared understandings of conflict systems, and eventually, collective action, by stepping through varying stages of deliberation, learning, and reflection (Schusler, Decker, and Pfeffer 2003; Pahl-Wostl 2006; Mostert et al. 2007).

Viewing this topic from a sociological angle, Barbara Gray (2003, 2004) discusses conflicts where failures to develop communication norms and jointly viewed facets of conflicts were major factors in the failure of conflict resolution efforts. Gray's work demonstrates the importance of *"frames"* (referring to risk attitudes, power relationships and roles, management and conflict styles) in understanding how stakeholders understand problems and engage in successful collaboration. Frames are often the product of social interactions, norms and cultures, and collective frames support the formation of group identities; for example, environmental groups form coalitions and proclaim, "we're in this together to fight the oil companies!" Gray's work suggests that conflict resolution must include processes of collaborative fact finding and problem understanding to enhance individual conflict reframing (Pahl-Wostl 2006). As part of this, however, she also recognizes the need to reframe perceived collective group identities, including perceptions of sinister coalitions formed by agents on the "other side" of a dispute.

Collaborative Learning and Participatory Processes/Modeling

Muro and Jeffrey (2012) explored the connection between participatory processes and collaborative learning, finding that the extent to which stakeholder platforms promote collaborative learning is shaped by organizational arrangements and time provided for the engagement process. Claudia Pahl-Wostl (2006) has observed that collaborative learning occurs at (1) short-to-medium timescales where learning processes occur between actors and at (2) medium-to-long timescales when changes in governance structure (e.g., changes to institutional settings, norms, values, cooperative agreements, or laws) are sought as means of improving environmental conditions. Collaborative learning efforts do not need to yield complete stakeholder consensus; rather, they act as a starting point for understanding and constructively managing differing perspectives of conflicts (Pahl-Wostl 2006, p. 3):

> The notion of social learning implies that the social capacity of a group is enhanced during the interaction process. Hence, a process in which the recognition of different perspectives leads to even larger conflict and polarization would be called a failure from the perspective of social learning. Social learning is an important factor needed to recognize why there are different perspectives, and it needs to be considered to constructively deal with these differences.

These ideas imply that modeling is an important component of social learning; the same actors who are supposed to eventually use models for planning, decision making, and conflict resolution can be instrumental in contributing to the modeling process that aims to represent their behavior or system. This assumes that the modeling process is just as important for gaining legitimacy and aiding decision support as the validity of both the assumptions and data driving the model, and model simulations that are finally produced.

System dynamicists have long argued that the best value from computer modeling does not arise when models help to confirm what we already know, but when models actually change the way we see the world (i.e., our mental models; Forrester 1987). For example, van Eeten, Loucks, and Roe (2002) propose the notion that a chief manifestation of learning in complex systems arises through "surprise" and discovery (Demchak 1991; Stave 2010), which are often the result of game playing during conflict modeling exercises (e.g., D'Aquino et al. 2002; Becu et al. 2008). Thomas Kuhn (1962) argued that surprises occur when we learn things that we do not expect to learn; they are unexplained inconsistencies in the way systems operate that are not explained by our mental models (Stave 2010).

One way to articulate the role of participatory modeling in the social learning context is to enmesh it within the framework developed by Muro and Jeffrey (2012), who create a compound model of social learning drawn from their review of the literature in this area (Figure 4.1). Participatory modeling can act as a conduit for various facets of stakeholder communication and interaction that enables the social learning process.

The Need for More Evidence

It is important to point out that while scholars and practitioners have called for increased collaborative learning throughout environmental management and conflict resolution efforts, there is a huge hole in the research in this area; little empirical evidence substantiates the link between participatory modeling and collaborative learning outcomes. In fact, Reed et al. (2010) argue that although "collaborative learning" has become a widespread strategy for environmental decision making and policy, the concept is commonly confused with other educational processes that induce pro-environmental behavior or enhance public participation. Instead, the authors rightly contend that more distinction needs to be made between learning at the individual level and learning on a wider social scale. That is, "evidence" of social learning must demonstrate that a change in individual understanding

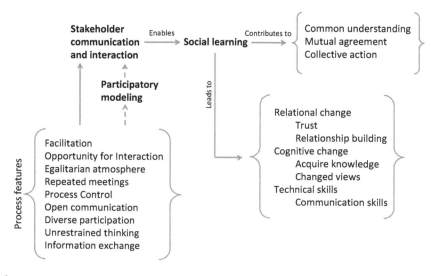

FIGURE 4.1

A compound model of social learning drawn from the literature (with participatory modeling processes embedded). Adapted from Muro and Jeffrey (2012).

has taken place, and that, through social processes and interactions between stakeholders, these individual changes become embedded within the broader stakeholder community.

Lessons Learned for Conducting Participatory Modeling Interventions

The vast majority of the literature on participatory modeling is composed of case studies that offer lessons for future modeling efforts based on their experiences. Lessons frequently relate to (1) the cognitive styles used for stakeholder engagement; (2) the manner of introducing and implementing modeling tools; (3) the process of model development, calibration, and scenario testing; and (4) the way that stakeholders apply the results of the modeling exercise to future education, decision making, and negotiation activities.

What is model *"usefulness?"* Here, it can mean that the model serves the purpose for which it was developed, (1) addressing the problem at the right scale and scope (correct model breadth and depth) and (2) functioning in a "valid" way by representing the system and its behavior correctly (Winz, Brierley, and Trowsdale 2009). Alternatively, Rouwette et al. (2011) consider usefulness as the "actionable insights" gleaned from the modeling process.

Antunes et al. (2009), Reed (2008), and the US National Research Council (2008) have identified best practice features in participation and deliberation. Important lessons can be derived from this research, including

1. Stakeholders need to be able to trust modelers and the modeling process.
2. Modeling processes need to acknowledge and manage historic stakeholder disagreements.
3. Modelers must remain flexible and respond to needs of the group (van den Belt 2004).
4. When models are seen to be practically *useful*, they are accepted and deemed to be successful.
5. The modeling process needs to be transparent and objective, and modelers must remain value-neutral.
6. Stakeholders need to remain interested in the process.
7. The modeling process must be professionally facilitated by an independent, third-party facilitator. This is a stance strongly advocated for many studies, as it is a crucial technique for enhancing collective learning processes (Castella, Trung, and Boissau 2005) and for eliciting numerous and diverging perceptions and interests of stakeholders. Andersen and Richardson (1997) recommend against lectures (by facilitators or modeling leaders) that deliver information to stakeholders lasting more than a couple minutes.
8. Knowledge solicitation and stakeholder understanding takes time. However, excessive demands for time or financial resources can make the modeling process inequitable.

9. Communication is key—including communication during model construction, rapid feedback during model testing, and interactive engagement with stakeholders while reporting model results (Repenning 2003; Prell et al. 2007).

10. Stakeholders must buy-in to the modeling process and feel that they are partial owners of the final model product (Sivakumar 2006; Cabrera, Breuer, and Hildebrand 2008).

11. Stakeholders must establish clear goals for the modeling process very early.

12. Group dynamics problems can derail the entire process (Moxey and White 1998; Nicolson et al. 2002; Cockerill et al. 2007).

These lessons demonstrate that the success of stakeholder participation in modeling processes is largely determined by the methods employed to engage and elicit local knowledge and other information from stakeholders, the ability of modelers to gain the trust of stakeholders and maintain transparency and objectivity in the creation of the model, the social relations among the stakeholders and with the modelers, the inclusion of diverse stakeholders into a representative group of proper size, the abilities of modelers and stakeholders to communicate and exchange information and knowledge, and the time available for the process to develop (e.g., Roberts 2004; Voinov and Brown Gaddis 2008).

Modeler and Methodological Transparency

Stakeholders' use of models for conflict resolution and environmental damage mitigation has, to date, largely depended on modelers that are willing to engage with stakeholders throughout the entire modeling process and provide tools that they need, including multiple types of models and modeling frameworks (Voinov and Bousquet 2010). This dedication to scientific openness and methodological transparency also requires that modelers be fully open to the participation of a wide range of stakeholders, including unsolicited (i.e., uninvited or unexpected) stakeholders (Metcalf et al. 2010).

True "bottom-up" approaches to model development also mean that stakeholders are allowed, from the beginning, to identify and prioritize conflict problems. In contrast to the typical, and usually unconscious, process of matching modeling techniques to the available expertise on the modeling team, this requires early stakeholder involvement for problem formulation to ensure that modeling aids in addressing useful conflict problems while producing relevant outputs (Prell et al. 2007). As an example of this dynamic, Prell et al. (2007) and Cockerill et al. (2007) describe situations where certain modeling techniques have become so dominant in hydrology and water resource management that the choice of a "pre-designed" model can often determine the problem to be solved, as opposed to the problem acting as a catalyst for selecting and using a modeling technique that addresses stakeholder needs while yielding useful insights into a system and its problems. This is a great example of the cognitive bias associated with overreliance on familiar tools, which is known as Maslow's "law of the hammer" (Maslow 1966): "If all you have is a hammer, everything soon looks like a nail."

In light of this, Bots and van Daalen (2008) theorize six "reasons to model" (Figure 4.2) that they hope will force researchers and stakeholders to determine the purpose of the modeling exercise, the modeling method and type to be used, and the approach to participation. They point to several additional questions that must be answered in selecting modeling tools, including, what are the objectives of the participatory process? What could modeling actually contribute to conflict resolution or stakeholder participation? Who is the client/process sponsor and what is the current phase of the decision process?

Stakeholder Selection

One critical part of the participatory modeling process involves the identification of a proper and helpful group of stakeholders to be included in the participatory modeling processes (Videira et al. 2010). However, while the importance of public participation in environmental decision making is almost universal (National Research Council 2008), there is less agreement about how and when to involve the public in conflict modeling. Korfmacher (2001) and Tidwell and Van Den Brink (2008) even explore the arguments against public participation altogether, identifying problems such as:

1. the public's lack of environmental expertise;
2. the high costs of participatory processes;
3. the potential for bias, including insufficient influence of subject matter expertise, underrepresentation of disenfranchised groups, and disproportionately large influence exercised by powerful groups (Coglianese 1999);
4. unrealistic expectations of success;
5. potential delegitimization of the scientific validity of modeling;
6. misrepresentation of group consensus as universal among stakeholders (even those not involved in modeling);
7. the overlegitimization of modeling decisions made on political grounds (combining technical and political rationales for modeling decisions and assumptions).

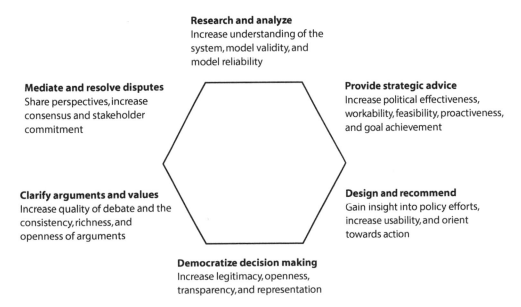

Research and analyze
Increase understanding of the system, model validity, and model reliability

Mediate and resolve disputes
Share perspectives, increase consensus and stakeholder commitment

Provide strategic advice
Increase political effectiveness, workability, feasibility, proactiveness, and goal achievement

Clarify arguments and values
Increase quality of debate and the consistency, richness, and openness of arguments

Design and recommend
Gain insight into policy efforts, increase usability, and orient towards action

Democratize decision making
Increase legitimacy, openness, transparency, and representation

FIGURE 4.2
Purposes for engaging in participatory modeling exercises. Adapted from Bots and van Daalen (2008).

Moreover, Lord (1986) and Ingram and Schneider (1998) point out that disagreements over values are often mistakenly portrayed as public disputes over facts, which become the focus as experts debate scientific information.

Given the complexity and often contentious nature of environmental conflicts, successful resolution requires a transparent and inclusive process for exchanging ideas (Tidwell and Van Den Brink 2008). This means that participatory modeling must assemble a wide-ranging set of representatives from different stakeholder groups (Pahl-Wostl and Hare 2004). As Prell et al. (2007) point out, by excluding stakeholders from conflict modeling, the process risks can become irrelevant to the needs and priorities of those who would implement and live within any identified solutions. Therefore, it is important to use careful approaches for eliciting a proper set of stakeholders for participation.

In some projects, stakeholders are invited to join the modeling process based on their "stake" in the conflict system. In other cases, involvement in the modeling effort may be open to any and all members of the public (Voinov and Brown Gaddis 2008). Marjan van den Belt (2004) described another common strategy, whereby steering committees or organizations tasked with administering modeling interventions have used "snowball" strategies (stakeholders referring other stakeholders for inclusion) for contacting and iteratively expanding the list of stakeholders involved. This can be problematic as stakeholders are obviously likely to refer others in organizations that they know, trust, or agree with (e.g., environmental stakeholders referring other environmentalists), thereby biasing the makeup of stakeholders in the modeling effort (Metcalf et al. 2010).

A final, more objective technique involves an institutional analysis (drawing on interviews and document analysis) prior to the modeling exercise, which identifies all stakeholder groups of relevance for the problem under consideration. An institutional analysis can also help to identify the organizations and roles underpinning stakeholder interactions. Ideally, the selection of stakeholders should be geared toward (1) promoting a diversity of values and approaches (thereby representing the tensions produced within the conflict), (2) producing and integrating knowledge, and (3) enhancing transparency (and the perception of transparency).

Approaches to Participatory Modeling

If participatory modeling is not simply intended to scientifically evaluate conflict, but rather to identify paths toward conflict resolution, then it suggests a spectrum of roles within conflict resolution processes (Tidwell and Van Den Brink 2008). Researchers and professionals around the world have developed and applied a variety of participatory modeling techniques. Although some are quite similar, many have methodological or theoretical twists that can make them especially suited for use in different types of conflict situations. In some cases, modeling approaches act as conceptual and illustrative vehicles for conflict visualization (Vennix et al. 1990; Andersen et al. 2007), while in others, models can be used as rigorous instruments for selecting among competing future courses of action (Ford and Sterman 1998; Sterman 2000). Between these divergent uses are additional modeling roles, such as problem scoping, outreach, and education. One primary difference between these uses involves the manner by which participants interact with the model; some modeling frameworks enable users to directly interact with formal mathematical models, testing assumptions and running scenarios in an independent manner. Others only allow users to interact with slick graphical interfaces, removing access to most specific model details and inner workings.

In this section, we discuss three participatory modeling approaches, including participatory modeling and system dynamics, participatory simulation and role playing, and decision analysis and support systems. We contend that these approaches, and their constituent practices, cover at least a majority of modeling interventions to date.

Participatory Modeling and System Dynamics

Group Model Building (GMB), the first application of system dynamics techniques (discussed in depth in the next chapter) to participatory modeling, originated from research in the Netherlands and in the "Decisions Techtronics Group" at SUNY-Albany's Rockefeller College of Public Affairs and Policy (New York; e.g., Andersen and Richardson 1997; Vennix 1999; Andersen et al. 2007). Growing out of this technique, *mediated modeling* was introduced in the late 1990s and is summarized in the seminal book of the same name by Marjan van den Belt (2004). GMB and mediated modeling are united in their common use of system dynamics modeling as a framework for articulating conflict systems. Perhaps, the primary difference between these techniques is their application areas: GMB has focused on organizational and business applications, particularly where stakeholders are business managers and client groups with divergent viewpoints, especially within the same firm (Rouwette and Vennix 2006). Mediated modeling, on the other hand, has primarily been applied to fostering consensus and collaborative team learning experiences in complex natural resource and environmental conflicts (van den Belt, Deutsch, and Jansson 1998; van den Belt et al. 2006; Metcalf et al. 2010).

GMB and mediated modeling are unique in their focus on structuring stakeholder discussions to understand and delineate problems in a sufficiently precise manner (Videira et al. 2010). Both techniques typically begin stakeholder meetings in a structured manner so as create boundaries around the conflict system (commonly referred to as problem "scoping" or "abstraction"; Videira, Antunes, and Santos 2009; Videira et al. 2010) and elicit stakeholder knowledge about problems, key model variables, individual cause–effect relationships (Rouwette, Vennix, and Mullekom 2002; van den Belt 2004), and system structure underlying the conflict (Vennix et al. 1990; Ford and Sterman 1998). This information is then used to create conceptual models, typically in the form of causal maps ("causal loop diagrams" in system dynamics parlance), which are often then transformed into full-fledged, quantitative system dynamics models. This process typically involves several, independently-facilitated stakeholder meeting sessions to build the conceptual model, and sometimes many more, to build any resulting quantitative model (Butler and Adamowski 2015).

In Figure 4.3, we present two conceptual frameworks that use system dynamics in a participatory setting and summarize a variety of different modeling efforts. While these frameworks encompass iterative model construction, they differ in the explicit ways in which stakeholders provide input and socially validate models.

Participatory Simulation and Role Playing

Perhaps one of the most promising and "hands-on" types of participatory conflict modeling is called "participatory simulation," which grew out of the system dynamics group at the Massachusetts Institute of Technology from the early 1960s through the 1990s. There, participatory modeling was primarily implemented through interactive role-playing games such as *Fish Banks*, a fishery optimization game, and *The Beer Game* (Van Ackere, Larsen, and Morecroft 1993), a supply chain simulation game.

In role-playing games, modeling and stakeholder exercises are linked as participant social interactions drive and validate simulations, and shed light on stakeholders' thought processes and their conceptual models of conflict (van Eeten, Loucks, and Roe 2002; Prell et al. 2007). When participants adopt the roles of other participants in a conflict system, they often improve their understanding of other conflicting actors, enabling reflection on changes in decision-making processes needed for conflict resolution (Pahl-Wostl 2006).

These games also provide valuable data and insights for researchers studying the social and decision processes that produce conflict, including the goals, actions, and behavior

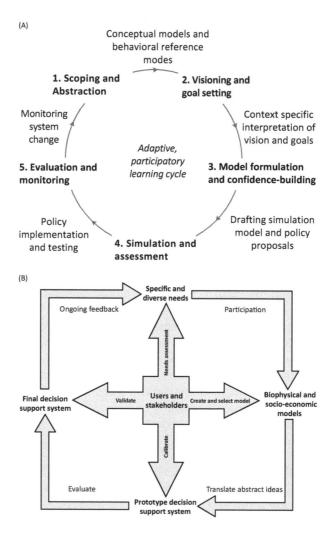

FIGURE 4.3
Alternative conceptual frameworks for participatory modeling. (A) Participatory modeling as an adaptive, social learning cycle, replete with system monitoring, process evaluation, and outcomes assessment. Adapted from Videira et al. (2010), which builds on work by Sterman (2000) and Weaver and Jordan (2008). (B) A stakeholder-centered framework that focuses more intensely on needs assessment and method selection than on process or outcomes evaluation. Modified from Cabrera, Breuer, and Hildebrand (2008) and based on a climate-related participatory modeling project.

of conflict participants. Understanding how stakeholders' conceptual models of conflict jointly function together is one of the keys to understanding the results of group decisions. Research using the *Companion Modeling* technique aims to do just this, especially when stakeholder games and simulation exercises act as experiments that learn about agent behavior and implicit decision rules (Bousquet and Le Page 2004; Barnaud, Bousquet, and Trébuil 2008; Becu et al. 2008).

Castella, Trung, and Boissau (2005) found that the role-playing approaches can overcome severe obstacles to data collection inherent in traditional ethnographic studies, such as surveys, questionnaires, or interviews, which require long-term and direct interactions between the interviewer and the stakeholder to build trust. The problem is that these interactions often only occur when the researcher shares the life of the local community for long periods. Trust issues are particularly apparent in conflicts where the researchers initiating conflict resolution or research are foreign or from a different social environment or ethnic or cultural group. For example, Castella, Trung, and Boissau (2005) point out problems in Vietnamese land use and water conflicts, where conflict researchers from the Kinh ethnic group (which dominates Vietnam's national research system) may inadvertently intimidate local villagers engaged in conflict.

Participatory simulation has more recently been articulated through agent-based modeling tools, including the ever-evolving "Logo" educational programming language. Logo was originally implemented for "constructivist" learning, where students could manipulate virtual or robotic "turtles" in a landscape (Watt 1983). Since then, it has morphed into a full-fledged modeling software language in several flavors, including Massachusetts Institute of Technology's StarLogo, and the more research-oriented NetLogo software that has been developed at Northwestern University (Wilensky and Rand 2015) and is the software that is used in the applications in Part III of this book.

While StarLogo is focused on classroom applications, NetLogo enables users to collaboratively run and experiment with models using "HubNet," a client–server feature that allows multiple users, connected through the internet, to control the behavior of individual agents within a model and to view the aggregated results on a central computer (Wilensky and Stroup 1999). However, although participatory simulation software has evolved to track and analyze all of the decisions and interactions made by each user, in most applications users have not been able to modify the models themselves, limiting the participatory nature of the simulations. In other words, in most participatory simulations, it is not possible for players to change the rules of the game.

This lack of user input has changed with a new approach, known as "companion modeling," where stakeholders are represented through a combination of agent-based modeling and role-playing games (e.g., D'Aquino et al. 2002; Becu et al. 2003; Gurung, Bousquet, and Trébuil 2006; Barnaud, Bousquet, and Trébuil 2008). Companion modeling has placed an emphasis on three primary principles, including stakeholder-led model construction, transparency in the modeling process, and adaptive modeling (i.e., models that can evolve throughout the process; Voinov and Bousquet 2010). Participants first lend their input in constructing models and then engage in role-playing games to test and analyze their behavior in a variety of situations.

Developed since the mid-1990s at CIRAD[1] a French research center that studies agricultural and development issues in developing countries, companion modeling interventions have occurred in many areas stricken by water and agricultural conflicts throughout Africa and Southeast Asia. The technique's priority objective is to raise the

---

[1] CIRAD is the Centre de coopération internationale en recherche agronomique pour le développement.

awareness of different stakeholders, commonly including farmers, scientists, and government officials, to the many other types of stakeholders in conflict systems, and their points of view. In this sense, companion modeling is linked to the concepts of organizational change and social learning (Pahl-Wostl 2006; Reed et al. 2010). If we now zoom out of participatory simulation techniques, we can explore a variety of other approaches that are inclusive of stakeholder ideas and local knowledge but do not necessarily involve model construction or simulation.

Decision Analysis and Decision Support

Like participatory simulation, *participatory decision analysis* is a fairly broad area that has evolved from a different set of background fields with different tools for involving stakeholders and capturing their understanding of problems. Buchholz, Volk, and Luzadis (2007, p. 6090) define decision analysis as the process of

> ...framing information as effectively as possible so that the decision maker will base the decision on the best possible knowledge of the outcomes....[I]ts goal is to reduce uncertainty and make the outcomes and consequences of a given decision more predictable, or at least help in the detection of possible consequences (i.e., reduce ignorance).

A class of participatory modeling frameworks function with a high level of interactivity are commonly known as *decision support systems* (Power 2002; Ackermann et al. 2010). These approaches require model-user interfaces, databases, decision-analysis tools, and integrated software environments, all of which contribute to a *level of abstraction* that masks the detailed and complex mathematical inner workings of the model behind user-friendly visualization and decision-support tools. Tidwell and Van Den Brink (2008) highlight the two scenarios where this is possible. The first involves relatively simple models with short computation times; the second includes models that have been developed previously, tested, and stored in such a way to mimic interactivity by dynamically displaying a variety of different scenarios.

Decision support systems are formalized approaches to structure the collection and evaluation of information about systems and conflicts (Buchholz, Volk, and Luzadis 2007). The concept of abstraction, or relative *"levels of abstraction,"* refers to the amount of complexity that is experienced when viewing or interacting with a system. Higher levels of abstraction mean that less detail is experienced (e.g., using voice commands to ask your smartphone for driving directions). Lower levels of abstraction mean that more detail is experienced (e.g., writing a program to calculate and display driving directions using cost-distance algorithms that you create using a programming language).

Along these lines, *negotiation support systems* have been created to help overcome cognitive limitations and move away from negotiating positions and toward the actual interests of negotiators (Fisher and Ury 1981; Eden and Ackermann 1996; Tidwell and Van Den Brink 2008). These systems include those that assemble information prior to negotiations (Starke and Rangaswamy 1999), systems that are focused on supporting group decisions during multiagent negotiations (Nandalal and Simonovic 2003), and autonomous negotiation agents that negotiate on behalf of stakeholders (Sheer, Baeck, and Wright 1989; Sierra, Faratin, and Jennings 1997).

Other forms of decision analysis include techniques like multicriteria decision analysis (MCDA) and Bayesian belief networks, which describe and rank stakeholder actions and choices (Voinov and Bousquet 2010). Buchholz, Volk, and Luzadis (2007) argue that MCDA is a tool that can help (1) identify and evaluate risks in dynamic environments, (2) integrate values and perceptions, and (3) gather large amounts of information from stakeholders, including system components and interactions. MCDA implementations begin with problem identification, definition of evaluation criteria, sensitivity analysis, and generation and evaluation of alternatives, followed by weighting and ranking of factors affecting the alternatives.

Multicriteria decision analysis (MCDA), is an area of operations research that seeks to evaluate multiple, conflicting criteria when making decisions. A *Bayesian belief network* is a type of statistical model that represents the conditional dependencies of random variables through a network (graph) structure (which we defined in Chapter 3).

Although many decision support techniques lead to single optimal solutions, MCDA can handle evolving processes that may lead to several optimal solutions (Petry 1990). MCDA can also help stakeholders evaluate decision feasibility in the context of trade-offs, including decision costs and sacrifices. For example, Bojórquez-Tapia, Sánchez-Colon, and Florez (2005) demonstrated the consensus-building use of MCDA in a controversial project: ranking two alternative sites for Mexico City's new airport based on its environmental impact. In fact, there have also been quite a few applications of MCDA to analyses of the suitability of land for different purposes (Malczewski 2006; Yu et al. 2011; Shaaban et al. 2017), although this technique has the potential to overlook many of the social and economic factors underlying land use planning (Yu et al. 2011).

Complexity and Participatory Modeling

"Complexity" is a term that is typically used to characterize a system with multiple parts, where those parts interact and create a higher-order, "emergent" behavior, which is greater than the sum of the interactions of the parts. The New England Complex Systems Institute (NECSI 2016) defines complex systems as "…a new field of science studying how parts of a system give rise to the collective behaviors of the system and how the system interacts with its environment."

In this chapter, we have gone to great lengths to describe the advantages and caveats of participatory modeling, focusing on ways of different types of modeling processes and ways of creating better models. In taking a step forward from this discussion, we need to think about how we can effectively represent the complexity of conflict systems in a manner that can satisfy many of the concerns articulated in so many previous efforts. Work by Alexey Voinov et al. (2008) offers an excellent starting point; they suggest that the origin of persistent conflicts starts with a lack of resources for dealing with complex problems. In doing so, they identify five factors that complicate environmental management and breed conflict:

1. Disputes between public and private interests create a zero-sum game of winners and losers based on any future course of action;
2. Political pressure forces rapid and aggressive changes in public policy;

3. Conflicts occur in scenarios that are irreversible, creating costly risks;

4. Uncertain technical information about ecological and sociological systems is prevalent;

5. Policy decisions can have wide-ranging effects throughout broader communities and ecosystems.

While we have shown that participatory interventions are important for addressing several of these issues, participants must also be empowered to deal with complex interactions that occur over space and time in conflict systems. Traditionally, dispute resolution professionals have attempted to simplify problems and reduce their cognitive complexity (Voinov et al. 2008). The issue is that problem definition is often complicated by conflicting stakeholder understanding of the complex operation of conflict systems. Moreover, feedback relationships between environmental and human systems complicate the direct or rapid understanding of the effects of decisions in the system. This is so important that Stave (2010, p. 2764) asserts that "[s]olving environmental problems can be as much about defining the problems to be solved as finding technical solutions, or negotiating between acceptable rather than optimal strategies."

Voinov and Bousquet (2010) point out that the introduction of social learning into environmental management processes can be directly linked to the increasing complexity of management regimes, which have become more adaptive to deal with uncertainties and nonlinearities. While stakeholder involvement alone is empowering from a social and psychological perspective, it is evident that cyberinfrastructure will become increasingly essential for empowering stakeholders that are mired in complex conflicts; clarifying and synthesizing complex conflict relationships and data have become essential to true stakeholder engagement (Voinov and Brown Gaddis 2008; Voinov et al. 2008). Fortunately, computational advances since the 1970s have consistently improved the effectiveness of participatory modeling, allowing individual stakeholders to better understand the complexity inherent in the conflicted system (Cockerill et al. 2007).

Integrated modeling systems have thus far been divided into two groups: user-friendly general-purpose modeling systems and participatory modeling systems that require expert assistance for their use. Unfortunately, while successful participatory modeling efforts typically require specialized modeling and data processing tools, they are often inaccessible for stakeholders, either due to their cost or due to the huge technical requirements (computer programming skills, computing power, and data collection; e.g., BenDor et al. 2011). For example, many role-playing games allow users to understand the complicated behavior that results from simple decisions by a great number of people. Some work has focused on using more sophisticated technology to increase the realistic nature of participatory role playing games—for example, using small, communicating computers to more quickly and completely convey feedback and emergence in complex dynamic systems (e.g., Colella 2000).

In perhaps one of the boldest attempts to democratize participatory modeling, Voinov et al. (2008) designed a modular framework that enables stakeholders to harness previously developed, "state-of-the-art" models and test policy alternatives. While we are not aware of any applications of their suggested modeling system, this view represents a shift away from the historical trade-offs—inclusion vs. realism—that have limited the effectiveness and use of participatory conflict modeling. We hope to further break down these trade-offs in our description of agent-based conflict modeling (Chapter 7) and applications of this technique (Chapters 8–10).

Summary

In this chapter, we introduced a wide range of research examining why models built to aid in conflict resolution often fail. We focused on the paramount importance of incorporating the perspectives of social actors whose decisions and actions can alter a system's trajectory (Gunderson and Holling 2002; Cockerill, Passell, and Tidwell 2006; Reed et al. 2010). We argued that conflict modeling efforts can fail when models are unable to adequately represent environmental or economic realities, or gain stakeholder trust and buy-in. We also explored social learning aspects of participatory modeling, as well as practical issues during the modeling process, including best practices and stakeholder selection. Finally, we reviewed different modeling approaches and tools and introduced concepts around complexity and participatory modeling.

The use of participation in modeling as a means of resolving disputes has extended into a variety of fields with a range of goals (Figure 4.4). Participatory modeling is now widely and favorably recognized as an important technique for engaging the public in the scientific process of developing sound descriptions of system behavior, and exploring system interactions (Pahl-Wostl 2006; Voinov and Brown Gaddis 2008). Context-dependent knowledge held by stakeholders is a fundamental prerequisite for competently handling

FIGURE 4.4
Progress during a participatory modeling workshop aimed at determining new management regimes for the Upper Mississippi River floodplain (author Todd BenDor shown sitting on right). This workshop was held at the National Mississippi River Museum and Aquarium in Dubuque, Iowa, and the results are described in Metcalf et al. (2010).

real-world complexity and creating buy-in to both the modeling process and any associated policies recommended to reduce conflict (Tidwell and Van Den Brink 2008). As a result, participatory modeling acts as a vehicle for identifying courses of action to resolve conflicts, creating stakeholder buy-in and providing a transparent and structured process for understanding and learning from conflict systems (Polhill, Gimona, and Aspinall 2011).

In the next two chapters, we will review the two major techniques that will inform our VIABLE modeling technique: system dynamics and agent-based modeling.

Questions for Consideration

1. The word *"collaborative"* can be a loaded term in that it connotes a cooperative stance, which *"communicative"* does not (see Shmueli, Kaufman, and Ozawa 2008). *What is the difference between "collaborative" and "communicative" in the context of participatory modeling or planning? Find some examples of this difference.* For help, see Laura Basco-Carrera et al. (2017) excellent discussion in the context of water resources management.

2. Substantial research has demonstrated the important role of collaborative learning in organizational settings. Collaborative learning is enabled by increased "social capital," which Krystyna Stave (2010, p. 2779) defines as "the social bonds and norms that enable and regulate the interactions of people in communities." *How does social capital get created in a participatory setting? How could we strengthen social capital? Are there other definitions of social capital different from Stave's?*

3. We briefly discussed "bottom-up" vs. "top-down" approaches to decision making. *What are the differences between these two approaches? Summarize each approach's strengths and weaknesses.* An extensive literature explores this idea, and we refer interested readers to National Research Council (2008) and Cockerill, Passell, and Tidwell (2006) for case studies and more information.

4. Laws in the 1970s made regulatory decision making more transparent and publicly accessible (e.g., Foerstel 1999). More recently, we have seen participatory directives contained within the *European Water Framework Directive* (Directive 2000/60/EC), which is now one of the most important components in European water legislation (Muro and Jeffrey 2012). *What is the relationship between government transparency and public participation? What other government transparency initiatives have you seen near where you live? How have those affected public participation?*

5. The literature argues that the participatory modeling process needs to be transparent and objective, and modelers must remain value neutral. For example, participatory system dynamics modeling sessions frequently use methods like the "nominal group technique" to generate ideas and form system boundaries (Andersen and Richardson 1997). *What are the other ways that modelers could remain neutral?*

6. In *After the Cap: Risk Assessment, Citizen Science and Disaster Recovery*, Sabrina McCormick (2012) used crowdsourcing to produce real-time assessments of health-related exposures from the 2010 Deepwater Horizon oil spill in the Gulf of Mexico. She found that crowdsourcing helped to transition from basic mapping toward a more precise, online data collection system that was both better at detecting impacts and gathering broader sets of data. *Review this (freely available,*

open access) paper and reflect on the use of citizen science and crowdsourcing as a form of public participation. What are the other avenues that you could use crowdsourcing to address environmental conflict? For more ideas on crowdsourcing and participatory "Web 2.0" techniques, see Alexey Voinov et al.'s (2016) retrospective to Voinov and Bousquet's (2010) seminal paper, *Modelling with Stakeholders.*

Additional Resources

Interested in learning more about collaboration and participatory processes? The value of collaboration between stakeholders and policy makers has been well documented in a variety of case studies. See Selin, Schuett, and Carr (2000), Wondolleck and Yaffee (2000), Connick and Innes (2003), Stave (2003, 2010), and Cockerill et al. (2007) for numerous examples and case studies.

While the importance of public participation in environmental decision making is almost universal, there is less agreement about how and when to involve the public in conflict modeling. The National Research Council (2008) and Wondolleck, Manring, and Crowfoot (1996) discuss important factors in determining who will be at the table during participatory conflict resolution efforts. Bryson (2004) and Marcuse (2005) also provide comprehensive overviews of stakeholder identification and analysis techniques.

Although Table 4.1 depicts a wide range of modes of public participation seen in participatory management and modeling (chronicled by Videira et al. 2003), there is an entire professional association, the International Association of Public Participation (IAP2; http://www.iap2.org/), devoted to this area.

Interested in social learning? Thomas Kuhn (1962) argued that surprises occur when we learn things that we do not expect to learn. Rouwette and Vennix (2006) discuss this concept in great detail, particularly with regard to the social–psychological literature on persuasion and motivation.

For more information on the organizational and communicative processes associated with social learning, we refer readers to Muro and Jeffrey (2012), who discuss these issues in detail and provide many resources for readers. For example, further studies of the cognitive and social aspects of collaborative learning have explored a wide variety of theoretical concepts that link learning to participatory modeling and planning, including *double-loop* and *triple-loop* learning and experiential learning (the process of turning direct experiences into meaning; Kolb 1984; Itin 1999). Double-loop and triple-loop learning are learning patterns where the goals, values, and assumptions about solutions become dynamic, and individuals gain insights about why a solution works (Argyris and Schön 1978). Muro and Jeffrey (2012) also discuss important caveats for social learning processes, including case-study research that connects the practice to reports of mistaken learning, intensifications of tensions or conflict, and failures to reach agreement or verifiable consensus.

A note on "participatory action research." Developed by sociologists and social psychologists, *participatory action research* is an umbrella concept that describes research that centers on the effects of a researcher's actions on the quality or performance of a participatory community (Brown Gaddis, Vladich, and Voinov 2007; Voinov and Bousquet 2010). However, action research usually involves community- and stakeholder-initiated investigation (Kemmis and McTaggart 1998) and has not historically involved modeling.

Interested in knowledge elicitation and participatory modeling? Check out http://participatorymodeling.org/, a website dedicated to creating a stronger community of modelers and researchers engaged in participatory modeling efforts.

We refer readers to Prell et al. (2007), Ford and Sterman (1998), and Vennix et al. (1990) for more information on stakeholder knowledge elicitation during the participatory modeling process. Additionally, please refer to Andersen and Richardson (1997), Luna-Reyes et al. (2006), and Ackermann et al. (2010) for resources around participatory modeling "scripts" as a means of knowledge elicitation.

Knowledge solicitation and stakeholder understanding takes time. However, excessive demands for time or financial resources can make the modeling process inequitable. Rouwette, Vennix, and Mullekom (2002), Beall and Ford (2007), and Stave (2010) survey the broad range of workshop designs, where time investment can vary from hours to months, and participant groups can vary from several to over 160.

One participatory modeling best practice that we can elaborate on is that *stakeholders must establish clear goals for the modeling process very early*. Arguing that these goal statements should be matched to outcomes assessments, Rouwette, Vennix, and Mullekom (2002) present a list of items that should be recorded before, during, and after an intervention, including premeeting interviews, sources of information, and processes of knowledge elicitation. For more on participatory modeling evaluation, see Natalie Jones et al.'s (2009) work, *Evaluating Participatory Modeling: Developing a Framework for Cross-Case Analysis*.

For more detailed advice on best practices in participatory modeling, we refer readers to work by Rebecca et al. (2018), Sterling et al. (in press), Gray et al. (2018) Brown Gaddis, Vladich, and Voinov (2007), Voinov and Brown Gaddis (2008), Prell et al. (2007), Tidwell and Van Den Brink (2008), Cockerill, Passell, and Tidwell (2006), Cockerill et al. (2007), Stave (2010), Korfmacher (2001), Beall and Ford (2007), Mendoza and Prabhu (2003), and Andersson et al. (2008). We also refer readers to Steven Gray et al.'s (2016) edited volume, *Environmental Modeling with Stakeholders: Methods, Theories and Applications*, for a great resource on all things related to environmental participatory modeling.

Interested in role-playing games? Some of the earliest role-playing games are also the most fun! Check out the two most well-known:

- *Fish Banks*, a fishery optimization game. Interested readers can play this game at http://forio.com/simulate/mit/fishbanks. More information is available at the MIT Sloan School of Management: https://mitsloan.mit.edu/LearningEdge/simulations/fishbanks/Pages/fish-banks.aspx.
- *The Beer Game* (Van Ackere, Larsen, and Morecroft 1993), a supply chain simulation game. The beer game classically demonstrates the emergence of the "bull-whip effect," a troubling pattern of increasingly erratic swings throughout the supply chain due to small changes in supply or demand (Lee, Padmanabhan, and Whang 1997).

Additionally, you can learn more about the companion modeling (ComMod) technique at the group's website (http://cormas.cirad.fr/ComMod/en/), where you will find a large number of applications as well as more information on the Cormas software package that facilitates the modeling technique and the role-playing games it often uses.

Interested in different or new techniques for participatory modeling? There have been several surveys of participatory modeling techniques to understand how each involves stakeholders and enhances the dispute resolution process (Scheffran and Stoll-Kleemann 2003; Scheffran 2006b; Lynam et al. 2007; Renger, Kolshoten, and De Vreede 2008; Voinov and Bousquet 2010). Voinov and Bousquet (2010) discuss a wide variety of qualitative and quantitative techniques, while Mendoza and Prabhu (2003) elaborate on three general types of "soft" system dynamics frameworks, including cognitive mapping, qualitative system dynamics, and "fuzzy" cognitive mapping, each of which entail varying degrees of complexity, modeling rigor, and information requirements.

- Integrated modeling systems have historically been divided into two groups: user-friendly general-purpose modeling systems (e.g., STELLA, http://www.iseesystems.com/; Costanza and Voinov 2001) and participatory modeling systems that require expert assistance for their use (e.g., Cormas, http://cormas.cirad.fr/indexeng.htm). However, this is changing; check out Steven Gray's (2013) easy-to-use fuzzy cognitive mapping tool at http://www.mentalmodeler.org/.
- Buchholz, Volk, and Luzadis (2007) argue that MCDA is a useful tool. We refer readers to Belton and Stewart (2002) and Mendoza and Martins (2006) for excellent in-depth discussions of MCDA approaches and applications to environmental problems.
- Gordon, Schirra, and Hollander (2011) describe a technique they call "immersive planning," whereby digital technologies like computer-aided design, geographic information systems (GIS), and virtual and gaming environments can be used to enhance user experiences in public participatory processes for public decision making.
- Kolagani and Ramu (2017) provide a great overview of participatory GIS and its use in resource management. Participatory GIS can be particularly useful for eliciting and communicating high-resolution spatial information from stakeholders and the public (e.g., see Jankowski et al. 2016).
- "Shared vision planning," also called *Computer-Aided Dispute Resolution* (CADRe) or referred to as "collaborative modeling for decision support," is a research program headed by the US Army Corps of Engineers that aims to integrate collaborative modeling with participatory processes to inform natural resources management decisions. CADRe was created by system dynamicists at the US Army Corps of Engineers to assist the agency's planning activities and conflict interventions, which primarily focus on domestic US water management issues (Tidwell and Van Den Brink 2008). The technique draws on techniques from planning, engineering, systems analysis, public participation, and dispute resolution, making it a particularly self-reflective modeling technique (Stephenson et al. 2007; Lorie 2009). Bourget (2011) explores numerous U.S. applications of CADRe, including water supply planning in the Potomac River basin, drought preparedness in Northern California, and interstate collaboration and water management planning on the Lower Susquehanna and Rio Grande rivers.
- Vrana et al. (2012) address situations where decision-making processes are uncertain, multidimensional, and decisions are to be based on the diverse opinions

of experts with different viewpoints. The authors introduce a technique called *MaxAgr*, which is aimed at aggregating experts' opinions to optimize agreement between expert's proposals.

References

Ackermann, Fran, David F. Andersen, Colin Eden, and George P. Richardson. 2010. "Using a Group Decision Support System to Add Value to Group Model Building." *System Dynamics Review* 26 (4): 335–46.

Andersen, David F., and George P. Richardson. 1997. "Scripts for Group Model Building." *System Dynamics Review* 13 (2): 107–29.

Andersen, David F., Jac A. M. Vennix, G. P. Richardson, and E. A. J. A. Rouwette. 2007. "Group Model Building: Problem Structuring, Policy Simulation and Decision Support." *The Journal of the Operational Research Society* 58 (5): 691–94.

Andersson, L., J. A. Olsson, B. Arheimer, and A. Jonsson. 2008. "Use of Participatory Scenario Modelling as Platforms in Stakeholder Dialogues." *Water SA* 34 (4): 3.

Antunes, P., G. Kallis, N. Videira, and R. Santos. 2009. "Participation and Evaluation for Sustainable River Basin Governance." *Ecological Economics* 68: 931–39.

Antunes, Paula, Rui Santos, and Nuno Videira. 2006. "Participatory Decision Making for Sustainable Development – The Use of Mediated Modelling Techniques." *Land Use Policy* 23 (1): 44–52.

Argyris, C., and D. Schön. 1978. *Organizational Learning*. Reading, MA: Addison-Wesley.

Bandura, A. 1977. *Social Learning Theory*. Englewood Cliffs, NJ: Prentice-Hall.

Bandura, A. 1986. *Social Foundations of Thought and Action: A Social Cognitive Theory*. Englewood Cliffs, NJ: Prentice Hall.

Barnaud, Cecile, Francois Bousquet, and Guy Trébuil. 2008. "Multi-Agent Simulations to Explore Rules for Rural Credit in a Highland Farming Community of Northern Thailand." *Ecological Economics* 66 (4): 615–27.

Basco-Carrera, Laura, Andrew Warren, Eelco van Beek, Andreja Jonoski, and Alessio Giardino. 2017. "Collaborative Modelling or Participatory Modelling? A Framework for Water Resources Management." *Environmental Modelling & Software* 91 (May): 95–110.

Beall, Allyson, and Andrew Ford. 2007. "Participatory Modeling for Adaptive Management: Reports from the Field II." In *25th International Conference of the System Dynamics Society*. Albany, NY: International System Dynamics Society.

Becu, Nicolas, Andreas Neef, Pepijn Schreinemachers, and Chapika Sangkapitux. 2008. "Participatory Computer Simulation to Support Collective Decision-Making: Potential and Limits of Stakeholder Involvement." *Land Use Policy* 25 (4): 498–509.

Becu, Nicolas, P. Perez, A. Walker, O. Barreteau, and C. Le Page. 2003. "Agent Based Simulation of a Small Catchment Water Management in Northern Thailand: Description of the CATCHSCAPE Model." *Ecological Modelling* 170: 319–31.

Belt, Marjan van den. 2004. *Mediated Modeling: A System Dynamics Approach To Environmental Consensus Building*. Washington, DC: Island Press.

Belt, Marjan van den, Oscar A. Bianciotto, Robert Costanza, Serge Demers, Susana Diaz, Gustavo A. Ferreyra, Evamaria W. Koch, Fernando R. Momo, and Maria Vernet. 2006. "Mediated Modeling of the Impacts of Enhanced UV-B Radiation on Ecosystem Services." *Photochemistry and Photobiology* 82 (4): 865–77.

Belt, Marjan van den, Lisa Deutsch, and Asa Jansson. 1998. "A Consensus-Based Simulation Model for Management in the Patagonia Coastal Zone." *Ecological Modelling* 110 (1): 79–103.

Belton, V., and T. J. Stewart. 2002. *Multiple Criteria Decision Analysis. An Integrated Approach*. Boston, MA: Kluwer Academic Publishers.

BenDor, Todd, Philip Berke, David Salvesen, Yan Song, and Nora Lenahan. 2011. "Assessing Local Government Capacity to Manage and Model Military-Induced Growth in Eastern North Carolina." *Planning Practice and Research* 26 (5): 531–53.

Bojórquez-Tapia, Luis, Salvadur Sánchez-Colon, and Arturo Florez. 2005. "Building Consensus in Environmental Impact Assessment through Multicriteria Modeling and Sensitivity Analysis." *Environmental Management* 36 (3): 469–81.

Bots, Pieter, and C. van Daalen. 2008. "Participatory Model Construction and Model Use in Natural Resource Management: A Framework for Reflection." *Systemic Practice and Action Research* 21 (6): 389–407.

Bourget, Lisa, ed. 2011. *Converging Waters: Integrating Collaborative Modeling with Participatory Processes to Make Water Resources Decisions.* Maass-White Series. Washington, DC: Institute for Water Resources, U.S. Army Corps of Engineers.

Bousquet, F., and C. Le Page. 2004. "Multi-Agent Simulations and Ecosystem Management: A Review." *Ecological Modelling* 176 (3–4): 313–32.

Brown Gaddis, J. Erica, Helena Vladich, and Alexey Voinov. 2007. "Participatory Modeling and the Dilemma of Diffuse Nitrogen Management in a Residential Watershed." *Environmental Modelling & Software* 22 (5): 619–29.

Bryson, John M. 2004. "What to Do When Stakeholders Matter: Stakeholder Identification and Analysis Techniques." *Public Management Review* 6 (1): 21–53.

Buchholz, Thomas S., Timothy A. Volk, and Valerie A. Luzadis. 2007. "A Participatory Systems Approach to Modeling Social, Economic, and Ecological Components of Bioenergy." *Energy Policy* 35 (12): 6084–94.

Butler, Cameron, and Jan Adamowski. 2015. "Empowering Marginalized Communities in Water Resources Management: Addressing Inequitable Practices in Participatory Model Building." *Journal of Environmental Management* 153: 153–162.

Cabrera, Victor, Norman Breuer, and Peter Hildebrand. 2008. "Participatory Modeling in Dairy Farm Systems: A Method for Building Consensual Environmental Sustainability Using Seasonal Climate Forecasts." *Climatic Change* 89 (3): 395–409.

Cardenas, J. C., J. Stranlund, and C. Willis. 2000. "Local Environmental Control and Institutional Crowding-Out." *World Development* 28: 1719–33.

Castella, Jean-Christophe, Tran Ngoc Trung, and Stanislas Boissau. 2005. "Participatory Simulation of Land-Use Changes in the Northern Mountains of Vietnam: The Combined Use of an Agent-Based Model, a Role-Playing Game, and a Geographic Information System." *Ecology and Society* 10 (1): 27.

Cockerill, Kristan, Howard Passell, and Vince Tidwell. 2006. "Cooperative Modeling: Building Bridges between Science and the Public." *JAWRA Journal of the American Water Resources Association* 42 (2): 457–71.

Cockerill, Kristan, Vincent C. Tidwell, Howard D. Passell, and Leonard A. Malczynski. 2007. "Cooperative Modeling Lessons for Environmental Management." *Environmental Practice* 9 (1): 28–41.

Coglianese, C. 1999. "The Limits of Consensus." *Environment* 41 (3): 28–33.

Colella, Vanessa. 2000. "Participatory Simulations: Building Collaborative Understanding through Immersive Dynamic Modeling." *The Journal of the Learning Sciences* 9 (4): 471–500.

Connick, S., and Judith E. Innes. 2003. "Outcomes of Collaborative Water Policy Making: Applying Complexity Thinking to Evaluation." *Journal of Environmental Planning and Management* 46 (2): 177–97.

Costanza, Robert, and Matthias Ruth. 1998. "Using Dynamic Modeling to Scope Environmental Problems and Build Consensus." *Environmental Management* 22 (2): 183–95.

Costanza, Robert, and Alexey Voinov. 2001. "Modeling Ecological and Economic Systems with STELLA: Part III." *Ecological Modelling* 143: 1–7.

Daniels, S. E., and G. B. Walker. 2001. *Working through Environmental Conflict: The Collaborative Learning Approach.* Westport, CT: Praeger.

D'Aquino, Patrick, Olivier Barreteau, Michel Etienne, Stanislas Boissau, Sigrid Aubert, François Bousquet, Christophe Le Page, and William S. Daré. 2002. "The Role Playing Games in an ABM Participatory Modeling Process: Outcomes from Five Different Experiments Carried out in the Last Five Years." In *Integrated Assessment and Decision Support*, 1st Meeting iEMSs, edited by A. E. Rizzoli and A. J. Jakeman, 275–80. Lugano, Switzerland: International Environmental Modelling and Software Society.

Demchak, C. C. 1991. *Military Organizations, Complex Machines: Modernization in the U.S. Armed Services*. Ithaca, NY: Cornell University Press.

Durant, W., and A. Durant. 1961. *The Story of Civilization: The Age of Reason Begins, 1558–1648*. The Story of Civilization. New York: Simon and Schuster.

Eden, Colin, and Fran Ackermann. 1996. "'Horses for Courses': A Stakeholder Approach to the Evaluation of GDSSs." *Group Decision and Negotiation* 5 (4): 501–19.

Eeten, Michel J. G. van, Daniel P. Loucks, and Emery Roe. 2002. "Bringing Actors Together Around Large-Scale Water Systems: Participatory Modeling and Other Innovations." *Knowledge, Technology, & Policy* 14 (4): 94–108.

Fisher, R., and William Ury. 1981. *Getting to Yes: Negotiating Agreements without Giving In*. New York: Penguin Books.

Foerstel, Herbert N. 1999. *Freedom of Information and the Right to Know: The Origins and Applications of the Freedom of Information Act*. Westport, CT: Greenwood Press.

Ford, David N., and John D. Sterman. 1998. "Expert Knowledge Elicitation to Improve Formal and Mental Models." *System Dynamics Review* 14 (4): 309–40.

Forester, John. 1999. *The Deliberative Practicioner. Encouraging Participatory Planning Processes*. Cambridge, MA: MIT Press.

Forrester, Jay W. 1987. "Lessons from System Dynamics Modeling." *System Dynamics Review* 3 (2): 136–49.

Funtowicz, Silvio, and Jerome R. Ravetz. 1994. "The Worth of a Songbird: Ecological Economics as a Post-Normal Science." *Ecological Economics* 10 (3): 197–207.

Funtowicz, Silvio, Jerome Ravetz, and Martin Connor. 2004. "Challenges in the Use of Science for Sustainable Development." *International Journal of Sustainable Development* 1 (1): 99–107.

Geurtz, J. L. A., and F. Joldersma. 2001. "Methodology for Participatory Policy Analysis." *European Journal of Operational Research* 128: 300–310.

Giordano, R., G. Passarella, V. F. Uricchio, and M. Vurro. 2007. "Integrating Conflict Analysis and Consensus Reaching in a Decision Support System for Water Resource Management." *Journal of Environmental Management* 84 (2): 213–28.

Gordon, Eric, Steven Schirra, and Justin Hollander. 2011. "Immersive Planning: A Conceptual Model for Designing Public Participation with New Technologies." *Environment and Planning-Part B* 38 (3): 505.

Gray, B. 2003. "Framing Environmental Disputes." In *Making Sense of Intractable Environmental Conflicts: Concepts and Cases*, edited by R. J. Lewicki, B. Gray, and M. Elliott, 11–34. Washington, DC: Island Press.

Gray, B. 2004. "Strong Opposition: Frame-Based Resistance to Collaboration." *Journal of Community and Applied Social Psychology* 14: 166–76.

Gray, Steven A., Stefan Gray, Linda J. Cox, and Sarah Henly-Shepard. 2013. "Mental Modeler: A Fuzzy-Logic Cognitive Mapping Modeling Tool for Adaptive Environmental Management." In *System Sciences (HICSS), 2013 46th Hawaii International Conference On*, 965–73. IEEE.

Gray, Steven, Michael Paolisso, Rebecca Jordan, and Stefan Gray. 2016. *Environmental Modeling with Stakeholders: Theory, Methods, and Applications*. New York: Springer.

Gray, Steven, Alexey Voinov, Michael Paolisso, Rebecca Jordan, Todd BenDor, Pierre Bommel, Pierre Glynn, et al. 2018. "Purpose, Processes, Partnerships, and Products: Four Ps to Advance Participatory Socio-Environmental Modeling." *Ecological Applications* 28 (1): 46–61.

Gunderson, L. H., and C. S. Holling, eds. 2002. *Panarchy: Understanding Transformations in Human and Natural Systems*. Washington, DC: Island Press.

Gurung, Tayan Raj, Francois Bousquet, and Guy Trébuil. 2006. "Companion Modeling, Conflict Resolution, and Institution Building: Sharing Irrigation Water in the Lingmuteychu Watershed, Bhutan." *Ecology and Society* 11 (2): 36.

Ingram, H., and A. Schneider. 1998. "Science, Democracy, and Water Policy." *Water Resources Update* 113 (Autumn): 21–28.

Itin, C. M. 1999. "Reasserting the Philosophy of Experiential Education as a Vehicle for Change in the 21st Century." *The Journal of Experiential Education* 22 (2): 91–98.

Jankowski, Piotr, Michał Czepkiewicz, Marek Młodkowski, and Zbigniew Zwoliński. 2016. "Geo-Questionnaire: A Method and Tool for Public Preference Elicitation in Land Use Planning." *Transactions in GIS*, 20 (6): 903–24.

Jones, Natalie A., Pascal Perez, Thomas G. Measham, Gail J. Kelly, Patrick d'Aquino, Katherine A. Daniell, Anne Dray, and Nils Ferrand. 2009. "Evaluating Participatory Modeling: Developing a Framework for Cross-Case Analysis." *Environmental Management* 44 (6): 1180–195.

Jordan, Rebecca, Steven Gray, Moira Zellner, Pierre D. Glynn, Alexey Voinov, Beatrice Hedelin, Eleanor J. Sterling, et al. 2018. "12 Questions for the Participatory Modeling Community." *Earth's Future* 6.

Kemmis, S., and R. McTaggart, eds. 1998. *The Action Research Planner* (3rd Edition). Geelong, VIC: Deakin University.

Kolagani, Nagesh, and Palaniappan Ramu. 2017. "A Participatory Framework for Developing Public Participation GIS Solutions to Improve Resource Management Systems." *International Journal of Geographical Information Science* 31 (3). 163–80.

Kolb, D. A. 1984. *Experiential Learning: Experience as the Source of Learning and Development*. Englewood Cliffs, NJ: Prentice-Hall.

Korfmacher, Katrina Smith. 2001. "The Politics of Participation in Watershed Modeling." *Environmental Management* 27 (2): 161–76.

Kuhn, Thomas. 1962. *The Structure of Scientific Revolutions*. Chicago, IL: University of Chicago Press.

Laird, Frank N. 1993. "Participatory Analysis, Democracy, and Technological Decision Making." *Science, Technology & Human Values* 18 (3): 341–61.

Lee, Hau L., V. Padmanabhan, and Seungjin Whang. 1997. "The Bullwhip Effect in Supply Chains." *Sloan Management Review* 38 (3): 93–102.

Lord, W. 1986. "Social and Environmental Objectives in Water Resources Planning and Management." In *An Evolutionary Perspective on Social Values*, edited by W. Viessman, and K. Schilling, 1–11. New York: American Society of Civil Engineers.

Lorie, Mark. 2009. "Computer-Aided Dispute Resolution: 2nd Workshop Summary and Strategic Plan (October 20–21, 2009)." www.iwr.usace.army.mil/Portals/70/docs/iwrreports/10-R-5.pdf

Luna-Reyes, Luis Felipe, Ignacio J. Martinez-Moyano, Theresa A. Pardo, Anthony M. Cresswell, David F. Andersen, and George P. Richardson. 2006. "Anatomy of a Group Model-Building Intervention: Building Dynamic Theory from Case Study Research." *System Dynamics Review* 22 (4): 291–320.

Lynam, T., W. de Jong, D. Shell, T. Kusumanto, and K. Evans. 2007. "A Review of Tools for Incorporating Community Knowledge, Preferences, and Values into Decision Making in Natural Resources Management." *Ecology and Society* 12 (1): 5.

Malczewski, J. 2006. "Ordered Weighted Averaging with Fuzzy Quantifiers: GIS-Based Multicriteria Evaluation for Land-Use Suitability Analysis." *International Journal of Applied Earth Observation and Geoinformation* 8 (4): 270–77.

Marcuse, Peter. 2005. "Study Areas Sites and the Geographic Approach to Public Action." In *Site Matters: Design Concepts, Histories, and Strategies*, 249–80. New York: Routledge.

Maslow, Abraham Harold. 1966. *The Psychology of Science: A Reconnaissance*. New York: Harper & Row.

Mason, E. 1970. *Collaborative Learning*. London, England: Ward Lock Educational.

McCormick, Sabrina. 2012. "After the Cap: Risk Assessment, Citizen Science and Disaster Recovery." *Ecology and Society* 17 (4): 31.

Mendoza, G. A., and H. Martins. 2006. "Multi-Criteria Decision Analysis in Natural Resource Management: A Critical Review of Methods and New Modelling Paradigms." *Forest Ecology and Management* 230: 1–22.

Mendoza, G. A., and R. Prabhu. 2003. "Qualitative Multi-Criteria Approaches to Assessing Indicators of Sustainable Forest Resource Management." *Forest Ecology and Management* 174: 329–43.

Metcalf, Sara S., Emily Wheeler, Todd K. BenDor, Kenneth S. Lubinski, and Bruce M. Hannon. 2010. "Sharing the Floodplain: Mediated Modeling for Environmental Management." *Environmental Modelling and Software* 25 (11): 1282–90.

Mostert, E., Claudia Pahl-Wostl, Y. Rees, B. Searle, D. Tabara, and J. Tippett. 2007. "Social Learning in European River-Basin Management: Barriers and Fostering Mechanisms from 10 River Basins." *Ecology and Society* 12 (1): 19.

Moxey, A., and B. White. 1998. "NELUP: Some Reflections on Undertaking and Reporting Interdisciplinary River Catchment Modelling." *Journal of Environmental Planning and Management* 41 (3): 397–402.

Muro, Malanie, and Paul Jeffrey. 2012. "Time to Talk? How the Structure of Dialo Processes Shapes Stakeholder Learning in Participatory Water Resources Management." *Ecology and Society* 17 (1): 3.

Nandalal, K. D. W., and S. P. Simonovic. 2003. "Resolving Conflicts in Water Sharing: A Systemic Approach." *Water Resources Research* 39 (12): 1362–73.

National Research Council. 2008. *Public Participation in Environmental Assessment and Decision Making.* Washington, DC: The National Academies Press.

NECSI. 2016. "About Complex Systems." New England Complex Systems Institute. http://necsi.edu/ guide/

Nicolson, C. R., A. M. Starfield, G. P. Kofinas, and J. A. Kruse. 2002. "Ten Heuristics for Interdisciplinary Modeling Projects." *Ecosystems* 5 (4): 376–84.

Nonaka, Ikujiro. 1994. "A Dynamic Theory of Organizational Knowledge Creation." *Organization Science* 5 (1): 14–37.

Pahl-Wostl, Claudia. 2006. "The Importance of Social Learning in Restoring the Multifunctionality of Rivers and Floodplains." *Ecology and Society* 11 (1): 10.

Pahl-Wostl, Claudia, and M. Hare. 2004. "Processes of Social Learning in Integrated Resources Management." *Journal of Applied and Community Psychology* 14: 193–206.

Palmer, R. 1999. "Modeling Water Resources Opportunities, Challenges, and Tradeoffs: The Use of Shared Vision Modeling for Negotiation and Conflict Resolution." In *Proceedings of the ASCE's 26th Annual Conference on Water Resources Planning and Management (Tempe, Arizona).* Reston, VA: American Society of Civil Engineers.

Petry, F. 1990. "Who Is Afraid of Choices? A Proposal for Multi-Criteria Analysis as a Tool for Decision-Making Support in Development Planning." *Journal of International Development* 2 (2): 209–31.

Polhill, J. Gary, Alessandro Gimona, and Richard J. Aspinall. 2011. "Agent-Based Modelling of Land Use Effects on Ecosystem Processes and Services." *Journal of Land Use Science* 6 (2–3): 75–81.

Power, Daniel J. 2002. *Decision Support Systems: Concepts and Resources for Managers.* Westport, CT: Praeger.

Prell, Christina, Klaus Hubacek, Mark S. Reed, Claire Quinn, Nanlin Jin, Joe Holden, Tim Burt, et al. 2007. "If You Have a Hammer Everything Looks like a Nail: Traditional versus Participatory Model Building." *Interdisciplinary Science Reviews* 32 (3): 263–82.

Quick, Kathryn S., and Martha S. Feldman. 2011. "Distinguishing Participation and Inclusion." *Journal of Planning Education and Research* 31 (3): 272–90.

Reed, Mark S. 2008. "Stakeholder Participation for Environmental Management: A Literature Review." *Biological Conservation* 141 (10): 2417–31.

Reed, Mark S., A. C. Evely, G. Cundill, I. Fazey, J. Glass, A. Laing, and J. Newig, et al. 2010. "What Is Social Learning?" *Ecology and Society* 15 (4): r1.

Renger, M., G. Kolshoten, and G. De Vreede. 2008. "Challenges in Collaborative Modelling: A Literature Review and Research Agenda." *International Journal of Simulation and Process Modelling* 4: 248–63.

Repenning, Nelson P. 2003. "Selling System Dynamics to (Other) Social Scientists." *System Dynamics Review* 19 (4): 303–27.

Rhoads, B. L., D. Wilson, M. Urban, and E. E. Herricks. 1999. "Interaction between Scientists and Nonscientists in Community-Based Watershed Management: Emergence of the Concept of Stream Naturalization." *Environmental Management* 24 (3): 297–308.

Roberts, N. 2004. "Public Deliberation in an Age of Direct Citizen Participation." *American Review of Public Administration* 34 (4): 315–53.

Rouwette, Etiënne A. J. A., Hubert Korzilius, Jac A. M. Vennix, and Eric Jacobs. 2011. "Modeling as Persuasion: The Impact of Group Model Building on Attitudes and Behavior." *System Dynamics Review* 27 (1): 1–21.

Rouwette, Etiënne A. J. A., and Jac A. M. Vennix. 2006. "System Dynamics and Organizational Interventions." *Systems Research and Behavioral Science* 23 (4): 451–66.

Rouwette, Etiënne A. J. A., Jac A. M. Vennix, and Theo van Mullekom. 2002. "Group Model Building Effectiveness: A Review of Assessment Studies." *System Dynamics Review* 18 (1): 5–45.

Scheffran, Jürgen. 2006a. "The Formation of Adaptive Coalitions." In *Advances in Dynamic Games*, edited by A. Haurie, S. Muto, L. A. Petrosjan, and T. E. S. Raghavan, 163–78. Berlin, Germany: Birkhäuser.

Scheffran, Jürgen. 2006b. "Tools in Stakeholder Assessment and Interaction." In *Stakeholder Dialogues in Natural Resources Management and Integrated Assessments: Theory and Practice*, edited by S. Stoll-Kleemann and M. Welp, 153–85. Berlin, Germany: Springer.

Scheffran, Jürgen, and S. Stoll-Kleemann. 2003. "Participatory Governance in Environmental Conflict Resolution: Developing a Framework of Sustainable Action and Interaction." In *Transition Towards Sustainable Development in South Asia*, edited by K. Deb and L. Srivastava, 307–27. New Delhi, India: The Energy and Resources Institute.

Schusler, T. M., D. J. Decker, and M. J. Pfeffer. 2003. "Social Learning for Collaborative Natural Resource Management." *Society and Natural Resources* 15: 309–26.

Selin, S. W., M. A. Schuett, and D. Carr. 2000. "Modeling Stakeholder Perceptions of Collaborative Initiative Effectiveness." *Society and Natural Resources* 13: 735–45.

Sheer, D. P., M. L. Baeck, and J. R. Wright. 1989. "The Computer as Negotiator." *Journal of the American Water Works Association* 81 (2): 68–73.

Shmueli, Deborah F., Sanda Kaufman, and Connie Ozawa. 2008. "Mining Negotiation Theory for Planning Insights." *Journal of Planning Education and Research* 27 (3): 359–64.

Sierra, Carles, Peyman Faratin, and Nick Jennings. 1997. "A Service-Oriented Negotiation Model between Autonomous Agents." In *Multi-Agent Rationality*, edited by Magnus Boman and Walter Van de Velde, 1237: 17–35 Lecture Notes in Computer Science. Berlin/Heidelberg, Germany: Springer.

Sivakumar, M. V. K. 2006. "Climate Prediction Agriculture: Current Status and Future Challenges." *Climate Research* 33: 3–17.

Slavin, R. E. 1983. *Cooperative Learning*. New York: Longman.

Starfield, Anthony M. 1997. "A Pragmatic Approach to Modeling for Wildlife Management." *The Journal of Wildlife Management* 61 (2): 261–70.

Starke, K., and A. Rangaswamy. 1999. "Computer-Mediated Negotiations: Review and Research Opportunities." In *Encyclopedia of Microcomputers*, Vol. 25 (Supplement 4), edited by A. Kent and J. G. Williams, 47–71. New York: Marcel Dekker, Inc.

Stave, Krystyna. 2003. "A System Dynamics Model to Facilitate Public Understanding of Water Management Options in Las Vegas, Nevada." *Journal of Environmental Management* 67: 303–13.

Stave, Krystyna. 2010. "Participatory System Dynamics Modeling for Sustainable Environmental Management: Observations from Four Cases." *Sustainability* 2 (9): 2762–84.

Stephenson, Kurt, Leonard Shabman, Stacy Langsdale, and Hal Cardwell. 2007. *Computer Aided Dispute Resolution: Proceedings from the CADRe Workshop (Albuquerque, New Mexico (September 13–14, 2007)*. Washington, DC: Institute for Water Resources, U.S. Army Corps of Engineers.

Sterling, Eleanor J., Moira Zellner, Kirsten Leong, Karen E. Jenni, Steven Gray, Rebecca Jordan, Todd K. BenDor, Antonie J. Jetter, Laura Schmitt Olabisi, Michael Paolisso, Klaus Hubacek, Pierre Bommel, and Gabriele Bammer. Try, try again: Lessons learned from success and failure in participatory modeling. Elementa Sustainability Transitions (In Press).

Sterman, John D. 2000. *Business Dynamics: Systems Thinking and Modeling for a Complex World*. New York: Irwin/McGraw-Hill.

Tesh, S. N. 1999. "Citizen Experts in Environmental Risk." *Policy Sciences* 32: 39–58.

Tidwell, Vincent C., and Cors Van Den Brink. 2008. "Cooperative Modeling: Linking Science, Communication, and Ground Water Planning." *Ground Water* 46 (2): 174–82.

Van Ackere, Ann, Erik Reimer Larsen, and John D. W. Morecroft. 1993. "Systems Thinking and Business Process Redesign: An Application to the Beer Game." *European Management Journal* 11 (4): 412–23.

Vennix, Jac A. M. 1999. "Group Model-Building: Tackling Messy Problems." *System Dynamics Review* 15 (4): 379–401.

Vennix, Jac A. M., Jan W. Gubbels, Doeke Post, and Henk J. Poppen. 1990. "A Structured Approach to Knowledge Elicitation in Conceptual Model Building." *System Dynamics Review* 6 (2): 194–208.

Videira, Nuno, Paula Antunes, and Rui Santos. 2009. "Scoping River Basin Management Issues with Participatory Modelling: The Baixo Guadiana Experience." *Ecological Economics* 68 (4): 965–78.

Videira, Nuno, Paula Antunes, Rui Santos, and Sofia Gamito. 2003. "Participatory Modelling in Environmental Decision-Making: The Ria Formosa Natural Park Case Study." *Journal of Environmental Assessment Policy and Management* 5 (3): 421–47.

Videira, Nuno, Paula Antunes, Rui Santos, and Rita Lopes. 2010. "A Participatory Modelling Approach to Support Integrated Sustainability Assessment Processes." *Systems Research and Behavioral Science* 27 (4): 446–60.

Voinov, Alexey, David Arctur, Ilya Zaslavskiy, and Saleem Ali. 2008. "Community-Based Software Tools to Support Participatory Modelling: A Vision." In *IEMSs 2008: International Congress on Environmental Modelling and Software: Integrating Sciences and Information Technology for Environmental Assessment and Decision Making (4th Biennial Meeting of IEMSs)*, edited by M. Sànchez-Marrè, J. Béjar, J. Comas, A. Rizzoli, and G. Guariso. Barcelona, Spain: International Environmental Modelling and Software Society.

Voinov, Alexey, and Francois Bousquet. 2010. "Modelling with Stakeholders." *Environmental Modelling & Software* 25 (11): 1268–81.

Voinov, Alexey, and Erica J. Brown Gaddis. 2008. "Lessons for Successful Participatory Watershed Modeling: A Perspective from Modeling Practitioners." *Ecological Modelling* 216 (2): 197–207.

Voinov, Alexey, Nagesh Kolagani, Michael K. McCall, Pierre D. Glynn, Marit E. Kragt, Frank O. Ostermann, Suzanne A. Pierce, et al. 2016. "Modelling with Stakeholders – Next Generation." *Environmental Modelling & Software* 77 (March): 196–220.

Vrana, Ivan, Jiří Vaníček, Pavel Kovář, Jiří Brožek, and Shady Aly. 2012. "A Group Agreement-Based Approach for Decision Making in Environmental Issues." *Environmental Modelling & Software*, Thematic issue on Expert Opinion in Environmental Modelling and Management, 36 (October): 99–110.

Watt, Daniel. 1983. *Learning with Logo*. New York: McGraw-Hill, Inc.

Weaver, P., and A. Jordan. 2008. "What Roles Are There for Sustainability Assessment in the Policy Process?" *International Journal of Innovation and Sustainable Development* 3 (1/2): 9–32.

Wilensky, Uri, and William Rand. 2015. *An Introduction to Agent-Based Modeling: Modeling Natural, Social, and Engineered Complex Systems with NetLogo*. Cambridge, MA: MIT Press.

Wilensky, Uri, and W. Stroup. 1999. "HubNet." http://ccl.northwestern.edu/netlogo/hubnet.html

Winz, Ines, Gary Brierley, and Sam Trowsdale. 2009. "The Use of System Dynamics Simulation in Water Resources Management." *Water Resources Management* 23 (7): 1301–23.

Wondolleck, Julia M., Nancy J. Manring, and James E. Crowfoot. 1996. "Teetering at the Top of the Ladder: The Experience of Citizen Group Participants in Alternative Dispute Resolution Processes." *Sociological Perspectives* 39 (2): 249–62.

Wondolleck, J. M., and S. L. Yaffee. 2000. *Making Collaboration Work: Lessons From Innovation in Natural Resource Management*. Washington, DC: Island Press.

Yu, J., Y. Chen, J. Wu, and S. Khan. 2011. "Cellular Automata-Based Spatial Multi-Criteria Land Suitability Simulation for Irrigated Agriculture." *International Journal of Geographical Information Science* 25 (1): 131–48.

Part II

Modeling Environmental Conflict

In Part II of this book, we first introduce readers to the system dynamics modeling approach and the system-focused participatory techniques that have been used for intervening in conflicts (Chapter 5). In Chapter 6, we present a practical introduction to agent-based modeling, its application to conflict research, generally, and environmental conflict, specifically. We also discuss complex systems and multiagent modeling concepts and theory (stability, chaos, emergence), comparing them with the dynamic modeling approaches more frequently used in conflict modeling. Finally, in Chapter 7, we introduce the "VIABLE" modeling framework, which simulates Values and Investments for Agent-Based interaction and Learning for Environmental systems, leveraging aspects of agent-based modeling, evolutionary game theory, system dynamics modeling, and network theory. Using this framework, we explore alternative realities that conflicting parties construct and experience during the course of disputes. We will interpret conflict as a dynamic and complex form of human interaction, which often emerges from incompatible actions, values, and goals, and which consumes a considerable amount of resources.

5

System Dynamics and Conflict Modeling

[O]ne of the great revelations of systems theory is that the behavior of the system comes from the system. Most people prefer to look for causes of problems somewhere else, outside, over there.... Systems theory almost always reveals that a stress may be coming from outside, but the unproductive reaction of the system to that stress comes from the way the system is structured.

—*Dana Meadows (1991), The Global Citizen, "System Dynamics Meets the Press"*

...The system dynamics perspective can stretch [our] conception of how the system works. It can help sort out the things that are predetermined from those that can be chosen.

—*Krystyna Stave (2002)*

Introduction

In the last chapter, we looked at the ability of participatory modeling, broadly defined, to improve decision making and promote social learning. In this chapter, we will focus more closely on system dynamics (SD) modeling, an approach that focuses on the changing nature of problems and the manner by which different elements of systems interact and evolve over time to create complex systems (Sterman 2000). In this context, SD modeling is an important technique for improving our individual abilities to understand the complexity of human–human and human–environmental interactions.

In this chapter, we will begin by introducing system concepts and SD modeling, including qualitative and quantitative approaches to modeling dynamic, complex systems. We will introduce the philosophy of SD, contrasting it with statistical modeling techniques that may be more familiar for some readers. We will contrast last chapter's general look at participatory modeling with a more focused discussion of participatory SD modeling, particularly the wide world of modeling interventions that have focused on conflict. Finally, we will discuss several cautions and criticisms associated with SD modeling interventions, including the need for systematic process and outcome evaluations, issues in addressing conflict among stakeholders that occur due to the modeling process, and finally, the "tyranny of expertise" that can arise from SD modeling interventions.

What Is a System?

A system is an interconnected set of components (Buchholz, Volk, and Luzadis 2007). When we look at both the components and the boundaries of any system—and interactions across those boundaries—we see that components are themselves composed of smaller sets of subsystems. Likewise, systems are components of larger systems, creating a nested pattern of relationships that generates *system hierarchies* and *emergent behavior* (Allen and Hoekstra 1993).

> *"Emergence,"* a term originating from complex systems research, refers to the idea that the behavior of interacting components cannot be explained entirely by referring to the properties of individual components themselves.

Many emergent behaviors are nonintuitive, leading to outcomes that no one expected. A common example is the "murmuration" behavior of huge flocks of starlings, as thousands of birds create intricate twisting and turning flight patterns in the sky. Even though individual birds' behavior is quite simple, the reactions of many individual birds to each of their neighbors' actions compound, thereby creating complex and unpredictable feedbacks that shape the behavior of the flock.

Efforts to combat emergent outcomes that are negative can often make the problem worse; perhaps one of the most widely experienced examples of this is seen in Braess's Paradox (Braess, Nagurney, and Wakolbinger, 2005 [1968]), which finds that in certain instances, when moving entities selfishly choose their own routes, adding extra capacity to a network can actually reduce overall performance. In practice, this means that adding new lanes to a highway often actually makes traffic congestion worse, a counterintuitive result that is a product of the *independent* choices of transportation agencies and motorists, which are actually *interdependent* (Downs 2005).

The Philosophy of SD

SD encompasses a set of qualitative (*systems thinking*) and quantitative tools (computer-aided modeling) for understanding, managing, and problem solving in highly complex and dynamic contexts of all types. For example, possibly, the most famous SD model was the *Limits to Growth*, model developed by Dennis and Dana Meadows and Jørgen Randers (1972), which aimed to understand and predict patterns of worldwide natural resource use. Originally developed by Jay Forrester while exploring the causes of industrial growth and decline in his seminal 1961 book, *Industrial Dynamics*, SD is grounded in electrical engineering control theory and nonlinear dynamics theories originating in physics (Antunes, Santos, and Videira 2006). Models created using this technique can function as powerful decision support systems that provide a flexible, and rigorous way of characterizing complex, nonlinear systems while fully capturing *interconnections, feedback loops*, and *delays* (Costanza and Ruth 1998; Sterman 2000; Otto and Struben 2004).

SD modeling distinguishes itself from statistical analysis or modeling by explicitly seeking to represent the *cause and effect* relationships of systems, particularly the systems for which modelers have detailed knowledge.[1] Key insights from SD modeling arise as we recognize that system *behavior*, which is observed in collected data (output), is the direct

---

1 Although statistical advances around causal inference over the last decade have improved tests for causality and goodness-of-fit of different regression model specifications (Granger 1969, 1993; Pearl 2009), Gelman and Imbens (2013) argue that statistical models still provide little transparency in separating the "effects of causes" from the "causes of effects," while focusing too heavily on the former.

result of that system's *structure*. This is a concept which closely parallels theories connecting behavior and structure of ecological systems (Allen and Hoekstra 1993).1 Therefore, rather than solely harnessing data (system behavior) to infer relationships (system structure), modelers attempt to explicitly represent the relationships that they know exist (whether by theory or observation; Costanza and Ruth 1998; Ruth and Hannon 2012). The SD modeling process requires (1) that modelers be able to estimate parameter values and formulate mathematical relationships connecting variables together causally (Videira et al. 2003) and (2) that simulating the model will demonstrate the effect of the system structure on policy interventions (and vice versa; Stave, 2003).

As part of this modeling process, model "validation" determines whether the model can mimic real-world behavior (Kleindorfer, O'Neill, and Ganeshan 1998); can the modeled structure of the system mimic the behavior that has been observed? In some SD models, validation processes can require consensus of many parties that a model does, in fact, adequately replicate reality. This is particularly true when many modelers or stakeholders are engaged, thereby making model validation a social process (Barlas and Carpenter 1990), which requires transparency in both the model and its development process. Winz, Brierley, and Trowsdale (2009) argue that statistical forecasting models do not provide this level of transparency; although statistical advances over the last several decades (Pearl 2009) have improved tests for causality and goodness of fit of different regression model specifications (e.g., Granger 1969, 1993), model results are still articulated by fitting parameters to data using convenient estimation techniques, which do not necessarily ensure identification of the fundamental drivers of system change (Leontief 1982; Leamer 1983; Costanza and Ruth 1998).

> *"Policy levers"* in a system are the variables that decision makers can control to affect major change over an entire system (Meadows 2008). We can identify policy levers by looking for ways in which we can intervene in a system and enact substantial change.

SD approaches problems with the goal of understanding how systems generate dynamic behavioral patterns, and where we can find *policy levers* (sometimes called "leverage points") within the system that can convert problematic trends into desirable ones (Stave 2010). As a result, the usefulness of SD models often arises from their use as "virtual worlds" or "learning laboratories" (Winch 1993), where experimentation, tests of assumptions, and comparisons of alternative management regimes or policies, aid users in understanding how systems work, why problems or conflicts emerge, and how deliberative analytical processes can be managed in effective ways (Sterman 1994; Stave 2002; Videira et al. 2003; Winz, Brierley, and Trowsdale 2009).

These "laboratories" provide modelers with rapid feedback, allowing them to test ideas about how their actions affect complex systems that are complicated by feedbacks, delays, imperfect information, and misperceptions (Sterman 1994; Stave 2002). Therefore, the stated objective of many SD models is not necessarily to act as a predictive tool that generates accurate projections of future conditions. Rather, SD models are often designed as policy analysis tools that enable stakeholders to explore and assess alternative system configurations and policy designs. Drawing on information about how systems actually fit together and function as well as how decisions are made, models are designed to simulate information flows and decision-making processes (Morecroft 1985; Sterman 1987).

---

1 When we discuss agent-based modeling in the next chapter, will see that this structure–behavior relationship is complicated by the idea that *individual behaviors* can aggregate to create emergent *system structure*.

The real innovation of SD is in being able to represent the complex mathematical underpinnings of nonlinear, complex systems in a comprehensive, nonreductionist manner; systems are viewed as a collection of interconnected elements that accumulate material or information, flow and change, and interact in complex ways (Voinov and Bousquet 2010). This systems approach stands in stark contrast to the reductionist approach common to "traditional" scientific studies, where systems are broken into their component parts and understood in isolation. This traditional approach is a practice that some researchers (e.g., Checkland 1988) have argued promotes mechanistic and reductionist views of systems, which does not necessarily conform to dealing with ill-defined problems like coupled human–natural system interactions (Mendoza and Prabhu 2006). Instead, SD understands that system elements jointly act to create emergent behavior and that general system principles act analogously across many different types of systems (Ruth and Hannon 2012). In doing so, SD models can easily capture feedbacks, nonlinearities, and time lags that are too often ignored in assessing problems, developing management practices, and resolving conflicts (Costanza and Ruth 1998).

Systems Thinking

Before quantitative models are constructed, modelers commonly begin by relying on systems thinking techniques to create a "dynamic hypothesis" of how a system operates. Dynamic hypotheses help to explain system structures that create the type of behavior observed (usually through data) in the past, while making assumptions explicit (Otto and Struben 2004). In this sense, dynamic hypotheses represent theories that a certain structure or process will contribute to certain behavior patterns (Sterman 2000).

The most common way of articulating a dynamic hypothesis is through the creation of a "causal loop diagram" (CLD; see example in Figure 5.1), which diagrammatically maps important variables that characterize dynamic problems and the cause-and-effect relationships linking them together (BenDor 2012). CLDs allow modelers and diverse stakeholders to easily visualize complexity in a system, including nonlinear relationships and feedbacks (Cockerill, Passell, and Tidwell 2006).

However, although many systems thinking interventions are successful as first steps toward a dynamic understanding of complex problems and conflicts, Forrester (2007)

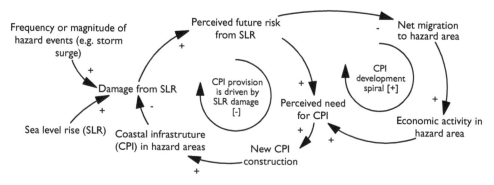

FIGURE 5.1

Example of a causal loop diagram (CLD) describing the interplay between sea level rise (SLR) and the construction of coastal protection infrastructure (CPI). Positive (+) arrowheads indicate positive (proportional) relationships and negative (–) arrowheads indicate inverse relationships.

argues that systems thinking methods represent only a small step toward understanding systems; the value and learning opportunities for future decision making occurs during the development of rigorous quantitative models, whose simulations demonstrate the inconsistencies within our own mental models.

Quantitative SD Modeling

In a quantitative SD model, system structure "...consists of the feedback loops, stocks and flows, nonlinearities, and delays created by the interaction of the physical and institutional structure of the system with the decision-making processes of the agents acting within it (Sterman, 2000, p. 107)." *Stocks* represent accumulations of material or information, flows represent rates of change affecting stocks (rates of inflow and outflow are defined using auxiliary variables), and feedbacks are represented as loops of information links between variables that enable representation of cause and effect (Ford 2009). Modelers attempt to represent all of the components of a model separately, while defining a structure for the model that links components with each other (Dudley 2008; BenDor 2009). Together, these components combine to create equations that describe how the system changes over time, whereby system state conditions at one point in time are used to compute system states at the next, continuing step-by-step over any desired time horizon (Winz, Brierley, and Trowsdale 2009).

In cases where systems are represented formally, the stock-and-flow structure of an SD model is built in user-friendly software packages, such as STELLA/iTHINK, Vensim, or Powersim, among others, which represent models as multiple communicating layers that contain progressively more detailed information on the structure and functioning of the model (Figure 5.2). These software packages are also graphically based, making them excellent for use with participant groups with varying levels of technical proficiency (Langsdale et al. 2007).

The range of abstraction levels[1] presented in these software packages allows modelers and users to tuck away mathematical intricacies into the background, presenting both groups with choice over the level of detail at which they wish to view and understand the model. By allowing users to avoid the mathematical intricacies of models, while still making all assumptions explicit without any hidden "black boxes," SD enables anyone with knowledge of a problem or system to begin constructing internally consistent model structures for understanding and solving problems.

Numerical methods (sometimes called *numerical integration*) are techniques for finding approximate solutions of ordinary differential equations. In many situations, these differential equations are complex enough to prohibit us from determining "closed-form" or precise solutions using mathematical rules. As a result, we can approximate these solutions numerically. In the context of SD, numerical integration algorithms are used to calculate the net change and resulting accumulation of multiple, connected stocks over time.

---

[1] As we defined in Chapter 4, *abstraction* refers to the amount of complexity that is experienced when viewing or interacting with software. Higher levels of abstraction mean that less detail is experienced (e.g., using voice commands to ask your smartphone for driving directions). Lower levels of abstraction mean that more detail is experienced (e.g., writing a program to calculate and display driving directions using cost-distance algorithms that you create in a programming language).

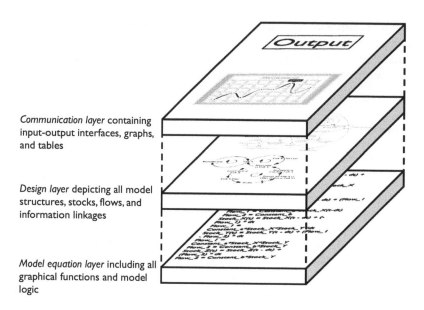

Communication layer containing input-output interfaces, graphs, and tables

Design layer depicting all model structures, stocks, flows, and information linkages

Model equation layer including all graphical functions and model logic

FIGURE 5.2
Structure of models in SD software packages, including the widely used VENSIM (http://www.vensim.com) and STELLA software (http://www.iseesystems.com/).

The Inner-Workings of SD

SD models are *aggregate* models in that they are not meant to capture the system at the level of an individual unit of analysis (e.g., the actions of an individual stock broker or a bear foraging for food) but rather the overall accumulations and rates of change of the properties of collections of those individuals (e.g., the behavior of the overall market or the consumption patterns of a population of bears). Mathematically, SD models are composed of systems of nonlinear, first-order differential equations, which describe of change with regard to time. During simulations, these differential equations are turned into *difference equations* that are solved through *numerical integration* techniques (e.g., Euler's method; Press et al. 2007) over a specified time horizon. It is important to note that because calculations are carried out using these numerical techniques, SD models are not limited by analytical tractability of the underlying differential equations; the model structure can be changed in almost any way to examine increasingly dynamic and complex situations (Dudley 2008). This gives SD conflict models a major advantage over many of the game-theoretical modeling techniques that rely entirely on analytically tractable solutions (e.g., Reuveny, Maxwell, and Davis 2011).

Participatory SD Modeling

As we discussed in the last chapter, a common role of SD in conflict resolution involves the inclusion of stakeholders in the modeling process through *participatory SD* approaches, including *mediated modeling* and *group model building* (GMB) techniques. Following Stave (2010, p. 2763, 2766), we will consider *participatory SD modeling* to be

> ...any use of a system dynamics model to structure group analysis of a problem, whether it is a conceptual or fully operational model, and whether or not the model

users are involved in model development....[Furthermore, this approach involves,] ... the use of a system dynamics perspective in which stakeholders or clients participate to some degree in different stages of the process, including problem definition, system description, identification of policy levers, model development and/or policy analysis.

The aim of participatory SD modeling is to integrate and structure available information about a conflict. Doing this allows users to avoid some of the pittfalls of traditional decision-making techniques, which often exacerbate discrepancies in power and status among stakeholders while failing to create shared languages and environments that facilitate open and supportive communication (Rouwette et al. 2011). Along these lines, Rouwette et al. (2011, p. 3) outline four sets of goals of participatory SD (GMB in particular):

1. Goals for individual participants, including positive reactions to the modeling intervention, learning and mental model refinement, and commitment to behavioral change;
2. Goals for the entire stakeholder group, including alignment of mental models, consensus, and group commitments to a decision;
3. Goals for the conflict as a whole, including systemic changes ("doing things differently") and improvement of the problematic conflict condition;
4. Goals for modeling itself, including improvements to the efficiency of the modeling process and further use of SD modeling in the conflict.

Like other participatory modeling techniques, mediated modeling is useful for promoting stakeholder understanding of the system as well as social learning and communication among participants (construction of social capital; Stave 2003). However, unlike other participatory modeling techniques, SD modeling introduces a robust framework for "analytic-deliberation" (i.e., formalized discussion) in dynamic environmental systems and conflicts (Stave 2010; Videira et al. 2010). This means that SD models incorporate a new type of technical analysis, while enforcing a level of internal consistency (a "structure for deliberation and education"; Stave 2002, p. 155), which is difficult to achieve in discussion or in other participatory modeling frameworks (Forrester 1987; Stave 2002). Participatory SD mediates the interface between complexity and participation; as we will show, the method helps in establishing and linking a jointly viewed conflict reality with jointly-defined, mutually compatible goals. Under this view, any changes in conflict systems are the direct result of changes in the management actions of conflict stakeholders and their organizations.

Since the earliest years of SD practice, professionals seeking to develop models of organizational problems attempted to involve clients in the model development. For example, Jay Forrester (1961, 1985, 1994) continually emphasized the need to "access the mental database" of managers in order to construct SD models of strategic problems in business (Voinov and Bousquet 2010). Vennix (1999) argued that this was important for three major reasons, including (1) capturing the needed information from the mental models of the client group, (2) increasing the likelihood of implementing model results (stakeholder/"client" buy-in), and (3) enhancing stakeholder learning processes (social/collaborative learning; Rouwette and Vennix 2006). Although relatively few metrics have been created to judge outcomes, many of these modeling interventions have been judged by involved parties to have been fairly successful (Rouwette et al. 2011), and the number of projects involving business clients or other types of organizational stakeholders has increased rapidly over the last several decades (Scott 2018; Scott, Cavana, and Cameron 2016).

What Makes Participatory SD Modeling Unique?

The participatory SD modeling process attempts to translate individual viewpoints into a shared, common language, which is an important element of any mediation effort. The idea is that individual stakeholders will work together to collectively guide dynamic model development, linking their different viewpoints about the system into a coherent whole, and developing theories around conflict structure and dynamics on the basis of conflict information and stakeholder experience (Luna-Reyes et al. 2006). Building a model allows participants to appreciate conflict complexity while developing a common understanding of the issue, which, hopefully, improves the odds that solutions coming out of the process will be implemented. The process requires intensive participation of stakeholders to make their assumptions about the system explicit and draw out the disputes over the actual system operation (Videira, Antunes, and Santos 2009).

Although stakeholder views are often very complex, wide-ranging, and difficult to articulate, creating a common understanding of system structure in dynamics has been shown to be essential for management teams and other organizations (Winch 1993). By constructing a causal model together, stakeholders can better appreciate the difficulties of explicitly representing relevant variables and backing up their arguments—which are created by their mental models—with accurate data (Cockerill et al. 2007). Work on participatory SD modeling (e.g., Videira et al. 2003; Antunes et al. 2006) has suggested that the practice can facilitate quality decision making and conflict resolution by:

1. Creating a collaborative learning environment for stakeholders, where models are built on shared assumptions and a common understanding of systems (Stave 2003);

2. Internalizing the public's values and preferences (Videira et al. 2003);

3. Promoting a multidimensional understanding of relevant systems (ecological, economic, and social) while clarifying resource management problems;

4. Forcing all modeler and stakeholder assumptions about the way the system works to become explicit, thereby making the modeling process more open and tackling some of the transparency problems inherent in participatory modeling (Meadows and Robinson 1985);

5. Supporting problem definition and structuring, as well as boundary setting around conflicts (Woltsenholme 1990; Lane 1993; Vennix 1996, 1999), particularly when working with ill-defined, badly structured, or "wicked" problems (i.e., situations where stakeholders disagree on problem existence or definition;

6. Aiding communication about system structure and depicting the consequences of management actions visually, with minimal technical language or jargon (Stave 2003). In this sense, SD can promote public participation by easily and explicitly showing that stakeholder choices affect the direction that their own futures can take (Stave 2002);

7. Increasing shared understanding of the problems, which is likely to drive consensus and commitment toward action (creating buy-in and fostering trust in institutions);

8. Creating an environment to test the long-term effects of decisions or strategies in complex systems with high uncertainty (Winz, Brierley, and Trowsdale 2009); and

9. Enabling exploration of large systems and incorporating extensive input from a wide range of stakeholders (Costanza and Ruth 1998; Pahl-Wostl 2006; Videira, Antunes, and Santos 2009).

It is important to note that there is often a discrepancy found between the long-term thinking endangered in the SD paradigm and the short-term pressures faced by most decision makers (Meadows and Robinson 1985; Stave 2002). For example, when modelers recommend solutions that require long-term, structural change, which could be either simple (e.g., using new sources of information) or more complex (e.g., revising authority and incentive structures, changing goals), these changes are sometimes politically unpalatable. Even worse, when SD modeling is used to search for the root causes of problems, it can discover that problems are frequently caused by the stakeholders themselves. This can create tensions between clients and modelers that can be difficult to overcome.

The Participatory SD Modeling Process

SD modelers and researchers have developed techniques for "knowledge elicitation," which aims to obtain the necessary knowledge, including information about system structure and governing policies, from a group of people to inform models (Vennix et al. 1992). Traditional methods for eliciting this information include interviews and focus groups (Vennix et al. 1990), which often assist initial modeling phases (prior to formal model building), such as problem definition and qualitative causal mapping. Researchers have even developed scripts that function as recipes for participatory modeling workshops (Andersen and Richardson 1997; Luna-Reyes et al. 2006).

However, it is important to stress that participatory SD interventions involve much more than merely eliciting knowledge from stakeholders about the problem and the system; interventions involve efforts to build shared ownership of the analysis and the model, a shared understanding of the conflict and its description, and a shared appreciation of the trade-offs among different decisions or solutions (Stave 2010). As a result, participatory SD modeling efforts usually include a series of workshops covering different modeling stages, (Videira et al. 2003; van den Belt 2004).

During these workshops, stakeholders work through a number of model development stages, including problem definition, system conceptualization (i.e., dynamic hypothesis construction), model parameterization, and policy analysis. Stave (2003) and Beall and Ford (2007, 2010) discuss different designs of these workshop series, including flexibility in stakeholder involvement and delineation of modeling tasks into separate workshops. The length of each workshop, as well as the entire process, is dependent on stakeholder group size, model detail, problem complexity, and the relative magnitude of conflict antagonism (Vennix 1996; Videira et al. 2003; van den Belt 2004). Figure 5.3 compares the implementation of a participatory SD framework across four separate studies, each of which endeavored to create a formal, quantitative SD model.

The first stage of participatory SD modeling processes typically begins with the development of a preliminary conceptual model by stakeholders. In some projects, modeling leaders actually prepare this initial "scoping" (or "preliminary") model (Costanza and Ruth 1998) before interacting with stakeholders, as it may be easier to confront them with a preliminary model in hand rather than to approach them for the first time unprepared (Beall and Zeoli 2008). However, this approach can be controversial; work by Jac Vennix (1996) suggests that if stakeholders have no role in creating preliminary models, use of these models can inadvertently decrease participants' feelings of ownership and buy-in.

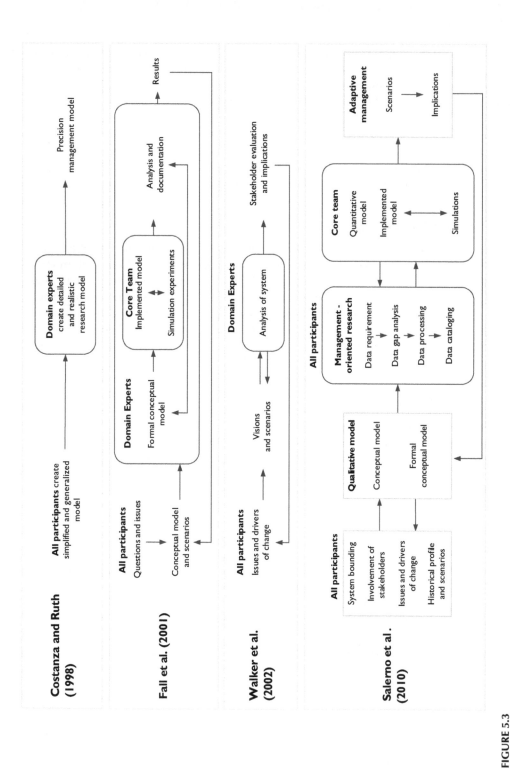

FIGURE 5.3
A comparison of participatory SD frameworks across four studies. Adapted from Salerno et al. (2010) and Videira et al. (2003).

(Continued)

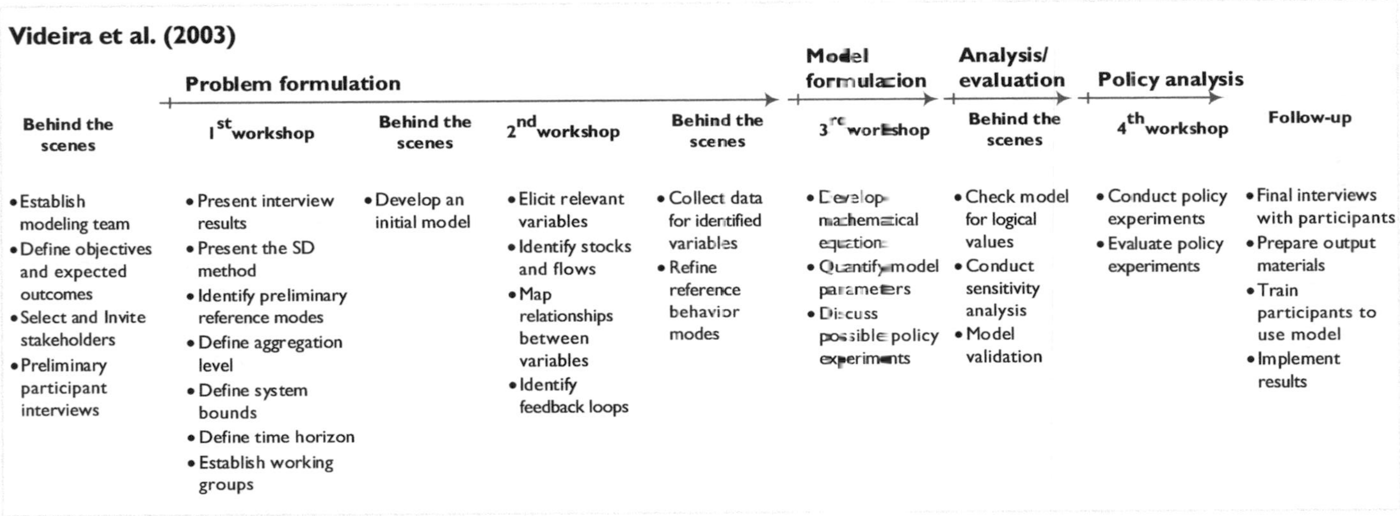

Videira et al. (2003)

Problem formulation					Model formulation	Analysis/ evaluation	Policy analysis	
Behind the scenes	1st workshop	Behind the scenes	2nd workshop	Behind the scenes	3rd workshop	Behind the scenes	4th workshop	Follow-up
• Establish modeling team • Define objectives and expected outcomes • Select and Invite stakeholders • Preliminary participant interviews	• Present interview results • Present the SD method • Identify preliminary reference modes • Define aggregation level • Define system bounds • Define time horizon • Establish working groups	• Develop an initial model	• Elicit relevant variables • Identify stocks and flows • Map relationships between variables • Identify feedback loops	• Collect data for identified variables • Refine reference behavior modes	• Develop mathematical equation • Quantify model parameters • Discuss possible policy experiments	• Check model for logical values • Conduct sensitivity analysis • Model validation	• Conduct policy experiments • Evaluate policy experiments	• Final interviews with participants • Prepare output materials • Train participants to use model • Implement results

FIGURE 5.3 (CONTINUED)

A comparison of participatory SD frameworks across four studies. Adapted from Salerno et al. (2010) and Videira et al. (2003).

During subsequent phases, in which qualitative or basic quantitative scoping models are made more sophisticated (e.g., see Costanza and Ruth 1998), modelers help to facilitate the process using either SD software (to help actually develop models during the workshops) or other types of systems thinking tools, including diagrammatic modeling using CLDs (Videira, Antunes, and Santos 2009). Although these conceptual models do not rely on computer simulation to precisely infer dynamic behavior, they can be particularly useful for understanding complex systems and determining the structure of underlying feedback loops.

System Dynamics and Conflict

Quite a bit of work has applied SD more specifically to conflict analysis and resolution. Cioni (2009) and Stave (2002) summarize five primary uses of SD in this domain:

1. As a normative tool for describing an ideal system; perhaps one that is conflict free with an idealized or improved resource distribution or management situation;
2. As a descriptive tool for describing current conflict or system states;
3. As a prescriptive tool for exploring scenarios or management regimes that would improve current conflict or system states;
4. As a cognitive tool for building knowledge and skills for better understanding problems and solutions; and
5. As a tool for the analysis of group decision processes.

In many cases, SD models can be used to fill several of these roles, including use in a participatory context, functioning as a cognitive and group decision-making tool, as a normative tool that describes a jointly agreed-upon representation of a system, and as an analytical tool for exploring management scenarios.

Quantitative vs. Qualitative SD Approaches to Conflict

Many GMB and other systems thinking interventions consider simulation to be the primary and significant contribution toward resolving group conflicts or enhancing problem solving capacity (Andersen and Richardson 1997; Winz, Brierley, and Trowsdale 2009). They argue that quantitative modeling (mathematically based SD simulation) is essential for (1) adding rigor to group analysis and discussions, (2) identifying feedback loops, (3) using the explicit and transparent nature of the model for observing and managing complexity (keeping track of feedback structure and many variables), and (4) serving as "group memory" of how models were developed, particularly when working with wicked, "messy," or complex problems.

Some relatively recent analysis has lent evidence to these claims; in an assessment of over 100 applications of GMB, Rouwette, Vennix, and Mullekom (2002) found that qualitative modeling (employing systems thinking and other nonsimulation techniques) was used only 21% of the time. However, when they dug deeper, they found that the data on qualitative modeling and mental mapping for conflict resolution were relatively nuanced. They surveyed 22 qualitative modeling studies, dividing them into two categories, whereby participants created either qualitative "demonstration" models (results of which participants never expected to implement) or qualitative models that were used for finding implementable results (usually involving a small group of stakeholders over a limited timespan). Rouwette, Vennix, and Mullekom (2002, p. 24) usefully summarized their observed differences between qualitative and quantitative participatory modeling interventions:

> Qualitative models seem to be less likely to lead to commitment, consensus or system changes than (small or large) quantitative models. The likelihood of positive results of system changes is equal for all ... types of models. In addition, qualitative models do not seem to differ substantially from quantitative models with respect to their capacity for generating insight, behavioral changes or communication. These outcomes should be considered in terms of the different contexts in which models are used. Qualitative modeling appears to be an intervention requiring relatively few organizational resources (time and participants) and is instrumental in clarifying intangible matters. Quantification adds substantial benefits in situations where the problem is more structured and justifies a larger investment in terms of participants' time; it is thus more effective in producing consensus and system changes.

History has shown that effective conflict negotiation uses multiple techniques to illuminate, clarify, and challenge hidden assumptions (Fisher and Ury 1981). While simulation remains the gold standard for examining these assumptions in truly explicit ways, like other SD applications that tackle problems at multiple levels of precision, participatory SD takes mental models and builds them into conceptual models and eventually into quantitative simulations (Vennix 1999). Like other participatory modeling techniques, SD models can be designed as communication and analysis platforms that connect to conflicts in a variety of ways and which embed public participation at different levels or during different stages of the modeling process (e.g., scoping, identification of environmental pressures, setting policy objectives, and evaluation of management alternatives; Videira et al. 2003).

For example, significant work has developed methods focusing entirely on mental models, attempting to elicit and make explicit the ways that various stakeholders view a conflict (Becu et al. 2003; Mendoza and Prabhu 2006; Lynam et al. 2007; Voinov and Bousquet 2010). Many other modeling processes are nearly entirely devoted to problem definition or characterizations of system structure (Langsdale et al. 2007). Qualitative models articulating collaborative problem visions or problem boundaries can be the most valuable product of the process and essential for group learning (Beall and Zeoli 2008).

Finally, if we think about the level of participant involvement in modeling workshops as a continuum, many models have been developed in an iterative process away from stakeholders, but with frequent and regular opportunities for them to contribute and get involved in the process (Langsdale et al. 2007). As we saw in the last chapter, there is evidence that even for simple forms of modeling, where stakeholders draw on completed models during hands-on analyses and discussions, participants continue thinking about problems and conflicts using dynamic, long-term, and interconnected frameworks (Winz, Brierley, and Trowsdale 2009).

Drawbacks to Conflict Modeling with System Dynamics

Conflict during and after the Modeling Process

Building an effective SD model in a client-group environment is a challenging task, particularly with a heterogeneous clientele and a variety of stakeholders influencing the boundaries of the model (Otto and Struben 2004). Winch (1993) divides disparities in the problem of stakeholder system conceptualization into two broad types: (1) stakeholder views that simply do not coincide, and therefore lead to different interpretations of the situation, and (2) stakeholder views that collectively lead to incomplete understanding of the full system or conflict problem.

While evidence has demonstrated that participatory modeling is a helpful tool for social and collaborative conflict resolution (Rouwette and Vennix 2006), some have suggested that participatory modeling is analogous to flight training using an aviation flight simulator; no matter how accurate the model is, or how much participants actually use it, it may not significantly improve the decisions of those using it (Vennix 1999). Persistent social resistance, like defensive routines and "face saving," often prevent real learning, particularly when participants are not actively involved in model construction (Sterman 1994). This is due to the reality that information can only act as a persuasive argument when it is relevant and new to the receiver (Rouwette et al. 2011). Hearing information in a modeling session that is already known to stakeholders does not act to change their beliefs; rather, the interpretation of modeling outcomes is entirely determined by the meaning that stakeholder participants attach to the findings.

Moreover, special attention must be paid to strengthening participants' feeling of control over the modeling process (Rouwette et al. 2011). Vennix (1999) has argued that two important issues prevent acceptance of models, including (1) differences in perceptions of conflict systems (multiple perceived "realities") and (2) ineffective communication among stakeholders and between modelers and stakeholders. Together, these barriers enable conflict and prevent convergence of these realities into one, commonly shared vision of the system structure.

Are SD Modeling Interventions Effective?

To the extent that data are available, there has only recently started to emerge unbiased evidence regarding the effectiveness of GMB (Scott, Cavana, Cameron, 2016a, b). Many modeling interventions provide little insight into the effectiveness of their efforts and most of the literature exists as individual case studies that fail to provide insight into how the outcomes of modeling efforts compare to different interventions' goals (Rouwette et al. 2011). Rouwette and Vennix (2006) note that there has not been a consistent theoretical framework for conducting or evaluating participatory SD modeling interventions; many participatory SD interventions are created and designed by gifted and experienced modelers who rely on insight and experience (and trial and error; Andersen et al. 2007), thereby limiting the spread of the practice considerably. Furthermore, from a traditional program and intervention viewpoint, there is no conceptual framework that links elements of the participatory SD modeling process to conflict resolution outcomes or goals (Rouwette et al. 2011). This makes it difficult to draw valid conclusions about the effectiveness of modeling interventions, as participatory model construction can be highly complex and draw on an enormous number of elements and processes (Rouwette, Vennix, and Mullekom 2002). The only way around this is to open up this black-box process and record, in meticulous detail, exactly how the interventions work and what happens.

Why then, after so many years of practice, has participatory modeling not developed these types of metrics? How can these models be viewed as valid with little hard evidence to support their effectiveness? Perhaps the most convincing answer is that many of the insights and intuitions that are the results of SD interventions are exceedingly difficult to test in any sort of rigorous or objective way; many mental models entail soft, intangible factors that are the basis of conflict complexity. The other major explanation is, as Winz, Brierley, and Trowsdale (2009, p. 1306) bluntly states, "...the concept of validity is flawed and models are never valid." Model "usefulness" is, at its core, a subjective concept that is difficult to objectively define. As conflict complexity and uncertainty increase, "objective" comparisons between observed data and predicted results become superficial. Although statistical verifications and fitting can help to increase our confidence in models, it also decreases the ability of models to adapt to new situations while increasing their cost (Winz, Brierley, and Trowsdale 2009). Videira et al. (2003) and Beierle (1998) devised numerous qualitative assessments of participatory processes that can also be used to evaluate the relative success of participatory SD interventions (Table 5.1).

TABLE 5.1

Example Criteria for Assessing the Success of Participatory System Dynamics Modeling

Criteria	Assessment Questions
Individual and collaborative learning	Did participants say that they learned from the process? What did they learn?
	Did participants recognize more interlinkages between the system's sectors than before?
	Were feedbacks, delays, nonlinearities, and uncertainties identified?
	Did the simulation of behavior help formulating and evaluating alternative policies at a strategic decision-making level?
Potential to involve the public and stakeholders	How many stakeholders were involved or benefited from the modelling intervention?
	Did stakeholders understand their role in the modeling process?
	Were all reasonably affected parties included or represented?
	Were there any efforts to accommodate potentially excluded stakeholders over different stages of the process?
Inclusion of stakeholder values, assumptions, and preferences	Was information from the stakeholders used to inform or review decisions?
	To what extent did stakeholders feel they had an impact on decisions?
	Where stakeholder input was not incorporated, was there a justification?
	Did the active participants feel that they had sufficient knowledge to contribute to the process?
Increasing decision quality	Did stakeholder involvement increase all parties' satisfaction?
	Was there an opportunity to generate new alternatives?
	Were new opportunities for trade-offs or compensation between parties identified?
	Was there sufficient time and money to obtain credible information?
Fostering trust in institutions	Was stakeholders' confidence in institutions increased?
	Did stakeholder involvement improve the image of the institution?
	Did the stakeholders feel that their interests are aligned or valued by the institution?
	Will stakeholders let the institution undertake decisions with less oversight?
Reducing stakeholder conflict	Did stakeholder involvement reduce opposition to the decisions?
	Did stakeholder involvement lead to less litigation?
	If an agreement was reached, will the agreement be stable over time? How long?
	Did stakeholder involvement improve or worsen cooperation among parties?
Cost-effectiveness	Was participation a cost-effective tool for achieving the proposed goals compared with other mechanisms?
	How much (time, money, other factors) did the participatory process cost the modeling team and to the participants in terms of time and money?
	What were the opportunity costs for all participants in terms of shifted resources and delayed actions?

Source: Adapted from Videira et al. (2003).

To focus this discussion, Rouwette and Vennix (2006) delineate modeling effectiveness into four different categories:

A. the extent to which participants contributed information during the modeling;

B. the impact of this information on other participants;

C. the impact of the modeling intervention on understanding decision-making biases and shortcomings; and

D. the impact of the modeling process on the broader organization (in the case of business or organizational interventions) or stakeholder organizations (in the case of multi-agent conflicts).

These measures build on previous work by Rouwette, Vennix, and Mullekom (2002), who analyzed case studies to assess modeling process outcomes, including documented reactions of individuals, insights or learning, commitment to results, behavioral changes, communication (exchange of viewpoints), reaching consensus, shared language developed between participants, system changes (organizational or physical), the results of system changes, further use of modeling methods, and the efficiency of the determined solution compared with other types of solutions. The authors used "metaanalysis" (i.e., reanalysis of previous studies) to evaluate 107 participatory SD interventions, finding a huge variety of the ways that modeling interventions are assessed and reported. They discovered that each case study establishes unique ways of assessing the effectiveness of the modeling exercise and that only 19 interventions (18%) actually assessed outcomes quantitatively. Perhaps most importantly, they found significant "publishing bias"; for the most part, unsuccessful modeling interventions are rarely reported.

Although, on the whole, Rouwette, Vennix, and Mullekom (2002) found positive results for all of the outcome variables that they tested, very few studies (about one quarter) actually included assessments of stakeholder reactions to the modeling process. Although stakeholders reported that GMB increased their insight into problems 95% of the time, much smaller effects were reported regarding:

1. development of consensus (50%);

2. system-level changes (46%);

3. intentions to actually implement results or changes in their behavior (42%);

4. improvement in communications between stakeholders (38%);

5. continued use of SD modeling after the initial project (38%); and

6. stakeholders belief that SD modeling was more efficient than more traditional models and/or processes used for problem solving or conflict resolution (32%).

In summary, an important outcome of participatory SD appears to be that stakeholders learn more about the underlying systems and problems generating conflict. However, although a number of interventions report increases in stakeholder insight into, and collaborative learning about, problems and conflicts, many of the outcome measures are simply too subjective (e.g., self-reports) to allow rigorous evaluation of the success of modeling interventions.

The Tyranny of SD

While most of this and the previous chapter expound on the advantages of participatory modeling for environmental conflict resolution, the fact remains that understanding and communicating the sheer complexity of many conflicts to a broad audience is extraordinarily difficult. Conflict dynamics may be incredibly complex, there may be vast disparities in the technical expertise or interest of the audience, and there may be deeply entrenched, value-based perspectives in conflict among stakeholders (Stave 2003).

Although true for any participatory modeling method, in some instances, SD can be a hindrance to conflict resolution, since the presumed objectivity and neutrality of models may have a cooling effect on discussion and eliminating further search for creative conflict solutions (Cioni 2007). SD, like other modeling techniques, is aimed at defining abstract representations of reality. However, literature on conflict resolution (e.g., Cobb 2001) does demonstrate that multiple realities can exist for different stakeholders. The participatory SD modeling process joins together the subjective, shared knowledge that stakeholders hold for a system. However, if the proper stakeholders are not represented during the modeling process, then this shared understanding of a system can have major deficiencies; this problem can compound the effects of power imbalances and inadequate stakeholder representation that occur during negotiations (Bryson 2004; Marcuse 2005; National Research Council 2008).

In this context, it is worth observing that SD models, with their focus on objectivity, transparency, and explicit cause–effect relationships, can produce premature solutions that, although based on incomplete information, can gain the stature of absolute truth and objectivity. If left unchecked, this stature can limit dissent, hide better solutions, and prevent legitimate inquiries into how and why the model is fundamentally incomplete. As John B. Robinson (1992), author of *Of Maps and Territories: The Use and Abuse of Socio-economic Modeling in Support of Decision Making*, wisely noted, "[b]y cloaking a policy decision in the ostensibly neutral aura of scientific forecasting, policy makers can deflect attention from the normative nature of that decision and the degree to which the future is being created (p. 148)." This concern points toward the surprising and unfortunate way that the "tyranny of expertise," can creep into the modeling process (Cioni 2007).

Summary

In this chapter, we have introduced SD modeling, a scientifically focused technique for modeling cause and effect among collections of interacting and changing variables. We have introduced the concept of a system, alluding to how we will amend and transform this idea in the next chapter on agent-based modeling. We have also introduced the philosophy of SD and inner-workings of the method distinguishing between qualitative systems thinking and quantitative SD modeling, as well as noting its mathematical basis in systems of ordinary different equations and numerical analysis. We have connected the SD method back to the last chapter, discussing participatory SD modeling in more detail, including aspects of the technique that distinguish it from other types of participatory modeling. We have also drawn distinctions between quantitative and qualitative SD approaches to conflict modeling and resolution, focusing on caveats, including the need for process and outcome evaluations, conflict during the modeling process, and the "tyranny of expertise" that can be brought on by SD modeling interventions.

In the next chapter, we will move on from SD to discuss agent-based modeling, a newer technique for characterizing the individual behaviors of actors engaged in conflict.

In presenting this contrasting framework for modeling conflicts, we will focus on the manner by which it places the SD technique on its head. While participatory SD efforts elicit stakeholder knowledge to describe system structure and explain system behavior, agent-based modeling describes individual stakeholder behavior, which explains the structure of conflicts as an emergent property of stakeholder interaction.

Questions for Consideration

1. Stave (2010, p. 2763) argued that "…[p]articipatory system dynamics modeling is more than simply eliciting knowledge from clients about the problem and the system. It involves building shared ownership of the analysis, problem, system description, and solutions or a shared understanding of the trade-offs among different decisions." *Given what you learned in the last chapter on participatory modeling, why would we want stakeholders to establish a shared ownership of a conflict?*

2. In his paper, *Of Maps and Territories: The Use and Abuse of Socio-economic Modeling in Support of Decision Making*, John B. Robinson (1992, p. 148) argued:

   > By cloaking a policy decision in the ostensibly neutral aura of scientific forecasting, policy-makers can deflect attention from the normative nature of that decision and the degree to which the future is being created.

   Discuss this criticism of modeling, and contrast it with the advantages of scientific forecasting as a decision support. How do you think that modelers can overcome this criticism?

3. The most famous SD model, called *Limits to Growth*, by Dennis and Dana Meadows and Jørgen Randers (1972), aimed to understand and predict patterns of worldwide natural resource use. The authors since updated their findings in *Beyond the Limits: Confronting Global Collapse, Envisioning a Sustainable Future* (1992) and *Limits to Growth: The 30-Year Update* (Meadows, Randers, Meadows 2004). Details for these books are listed in the references section of this chapter. *Check out these books and determine what has changed between these three models. Were the authors correct originally? Did their assumptions hold to be true? How did the newer models improve upon the original?*

4. Stave (2003) has argued that SD models are suited for systems steeped in feedback relationships and whose dynamics occur over a long-term time horizon. Therefore, SD can be badly suited for analyzing one-time decisions, such as single investment decisions or facility siting (Forrester 1987). *What techniques should we use for modeling these situations? How do those techniques compare to SD?*

5. Meadows and Robinson (1985) and, later, Stave (2002) highlight a common discrepancy between the long-term thinking provoked through the SD paradigm, and the short-term pressures faced by most decision makers. Modelers often recommend solutions requiring long-term, structural change. These solutions could be simple (e.g., use new sources of information) but are commonly more complex (e.g., revise authority and incentive structures, change goals) and, therefore, less politically palatable. When SD is used to search for root causes of problems, it is often "successful"—discovering that problems are caused by the stakeholders

themselves. Obviously, this can create tensions between clients and modelers that can be difficult to overcome (Rouwette, Vennix, and Mullekom 2002). *Can you think of an example where those suffering from an environmental problem have (perhaps unknowingly) created that problem themselves?*

Additional Resources

Interested in resources for learning SD online? MIT Sloan School of Management often has several of their SD graduate courses online (https://ocw.mit.edu/courses/sloan-school-of-management/), including *Introduction to System Dynamics* (Course number: 15.871), *System Dynamics II* (Course number: 15.872), *Applications of System Dynamics* (Course number: 15.875), and *Research Seminar in System Dynamics* (Course number: 15.879).

- The SD "Roadmaps" project created a collection of exercises and lessons for teaching yourself SD: http://web.mit.edu/sysdyn/road-maps/home.html.
- Check out the International System Dynamics Society's website (http://system dynamics.org/) to find many more opportunities to learn SD.
- Readers can experiment with and run lots of example simulations online using Forio's *Simulate* (requires Adobe Flash): https://forio.com/simulate/netsim/.
- The NetLogo software (discussed at length in Chapter 6 and used for models in Part III of this book) also contains a huge number of example models for reader experimentation: https://netlogoweb.org/.

Interested in SD software? Traditionally, the three main SD software packages have included Vensim (https://www.vensim.com/), STELLA/iThink (https://www.iseesystems.com/), and Powersim (http://www.powersim.com/). However, other packages have been developed in recent years, including Simile (http://www.simulistics.com/), NetLogo (https://ccl.northwestern.edu/netlogo/), and Anylogic, (https://www.anylogic.com/). Insight Maker (https://insightmaker.com/main) is a community aimed at creating and sharing SD models entirely online.

There have been efforts to create an open XML protocol to allow sharing of SD models and simulations, essentially making the various SD software packages interoperable. You can find out more about this XMILE project here: https://www.oasis-open.org/committees/xmile/.

One of the most interesting and powerful advances in SD software has been the development of pySD, a Python library facilitating the use of Vensim or XMILE files within the Python ecosystem, allowing for analysis, simulation, and linkages with the multitude of Python big data analytical capabilities, such as GIS [GRASS, ArcGIS] or statistical analysis [R]. Learn more about pySD here: https://github.com/JamesPHoughton/pysd.

Interested in SD applications? We refer readers to Sterman (2000) for more information on SD generally, and to Ford (2009), Videira et al. (2010), Deaton and Winebrake (1999), Stave (2010), and Beall and Ford (2010), for numerous examples of SD modeling for the study of environmental systems, many of which have included stakeholder groups. Additionally, Langsdale et al. (2007) describe SD applications in various case studies, including water resources management in Switzerland, Senegal, and Thailand; ecosystem management issues in Louisiana's coastal wetlands; South African fynbos ecosystems; Patuxent River watershed (Maryland, USA) management; vegetation management in Zimbabwe; water allocation issues in the Namoi River, Australia; transportation and air quality in Las Vegas, USA; Patagonian coastal zone management; and endangered species management in Illinois, USA. More details on best practices for using SD in socioecological modeling see Elsawah et al. (2017).

SD has also been extensively applied in environmental research, corporate planning and management, policy design, public management and policy, social science theoretical development, water resources, ecosystem services, and conflict resolution (e.g., see Videira et al. 2003; (Videira, Antunes, and Santos 2009). Given that SD modeling allows users to test the long-term effects of management decisions, policies, or strategies in uncertain and complex systems (Videira, Antunes, and Santos 2009; Winz, Brierley, and Trowsdale 2009), it is now frequently used as a part of corporate executive dialogs and policy analysis as a means of avoiding bias and misconceptions in business management decision making (Morecroft 1985; Sterman 2000; Saeed 2002).

Interested in making models simpler? Understanding conflicts and environmental systems can require a level of technical understanding, attention, and interest that may not be shared by all stakeholders. A major challenge in participatory SD concerns how to reduce conflict complexity during the modeling process, while clearly communicating how systems react to new policies or management regimes. One way to address this issue is to improve visualizations and general interactions with model output. A field of research in SD focuses on dynamic decision making, studying decision making in controlled environments often called "management flight simulators." These simulators can "abstract away" model complexity and present stakeholders with familiar measurements of the problem or system that they know well. Studies using these simulators attempt to mimic the types of decisions required in complex, dynamic environments, asking participants to manipulate decision variables, receive feedback, and test the relative effectiveness of certain decision aids (Rouwette and Vennix 2006). For more work in this area, see work by Bakken, Gould, and Kim (1992) and Rouwette, Grossler, and Vennix (2004).

Interested in the modeling process? While we have not discussed the significant literature that has studied the SD modeling process, a good starting point to better understanding this topic can be found in work by Costanza and Ruth (1998), who advocate the use of a three-step systems modeling process to enhance decision making regarding environmental problems, investments, and conflict resolution efforts. This process is composed of (1) a general, low-resolution scoping and consensus building model involving broad representation of stakeholder groups affected by the problem; (2) a research model containing more detailed and realistic attempts to replicate the dynamics of the particular system of interest; and finally, in the last stage, (3) a management model that is focused on producing a range of projected future scenarios that result from various management options and predicted environmental variations.

Costanza and Ruth (1998) test this participatory process for SD model construction using four case studies, including industrial systems (mining, smelting, and refining of iron and steel in the United States), ecosystems (Louisiana coastal wetlands and fynbos ecosystems in South Africa), and linked ecological–economic systems (Maryland's Patuxent River basin in the United States).

Interested in modeling bargaining with SD? In an interesting nonparticipatory conflict resolution application, Karamouz, Akhbari, and Moridi (2011) compare applications of bargaining models in the SD and Nash traditions (e.g., Karamouz, Akhbari, and Moridi 2011). *Nash* bargaining theory is a common modeling technique used in conflict resolution. It uses a utility function to represent different actor's preferences, as well as dissatisfaction points (agreements that will not be accepted by actors), and weights assigned based on the relative authority of each actor (Maskin 1999; Karamouz, Akhbari, and Moridi 2011).

Interested in participatory modeling with SD? A slew of participatory SD modeling interventions have been applied to environmental issues and conflicts, including work by Stave (2010), van den Belt (2004), Andersson et al. (2008), Beall and Zeoli (2008), Videira, Antunes, and Santos (2009), and Metcalf et al. (2010). Those interested in additional participatory modeling case studies and resources should check out http://participatorymodeling. org/, a website dedicated to creating a stronger community of modelers and researchers engaged in participatory modeling efforts.

You can also find out more about participatory modeling at *Integration and Implementation Insights* (also known as *I2Insights*; https://i2insights.org/), the blog of the Integration and Implementation Sciences team at the Australian National University, which is dedicated to pursuing better concepts and methods for understanding and acting on complex real-world problems.

Interested in Group Model Building (GMB)? Since the earliest years of the SD practice, professionals seeking to develop models of organizational problems have attempted to involve clients in the model development. GMB formally began in earnest in the late 1980s as Jac Vennix and his colleagues (1990) described procedures for developing dynamic models with the help of health-care professionals. Around the same period, Barry Richmond (1987, 1997) and many others began describing directed efforts for formally involving clients and stakeholders in model construction and analysis efforts.

Vennix (1996) and Andersen and Richardson (1997) consider GMB from the consultant–client viewpoint, while Korfmacher (2001) consider interventions from the government–public and scientist–public perspectives. GMB results have been fairly successful (Rouwette et al. 2011), and the number of projects involving business clients or other types of stakeholders has spread rapidly over the last several decades (Rouwette et al. 2002).

See Andersen and Richardson (1997) for more on the history of GMB, as well more information on notable work that involves groups for conceptualizing, formulating, and using models outside the SD field.

Interested in evaluating participatory modeling? Rouwette, Vennix, and Mullekom (2002) analyzed case studies to assess modeling process outcomes, discovering that during successful modeling interventions, stakeholders and modelers tended to take a broad perspective on problem identification, while jointly viewing learning as an important outcome. Building on this work, Rouwette and Vennix (2006) and Rouwette et al. (2011) developed measures for measuring modeling intervention outcomes.

The goals and techniques of these assessments can differ widely, as some researchers have called for evaluations of intervention outcomes (e.g., Andersen, Richardson, and

Vennix 1997; Rouwette, Vennix, and Mullekom 2002), while others have argued for evaluations of modeling processes (McCartt and Rohrbaugh 1989, 1995; Luna-Reyes et al. 2006). David Lane (2017) offers an insightful discussion of what he calls "behavioral system dynamics," which is intended to find ways of making better use of models and to study behavioral issues that need to be accounted for when using models for decision-making or decision-support purposes.

Interested in information elicitation? For more information on developing *scripts* for use during participatory modeling, which act as "recipes" for knowledge elicitation in modeling workshops, see Andersen and Richardson (1997) and Luna-Reyes et al. (2006). Additionally, Ford and Sterman (1998) describe a technique for eliciting formal modeling processes, including developing formal stock and flow diagrams, and specifying individual parameters and relationships.

Claudia Pahl-Wostl (2006) describes a sophisticated technique for eliciting personal perspectives about the interrelationships of variables in conflict systems. She introduces "hexagon modeling," a process in which participants describe key system concepts and how they are connected to problem or conflict situations. These concepts are then grouped and linked together to denote important relationships that connect different concepts together. By using these techniques, participants (1) reflect on their assumptions about neglected feedback effects and cause-and-effect relationships in the system and (2) discover differences between their mental models and those of other participants. Mental models derived in this manner act as a basis for the development of additional qualitative, conceptual, or quantitative computer models of the conflict.

References

Allen, Timothy, and Thomas W. Hoekstra. 1993. *Toward a Unified Ecology*. New York: Columbia University Press.

Andersen, David F., and George P. Richardson. 1997. "Scripts for Group Model Building." *System Dynamics Review* 13 (2): 107–29.

Andersen, David F., G. P. Richardson, and Jac A. M. Vennix. 1997. "Group Model Building: Adding More Science to the Craft." *System Dyanmics Review* 13 (2): 187–201.

Andersen, David F., Jac A. M. Vennix, G. P. Richardson, and Etiënne A. J. A. Rouwette. 2007. "Group Model Building: Problem Structuring, Policy Simulation and Decision Support." *The Journal of the Operational Research Society* 58 (5): 691–94.

Andersson, L., J. A. Olsson, B. Arheimer, and A. Jonsson. 2008. "Use of Participatory Scenario Modelling as Platforms in Stakeholder Dialogues." *Water SA* 34 (4): 3.

Antunes, Paula, Rui Santos, and Nuno Videira. 2006. "Participatory Decision Making for Sustainable Development – The Use of Mediated Modelling Techniques." *Land Use Policy* 23 (1): 44–52.

Bakken, B., J. Gould, and D. Kim. 1992. "Experimentation in Learning Organizations: A Management Flight Simulator Approach." *European Journal of Operational Research* 59 (1): 167–82.

Barlas, Y., and S. Carpenter. 1990. "Philosophical Roots of Model Validation." *System Dynamics Review* 6 (2): 148–66.

Beall, Allyson, and Andrew Ford. 2007. "Participatory Modeling for Adaptive Management: Reports from the Field II." In *25th International Conference of the System Dynamics Society*. Albany, NY: International System Dynamics Society. www.systemdynamics.org/assets/conferences/2007/proceed/papers/BEALL374.pdf

Beall, Allyson, and Andrew Ford. 2010. "Reports from the Field: Assessing the Art and Science of Participatory Environmental Modeling." *International Journal of Information Systems and Social Change* 1 (2): 72–89.

Beall, Allyson, and Len Zeoli. 2008. "Participatory Modeling of Endangered Wildlife Systems: Simulating the Sage-Grouse and Land Use in Central Washington." *Ecological Economics* 68 (1–2): 24–33.

Becu, Nicolas, P. Perez, A. Walker, O. Barreteau, and C. Le Page. 2003. "Agent Based Simulation of a Small Catchment Water Management in Northern Thailand: Description of the CATCHSCAPE Model." *Ecological Modelling* 170: 319–31.

Beierle, T. 1998. *Public Participation in Environmental Decisions: An Evaluation Framework Using Social Goals (Discussion Paper 99–06)*. Washington, DC: Resources for the Future.

Belt, Marjan van den. 2004. *Mediated Modeling: A System Dynamics Approach To Environmental Consensus Building*. Washington, DC: Island Press.

BenDor, Todd. 2009. "A Dynamic Analysis of the Wetland Mitigation Process and Its Effects on No Net Loss Policy." *Landscape and Urban Planning* 89 (1–2): 17–27.

BenDor, Todd. 2012. "The System Dynamics of U.S. Automobile Fuel Economy." *Sustainability* 4: 1013–42.

Braess, D., A. Nagurney, and T. Wakolbinger. 2005. "On a Paradox of Traffic Planning." *Transportation Science* 39: 446–50.

Bryson, John M. 2004. "What to Do When Stakeholders Matter: Stakeholder Identification and Analysis Techniques." *Public Management Review* 6 (1). 21–50.

Buchholz, Thomas S., Timothy A. Volk, and Valerie A. Luzadis. 2007. "A Participatory Systems Approach to Modeling Social, Economic, and Ecological Components of Bioenergy." *Energy Policy* 35 (12): 6084–94.

Checkland, P. B. 1988. "Soft Systems Methodology: Overview." *Journal of Applied Systems Analysis* 15: 27–30.

Cioni, Lorenzo. 2007. *How System Dynamics Can Be a Help or a Hindrance*. Pisa, Italy: Department of Computer Science, University of Pisa. www.di.unipi.it/~lcioni/papers/2007/systheory.pdf

Cioni, Lorenzo. 2009. "The Analysis and Resolution of Environmental Conflicts: Methods and Models." *Economia Aziendale Online* 1 (1): 17–41.

Cobb, S. 2001. "Dialogue and the Practice of Law and Spiritual Values: Creating Sacred Space: Toward a Second-Generation Dispute Resolution Practice." *Fordham Urban Law Journal* 28: 1017–2037.

Cockerill, Kristan, Howard Passell, and Vince Tidwell. 2006. "Cooperative Modeling: Building Bridges between Science and the Public." *JAWRA Journal of the American Water Resources Association* 42 (2): 457–71.

Cockerill, Kristan, Vincent C. Tidwell, Howard D. Passell, and Leonard A. Malczynski. 2007. "Cooperative Modeling Lessons for Environmental Management." *Environmental Practice* 9 (01): 28–41.

Costanza, Robert, and Matthias Ruth. 1998. "Using Dynamic Modeling to Scope Environmental Problems and Build Consensus." *Environmental Management* 22 (2): 183–95.

Deaton, Michael L., and James J. Winebrake. 1999. *Dynamic Modeling of Environmental Systems*. New York: Springer-Verlag.

Downs, Anthony. 2005. *Still Stuck in Traffic: Coping with Peak-Hour Traffic Congestion*. Washington, DC: Brookings Institution Press.

Dudley, Richard G. 2008. "A Basis for Understanding Fishery Management Dynamics." *System Dynamics Review* 24 (1): 1–29.

Elsawah, Sondoss, Suzanne A. Pierce, Serena H. Hamilton, Hedwig van Delden, Dagmar Haase, Amgad Elmahdi, and Anthony J. Jakeman. 2017. "An Overview of the System Dynamics Process for Integrated Modelling of Socio-Ecological Systems: Lessons on Good Modelling Practice from Five Case Studies." *Environmental Modelling & Software* 93 (July): 127–45.

Fisher, R., and William Ury. 1981. *Getting to Yes: Negotiating Agreements without Giving In*. New York: Penguin Books.

Ford, Andrew. 2009. *Modeling the Environment* (2nd Edition). Washington, DC: Island Press.

Ford, David N., and John D. Sterman. 1998. "Expert Knowledge Elicitation to Improve Formal and Mental Models." *System Dynamics Review* 14 (4): 309–40.

Forrester, Jay W. 1961. *Industrial Dynamics*. Cambridge, MA: MIT Press.

Forrester, Jay W. 1985. "'The Model' versus a Modelling 'Process.'" *System Dynamics Review* 1 (1): 133–34.

Forrester, Jay W. 1987. "Lessons from System Dynamics Modeling." *System Dynamics Review* 3 (2): 136–49.

Forrester, Jay W. 1994. "Policies, Decisions, and Information Sources for Modeling." In *Modeling for Learning Organizations*, edited by J. D. W. Morecroft and John D. Sterman, 51–84. Portland, OR: Productivity Press.

Forrester, Jay W. 2007. "System Dynamics—The Next Fifty Years." *System Dynamics Review* 23 (2–3): 359–70.

Gelman, Andrew, and Guido Imbens. 2013. *Why Ask Why? Forward Causal Inference and Reverse Causal Questions (Working Paper 19614)*. National Bureau of Economic Research.

Granger, C. W. J. 1969. "Investigating Causal Relations by Econometric Models and Cross-Spectral Methods." *Econometrica* 37 (3): 424–38.

Granger, C. W. J. 1993. "What Are We Learning about the Long Run?" *Economic Journal* 103: 307–17.

Karamouz, Mohammad, Masih Akhbari, and Ali Moridi. 2011. "Resolving Disputes over Reservoir-River Operation." *Journal of Irrigation and Drainage Engineering* 137 (5): 327–39.

Kleindorfer, George B., Liam O'Neill, and Ram Ganeshan. 1998. "Validation in Simulation: Various Positions in the Philosophy of Science." *Management Science* 44 (8): 1087–99.

Korfmacher, Katrina Smith. 2001. "The Politics of Participation in Watershed Modeling." *Environmental Management* 27 (2): 161–76.

Lane, David C. 1993. "The Road Not Taken: Observing a Process of Issue Selection and Model Conceptualization." *System Dynamics Review* 9 (3): 239–64.

Lane, David C. 2017. "'Behavioural System Dynamics': A Very Tentative and Slightly Sceptical Map of the Territory." *Systems Research and Behavioral Science* 34 (4): 414–23.

Langsdale, Stacy, Allyson Beall, Jeff Carmichael, Stewart Cohen, and Craig Forster. 2007. "An Exploration of Water Resources Futures under Climate Change Using System Dynamics Modeling." *Integrated Assessment* 7 (1): 51–79.

Leamer, E. 1983. "Let's Take the 'Con' out of Econometrics." *American Economic Review* 73: 31–43.

Leontief, W. 1982. "Academic Economics." *Science* 217: 104–7.

Luna-Reyes, Luis Felipe, Ignacio J. Martinez-Moyano, Theresa A. Pardo, Anthony M. Cresswell, David F. Andersen, and George P. Richardson. 2006. "Anatomy of a Group Model-Building Intervention: Building Dynamic Theory from Case Study Research." *System Dynamics Review* 22 (4): 291–320.

Lynam, T., W. de Jong, D. Shell, T. Kusumanto, and K. Evans. 2007. "A Review of Tools for Incorporating Community Knowledge, Preferences, and Values into Decision Making in Natural Resources Management." *Ecology and Society* 12 (1): 5.

Marcuse, Peter. 2005. "Study Areas Sites and the Geographic Approach to Public Action." In *Site Matters: Design Concepts, Histories, and Strategies*, edited by Carol J. Burns and Andrea Kahn, 249–80. New York: Routledge.

Maskin, Eric. 1999. "Nash Equilibrium and Welfare Optimality." *The Review of Economic Studies* 66 (1): 23–38.

McCartt, A. T., and J. Rohrbaugh. 1989. "Evaluating Group Decision Support System Effectiveness: A Performance Study of Decision Conferencing." *Decision Support Systems* 5 (2): 243–53.

McCartt, A. T., and J. Rohrbaugh. 1995. "Managerial Openness to Change and the Introduction of GDSS: Explaining Initial Success and Failure in Decision Conferencing." *Organization Science* 6 (5): 569–84.

Meadows, Donella H. 1991. *The Global Citizen*. Washington, DC: Island Press.

Meadows, Donella H. 2008. *Thinking in Systems: A Primer*. White River Junction, VT: Chelsea Green Publishing.

Meadows, Donella H., and J. M. Robinson. 1985. *The Electronic Oracle: Computer Models and Social Decisions.* Chichester, England: Wiley.

Meadows, Donella H., Dennis L. Meadows, and Jørgen Randers. 1972. *The Limits to Growth.* New York: Universe Books.

Meadows, Donella H., Dennis L. Meadows, and Jørgen Randers. 1992. *Beyond the Limits: Confronting Global Collapse, Envisioning a Sustainable Future.* White River Junction, VT: Chelsea Green Publishing.

Meadows, Donella H., Jørgen Randers, and Dennis Meadows. 2004. *Limits to Growth: The 30-Year Update.* White River Junction, VT: Chelsea Green Publishing.

Mendoza, Guillermo A., and Ravi Prabhu. 2006. "Participatory Modeling and Analysis for Sustainable Forest Management: Overview of Soft System Dynamics Models and Applications." *Forest Policy and Economics* 9 (2): 179–96.

Metcalf, Sara S., Emily Wheeler, Todd K. BenDor, Kenneth S. Lubinski, and Bruce M. Hannon. 2010. "Sharing the Floodplain: Mediated Modeling for Environmental Management." *Environmental Modelling and Software* 25 (11): 1282–90.

Morecroft, J. D. W. 1985. "Rationality in the Analysis of Behavioural Simulation Models." *Management Science* 31 (7): 900–16.

National Research Council. 2008. *Public Participation in Environmental Assessment and Decision Making.* Washington, DC: The National Academies Press.

Otto, P., and J. Struben. 2004. "Gloucester Fishery: Insights from a Group Modeling Intervention." *System Dynamics Review* 20 (1). 287–312.

Pahl-Wostl, Claudia. 2006. "The Importance of Social Learning in Restoring the Multifunctionality of Rivers and Floodplains." *Ecology and Society* 11 (1): 10.

Pearl, Judea. 2009. "Causal Inference in Statistics: An Overview." *Statistics Surveys* 3: 96–146.

Press, William H., Saul A. Teukolsky, William T. Vetterling, and Brian P. Flannery. 2007. *Numerical Recipes (3rd Edition): The Art of Scientific Computing.* Cambridge, England: Cambridge University Press.

Reuveny, Rafael, John W. Maxwell, and Jefferson Davis. 2011. "On Conflict over Natural Resources." *Ecological Economics* 70 (4): 698–712.

Richmond, Barry. 1987. *The Strategic Forum.* Hanover, NH: High Performance Systems.

Richmond, Barry. 1997. "The Strategic Forum: Aligning Objectives, Strategy and Process." *System Dynamics Review* 13 (2): 131–48.

Robinson, John B. 1992. "Of Maps and Territories: The Use and Abuse of Socio-economic Modeling in Support of Decision Making." *Technological Forecasting and Social Change* 42 (2): 147–64.

Rouwette, Etiënne A. J. A., A. Grossler, and Jac A. M. Vennix. 2004. "Exploring Influencing Factors on Rationality: A Literature Review of Dynamic Decision-Making Studies in System Dynamics." *Systems Research and Behavioral Science* 21: 351–70.

Rouwette, Etiënne A. J. A., Hubert Korzilius, Jac A. M. Vennix, and Eric Jacobs. 2011. "Modeling as Persuasion: The Impact of Group Model Building on Attitudes and Behavior." *System Dynamics Review* 27 (1): 1–21.

Rouwette, Etiënne A. J. A., and Jac A. M. Vennix. 2006. "System Dynamics and Organizational Interventions." *Systems Research and Behavioral Science* 23 (4): 451–66.

Rouwette, Etiënne A. J. A., Jac A. M. Vennix, and Theo van Mullekom. 2002. "Group Model Building Effectiveness: A Review of Assessment Studies." *System Dynamics Review* 18 (1): 5–45.

Ruth, Matthias, and Bruce Hannon. 2012. *Modeling Dynamic Economic Systems* (2nd Edition). Modeling Dynamic Systems. New York: Springer-Verlag.

Saeed, K. 2002. "System Dynamics: A Learning and Problem Solving Approach to Development Policy." *Global Business and Economic Review* 4 (1): 81–105.

Salerno, Franco, Emanuele Cuccillato, Paolo Caroli, Birendra Bajracharya, Emanuela Chiara Manfredi, Gaetano Viviano, Sudeep Thakuri, et al. 2010. "Experience With a Hard and Soft Participatory Modeling Framework for Social-Ecological System Management in Mount Everest (Nepal) and K2 (Pakistan) Protected Areas." *Mountain Research and Development* 30(2): 80–93.

Scott, Rodney J., Robert Y. Cavana, and Donald Cameron. 2016a. "Client Perceptions of Reported Outcomes of Group Model Building in the New Zealand Public Sector." *Group Decision and Negotiation* 25 (1): 77–101.

Scott, Rodney J., Robert Y. Cavana, and Donald Cameron 2016b. "Recent Evidence on the Effectiveness of Group Model Building." *European Journal of Operational Research* 249 (3): 908–918.

Scott, Rodney. 2018. *Group Model Building: Using Systems Dynamics to Achieve Enduring Agreement.* Singapore: Springer.

Stave, Krystyna. 2002. "Using System Dynamics to Improve Public Participation in Environmental Decisions." *System Dynamics Review* 18 (2): 139–67.

Stave, Krystyna. 2003. "A System Dynamics Model to Facilitate Public Understanding of Water Management Options in Las Vegas, Nevada." *Journal of Environmental Management* 67: 303–13.

Stave, Krystyna. 2010. "Participatory System Dynamics Modeling for Sustainable Environmental Management: Observations from Four Cases." *Sustainability* 2 (9): 2762–84.

Sterman, John D. 1987. "Testing Behavioural Simulation Models by Direct Experiment." *Management Science* 33 (2): 1572–92.

Sterman, John D. 1994. "Learning in and about Complex Systems." *System Dynamics Review* 10 (2–3): 291–330.

Sterman, John D. 2000. *Business Dynamics: Systems Thinking and Modeling for a Complex World.* New York: Irwin/McGraw-Hill.

Vennix, Jac A. M. 1996. *Group Model Building: Facilitating Team Learning Using System Dynamics.* New York: Wiley.

Vennix, Jac A. M. 1999. "Group Model-Building: Tackling Messy Problems." *System Dynamics Review* 15 (4): 379–401.

Vennix, Jac A. M., David F. Andersen, G. P. Richardson, and J. Rohrbaugh. 1992. "Model-Building for Group Decision Support: Issues and Alternatives in Knowledge Elicitation." *European Journal of Operational Research, Modelling for Learning* 59 (1): 28–41.

Vennix, Jac A. M., Jan W. Gubbels, Doeke Post, and Henk J. Poppen. 1990. "A Structured Approach to Knowledge Elicitation in Conceptual Model Building." *System Dynamics Review* 6 (2): 194–208.

Videira, Nuno, Paula Antunes, and Rui Santos. 2009. "Scoping River Basin Management Issues with Participatory Modelling: The Baixo Guadiana Experience." *Ecological Economics* 68 (4): 965–78.

Videira, Nuno, Paula Antunes, Rui Santos, and Sofia Gamito. 2003. "Participatory Modelling in Environmental Decision-Making: The Ria Formosa Natural Park Case Study." *Journal of Environmental Assessment Policy and Management* 5 (3): 421–47.

Videira, Nuno, Paula Antunes, Rui Santos, and Rita Lopes. 2010. "A Participatory Modelling Approach to Support Integrated Sustainability Assessment Processes." *Systems Research and Behavioral Science* 27 (4): 446–60.

Voinov, Alexey, and Francois Bousquet. 2010. "Modelling with Stakeholders." *Environmental Modelling & Software* 25 (11): 1268–81.

Winch, Graham W. 1993. "Consensus Building in the Planning Process: Benefits from a 'Hard' Modeling Approach." *System Dynamics Review* 9 (3): 287–300.

Winz, Ines, Gary Brierley, and Sam Trowsdale. 2009. "The Use of System Dynamics Simulation in Water Resources Management." *Water Resources Management* 23 (7): 1301–23.

Woltsenholme, Eric F. 1990. *System Enquiry, A System Dynamics Approach.* Chicester, England: Wiley.

6

Agent-Based Modeling and Environmental Conflict

We spend billions of dollars trying to understand the origins of the universe, while we still don't understand the conditions for a stable society, a functioning economy, or peace.
—*Dirk Helbing, "Sociophysicist" and Professor at ETH-Zurich (Cho 2009, p. 406)*

Introduction

In the previous two chapters, we have discussed approaches for using computer modeling to aid conflict resolution processes. These approaches help us to recognize that environmental conflicts also require us to understand the state of the ecosystem and its dynamics, the social and institutional dynamics prompting conflict, and the processes that might promote conflict resolution and long-term system viability. These modeling techniques are part of an expanding portfolio of approaches to understand the systems that underpin and either create, or prevent, conflicts. However, within this wide array of modeling techniques, there is an emerging necessity within the modeling world to represent the manners by which individual, interacting agents and institutions behave during conflict events (Beall and Ford 2007).

Agent based modeling is also often referred to as "individual-based modeling" in the ecology literature (Railsback and Grimm 2011) and as "multi-agent systems," a generalized term used in a diverse array of fields, including social sciences, robotics, interactive software design and many others (Bousquet and Le Page 2004; Gurung, Bousquet, and Trébuil 2006).

In this chapter, we will introduce agent-based modeling (ABM)[1] and explore how this technique can help researchers and stakeholders understand disputes in a much deeper way. ABMs have gained increasing interest in social and environmental modeling, including efforts to increase scientific understanding and improve environmental management and climate policy (Tango and Batiuk 2013; Berger and Troost 2014). Applications have also ranged from crowd management and traffic simulations, to urban, demographic, and environmental planning purposes, to studies of supermarket shoppers and the actions of stock market investors. Moreover, socially sophisticated ABMs have been used in environmental disciplines to describe and predict the way people ("social agents" or "stakeholders") are likely to behave in response to different stimuli given various decision rules (Billari et al. 2006).

ABMs represent a powerful tool for helping us to understand the complex interface between ecosystems and human actions. They have the abilities to couple and embed social interactions with environmental models, disaggregate decisions based on their origins and decision-making processes, and account for collective responses to changing

---

[1] We also refer to agent-based models as ABMs.

environments and management policies (Stillman, Wood, and Goss-Custard 2016). ABMs are also useful in situations where the future is unpredictable and traditional techniques for decision making are least effective.

In this chapter, we will begin by discussing the history and core elements of ABM as well as applications of the technique to conflicts. We will explore the links between ABM and complexity science, including contrasts between the approaches and philosophies of ABM vs. aggregate modeling (i.e., system dynamics [SD]). We will then delve into discussions of ABM development platforms and software environments, decision making and interactions between agents (including spatial interactions), and the incorporation of different paradigms of human behavior into models. Finally, we will discuss validation issues and the empirical basis of ABMs, along with the relatively recent movement toward participatory ABM.

What Is Agent-Based Modeling?

ABM is the computational study of autonomous agents that are capable of interacting with each other and their environment, according to behavioral rules. These rules can be simple or complex, deterministic or stochastic, fixed or adaptive, and heterogeneous or homogenous across the agents (Janssen and Ostrom 2006; Billari et al. 2006). ABMs explicitly represent the experience of discrete, individual entities—including their characteristics, decisions, and changes—and the dynamics of individual access to resources (Grimm and Railsback 2005).

ABM began with *cellular automata* (CA) concepts originated in John von Neumann's early work in the 1940s (von Neumann and Burks 1966). A CA model involves a set of uniform cells (most commonly squares), where each cell (automaton, agent) takes on one of a set of discrete states. For example, cells were either black or white in Schelling's (1971) famous racial segregation model, and in many urban growth models, cells represent either undeveloped or developed areas (e.g., Itami 1994; White and Engelen 1997). In a CA model, while each cell operates independently and maintains its own discrete state, all cells follow the same rules (i.e., agents act homogenously). These rules are based on the current and prior states of both the cell and its neighbors (called the "local" or "neighborhood" level; Moreno, Wang, and Marceau 2009). The idea is that these simple, homogeneous decision rules create emergent and complex patterns across the entire set of cells, which are also known as "global patterns." A variety of techniques, from areas including nonlinear dynamics and statistical physics, have been applied to understand cell interactions, state changes, and tendencies to replicate localized structures across the landscape ("self-organization"; Wolfram 2002; Billari et al. 2006).

An *"agent"* is any entity that is "...able to act according to its own set of rules and objectives (Billari et al. 2006, pp. 3–4)." The term "agent" in ABMs implies *agency* or the capacity of an entity to act in the world (Hitlin and Elder 2007).

By relaxing the homogeneity requirements of CA decision rules, ABMs have expanded beyond CA models to allow modelers to create heterogeneous agents that exhibit varying behaviors and decision-making rules. For example, Robert Axelrod's (1984) work on the repeated prisoner's dilemma (discussed in Chapter 3) modeled two agents with very different decision rules and conflict behavior.

The origin of ABM in the social sciences is often attributed to Thomas Schelling's (1971) landmark CA study on the rapid emergence of racial segregation in cities. Since that time,

ABM has become a popular technique in a wide range of fields, including economics, political science, sociology, geography, demography, ecology, and environmental sciences. Emerging social and environmental research areas are variously referred to as social simulation (Davidsson 2000: Gilbert and Troitzsch 2005), artificial societies (Etienne, Le Page, and Cohen 2003), individual-based modeling in ecology (see Grimm 1999; Grimm et al. 2005), agent-based computational economics (Batten 2000: Tesfatsion 2003), and agent-based computational demography.

The ABM modeling process can be described through six primary steps (Achorn 2004):

1. construct a conceptual model, built on initial knowledge of the conflict or problem, which defines agent types and behavioral rules, network relationships, and interaction mechanisms;

2. define the data to gather at each new time interval;

3. translate the "pseudocode" from the concept model into the computer language of the selected modeling framework (e.g., Repast, NetLogo, JAVA);

4. run the model;

5. calibrate the model (for conflict models, this often requires stakeholder involvement and/or role-playing games); and

6. validate the model to test robustness, including proposed policies, possible conflict solutions, and future conflict system changes.

In its modern form as a sophisticated modeling technique, ABMs begin with a set of synthetic, heterogeneous, and autonomous agents that represent individuals, groups, institutions, or any autonomous entity-making decisions. These agents' internal rules of behavior are defined by interactive principles that enable them to learn, adapt, and react to the actions of other agents and to the state of their ambient environment. These rules can depend on the state of the environment, the agent's location, the state of the agent, and the states of other agents (Voinov and Bousquet 2010). Contrasting many other modeling techniques, ABMs intuitively describe systems in terms of agents, objects, and the environment in which interactions occur (Barnaud, Bousquet, and Trébuil 2008). As a result, ABMs represent agent actions as a set of behaviors, including conditional decisions and concurrent subprocesses.

ABM has established itself as one of the most prominent approaches to modeling the complex dynamics observed in biological (DeAngelis and Gross 1992) and social systems (Epstein and Axtell 1996; Janssen and Ostrom 2006). The enormous number and variety of ABMs created over the years have drawn on a diverse set of theories and techniques. This diversity and magnitude of activity suggest that ABMs will not only produce new theories that connect emergent, system-wide behavior to individual-level processes in many different fields (Grimm et al. 2005; Smaldino, Calanchini, and Pickett 2015), but will also act as platforms for integrating theories across multiple social science and ecological perspectives (Robinson et al. 2007).

ABM Conflict Applications

While there are numerous examples where ABM has been used in environmental conflict (discussed later in this chapter's participatory modeling section), more extensive work has applied ABM to understanding decision making in violent conflicts, including civil wars or revolts, guerilla warfare, and autocratic governments. Here, we present and compare several representative examples of these applications.

We begin with Joshua Epstein's (2002) landmark modeling study of civil violence (unrest and revolts). This work explored the dynamics of decentralized civil unrest and interethnic violence, with the goal of developing improved policies for handling these types of conflict situations. Using an ABM, Epstein found unexpected deceptive behavior at the agent level, where "privately aggrieved agents hide their feelings when cops are near, but engage in openly rebellious activity when the cops move away (p. 7248)." This finding indicates thresholds that represent the low end of police efforts necessary to repress violence. When police efforts are below this tipping point, the model produces decentralized outbursts of violence, unrest, or full-scale revolts.

Along similar lines, Bhavnani, Backer, and Riolo (2008) used ABMs to understand how autocratic governments operate as a means of anticipating their domestic and foreign policy decisions. Drawing on extensive existing work on autocratic political regimes, the authors were able to use distinctions between different types of regimes and knowledge of how regimes transition among these types to create utility functions and risk profiles (again, drawing on this fundamental idea of agent-level optimizing behavior) to understand the internal dynamics of specific regimes.

Cioffi-Revilla and Gotts (2003) studied the evolution of nationalism and the effects of technology on war, using an ABM ("GeoSim") to explore how countries use resources for defensive and offensive purposes, eventually evolving into sovereign states that either survive (or thrive), collapse, or are conquered by other states. GeoSim represents the decision-making process of states in allocating resources to their territories (which is partly based on whether a province is at war), investigating ways that agents determine how they can use current resources to advance their interests.

Building on the groundbreaking conflict modeling work by Hirshleifer (1988, 1991) that we discussed in Chapter 3, Findley (2008) applied ABM to understand the complexity, evolution, and resolution (peace process) of civil wars, which pose important threats to international stability and peace. In this model, agents are explored as a heterogeneous mix of nonadaptive and adaptive combatants, the latter of which are capable of learning from history and their environment. Agent-based representations of civil conflicts allowed Findley (2008) to relax the common assumption that civil wars involve a fixed set of actors, whose allegiances and interests remain static throughout the conflict. Instead, civil war actors become dynamic, leaving and entering the conflict, and adapting new characteristics over time.

Findley's model explores the abilities of actors to adapt their use of resources (represented by ρ) in civil wars in order to optimally achieve their goals. Viewing battle as a way of increasing overall resources, adaptive agents select their investment in battle ("violence"; some proportion of ρ) in a game-theoretic fashion (after Wagner 2000; Reiter 2003; Slantchev 2003), where agent decision making is determined by the changing distribution of resources for fighting that results from battle outcomes. Findley (2008) finds that civil war dynamics, like we might expect for most conflicts, differ greatly when combatants have these important capabilities to adapt to and learn over time.

Another interesting application of conflict ABM can be found in Scott Wheeler's (2005) work studying the complex dynamics between Iraqi insurgents (guerilla warfare), United Nations peacekeepers, and the civilian populace in the armed conflict that had been raging since US intervention in the nation in 2003. In May 2003, following the occupation of Iraq, the Coalition Provisional Authority disbanded the Iraqi Army, dispersing trained soldiers into the civilian population while both inflicting loss of morale and removing financial support for individual soldiers. Wheeler (2005) set out to assess the feasibility of the NetLogo ABM software platform (also used in this book) to model several of the major behavioral, social, and physical interactions experienced in Iraqi guerilla warfare. He determined that the agent-based framework was useful for providing low-fidelity, preliminary insights into the nature of guerilla warfare, as well as simulating how agents behaved, acted, and reacted as they moved across the landscape, and detected and engaged with other agent types. In particular, the modeling study showed that fewer casualties could be sustained in situations where the local populace trusted the peacekeeping force enough to forewarn them of insurgent threats. This trust was likely heavily diminished when the Iraqi army was disbanded.

Finally, Taylor, Quist, and Hicken (2009) proposed a technique for capturing phenomena related to dynamic armed conflicts, using event data drawn from news reports. Lamenting the reality that ABMs are costly to design, program, and parameterize, the authors designed a system for simulating power dynamics among key actors. Based on a utility optimization model (rooted in the literature on rational choice theory; Coleman and Fararo 1992), this system simulates interactions among a configurable mix of political, military, economic, and social institutions, as they occur across conflict networks.

Taylor, Quist, and Hicken (2009) went on to experiment with this system, automatically producing ABMs from action-based event data in news reports. Using data on more than 41,000 events between 2001 and 2003 in the Philippines, their system linked actors together based on whether their interactions were positive or negative. They were able to successfully predict conflicts in 2004 at an accuracy equivalent to those achieved by experts, but with more than a 75% reduction in model construction time (and at substantially lower cost). While this approach requires substantially more research and input data, it represents a useful approach for aiding modelers and stakeholders (with specific conflict knowledge) in rapidly producing accurate models that can later be explored, validated, and improved upon by experts or stakeholders.

While there are many more applications of ABM to conflict research, the few that we have reviewed here are indicative of the growing field's wide range of goals, modeling approaches, and takeaways. In the next section, we will shift gears a bit to better understand the philosophical underpinnings of ABM and its relationship to other types of scientific research.

Complexity Science and ABM Philosophy

The methodologies that researchers and other professionals use to measure and assess social behavior have dramatically changed over the last 70 years (Achorn 2004). Between the 1940s and 1970s, computing and quantitative revolutions swept through many of the social sciences, including human geography (e.g., Keylock and Dorling 2004), psychology, economics, and political science. These revolutions produced many of the quantitative techniques (e.g., computer modeling and statistics) that we now rely upon for scientific analysis and measurement.

Developments in the field of ABM have been made possible partly by drastic improvements in information technology (including hardware and software; Billari et al. 2006). However, it is important to stress that ABM developments have also been made necessary in part by our inability to understand or resolve many problems through statistical analysis, alone. Many efforts to link theories of human behavior and empirical observations through statistical analysis have thus far missed the crucial notion that theoretically-grounded modeling techniques can be a useful aid in developing theories of human behavior, understanding and resolving conflict, and developing ways of dealing with alternative futures.

Although some view ABM as simply another quantitative research tool produced in parallel to the rapid growth in computing power since the 1970s, many others see ABM as a fundamentally new way of studying human behavior and interactions (Gilbert and Troitzsch 2005; Moss and Edmonds 2005; Janssen and Ostrom 2006). Along these lines, some have even called ABM, "...a new way of doing science..." (Achorn 2004), or a statement that echoes the title of Stephen Wolfram's (2002) magnum opus on CA and complexity, "A New Kind of Science."

"Complex systems" are defined as those systems, composed of many interacting parts, where nonlinearities give rise to properties such as feedback loops, emergence, adaptation, and *chaotic* behavior, which is seemingly random behavior that is based on the system's sensitivity to its initial conditions (Lewin 1992; Byrne 1998). "Emergence" is a phenomenon whereby global, system-level patterns or behavior arise as a result of interactions among the parts of the system. In ABMs, simple interactions between agents can yield complex behaviors for the group as a whole, and this system-level behavior can differ substantially from the behavior of individual agents (Holland 1998).

Understanding and predicting how systems of interacting entities work is now one of the most significant challenges confronting modern science, including social science (Grimm et al. 2005). This is one of the core challenges of modern complexity science and its focus on *complex systems*. Complex systems involve many elements interacting in ways that lead to phenomena like self-organization and emergent behavior, thereby making the whole of the system greater than simply the sum of the individual parts.

Cho (2009) uses the example of a car, which is complicated, but not complex: each part interacts with other parts in a predictable way. However, cars on a highway form a complex system as drivers pass each other, "rubberneck" (slowing to observe traffic stops or collisions), and react to the seemingly random shifts in speed and driving style of other drivers. Taken together, these individual behaviors lead to so-called "phantom" traffic jams (which have no specific cause), affecting the group of drivers and slowing traffic substantially. Complex systems are sensitive to both initial conditions and small changes in agent behaviors or interactions; tiny alterations can lead to wide swings in system structure and behavior, a phenomenon which is widely known as the "butterfly effect" (Hilborn 2004).

Techniques for studying complex systems include ABM, evolutionary game theory, network analysis, and many other techniques (Byrne 1998). Drawing on these techniques, modern complexity theory has attempted to arrange system behavior into categories based on linearity, knowledge, and dynamicism (Achorn 2004). This typology echoes the complex and adaptive nature of social and environmental systems, which often exist at the "edge of chaos," where small changes in initial conditions (e.g., resources historically available for conflicting entities) or individual agent whims can create completely new system-level behavior and feedbacks (Mora and Bialek 2011).

Feedbacks are inherent to the concept of *complex adaptive systems*, which have the additional attribute of having a goal to which the system can adapt (Grimm et al. 2005). The core concept is that individual "adaptive" agents can continuously learn and change to their surroundings, instead of following a fixed response behavior to external stimuli. This adaptation can be modeled in a more sophisticated manner by simulating individual agents' objectives, expectations, and delays in adaptation, e.g., delays in acquiring information and changing strategy (Billari et al. 2006). ABM can thus simulate nonlinear and adaptive processes that are initiated at the local ("micro" or individual) level and propagate to broader scales to generate complex and self-regulating structures, which often have feedbacks to agents to influence individual agent decision making.

"Resilience" is generally defined as the ability of a system to cope with disturbance and adapt under stress in order to maintain structure and function. The concept of resilience originated in ecology (Holling 1973) is now used widely in many fields, including urban and regional planning (Beatley 2009), economics (Briguglio et al. 2009; Vitro et al. 2017), and political science (Chu 1999).

As part of the complex adaptive systems' lexicon, *resilience*, has been promoted as a concept for integrating the management of social–ecological systems (Schlüter and Pahl-Wostl 2007). In complex environmental conflicts, there are often little-understood interactions linking the social and resource systems that determine their abilities to adapt, change, or collapse (Anderies, Walker, and Kinzig 2006; Perrings 2006). As a result, the need remains to understand conflicts from the point of view of complex, coupled socioecological systems. Modeling approaches, especially agent-based approaches, are valuable tools for exploring how individual agent behavior creates system structure and emergence of system characteristics like resilience (Schlüter and Pahl-Wostl 2007).

It has been difficult to establish the role of ABM in the pantheon of the philosophy of scientific research; many mention the logical paradox of ABMs, which are based on system assumptions, thus acting deductively, but fail to prove theorems as scientific deduction would suggest. Likewise, ABM data outputs ("simulated" data) can be analyzed in an inductive manner but do not necessarily represent the real world, thereby deviating from classic scientific inductive practices (Billari et al. 2006).

As a result, Achorn (2004, p. 3) and Boulanger and Br.chet (2005, p. 344) refer to the study of emergent behavior to be a "bottom-up" view of science. Additionally, Robert Axelrod (1997, 2006) has described ABM as a "third way of doing science" that augments techniques of induction and deduction. He observed that "[w]hereas the purpose of induction is to find patterns in data and that of deduction is to find consequences of assumptions, the purpose of ABM is to aid intuition."

A Departure from Previous Views of Structure and Behavior

The SD paradigm discussed in the last chapter parallels modern ecological theory in conceiving system behavior as a product of system structure (Allen and Hoekstra 1993; Sterman 2000). That is, the system output (behavior) that we witness and measure is the result of underlying system structures. When exposed to certain conditions (inputs), these structures create complex, often unexpected, and sometimes counterintuitive outcomes. An extensive literature now details the difficulties in connecting policy interventions to underlying system structure, including problems like policy resistance (e.g., BenDor 2012) and failures to intervene properly within feedback loops (Richardson 1991). In order to understand system structure, system dynamicists look toward historical system behavior ("reference modes;" Radzicki 1989) to create dynamic hypotheses of system behavior.

Contrasting this, ABM efforts do not begin with theories of system structure or even with observations of system behavior; given that re-creation of system-scale dynamics must begin at the individual level, ABMs require a different theory connecting behavior and structure together. Noted ABM practitioners, Volker Grimm et al. (2005), have argued that system structure—the arrangement of interacting entities—is the result of individual behavior taken in aggregate. That is, due to the behavior of many agents, system structure is actually an *emergent* phenomenon. Having observed that ABM structures are often chosen in an *ad hoc* fashion, without sufficient focus on analyzing and validating model applicability to real-world systems, Grimm et al. (2005, p. 987) propose a strategy for understanding how, why, and when individual behavior occurs, which they term "pattern-oriented modeling":

> Patterns are defining characteristics of a system and often, therefore, indicators of essential underlying processes and structures. Patterns contain information on the internal organization of a system, but in a "coded" form. The purpose of [pattern-oriented modeling] is to decode this information.

BOX 6.1   ABM DEVELOPMENT PLATFORMS AND SOFTWARE ENVIRONMENTS

There are numerous software platforms that have been developed over the last 30 years for creating ABMs, including AnyLogic (Borshchev 2013), RePast (North et al. 2007), NetLogo (Wilensky 1999), SWARM (Minar et al. 1996), and many others. Most platforms are based around object-oriented computer programming languages, which are based on the concept of "objects," which can have unique attributes (data fields) describing the object, and procedures (known as "methods") that describe the object's actions or behavior. Each object exists as a copy ("instance") of a larger defined "class," which defines the set of objects and their attributes. Objects then interact with each other in ways that make up the design of computer applications (Wu 2009). Many ABM platforms are based in JAVA, an efficient and popular object-oriented language, whose code can be compiled to run on any Java virtual machine, independent of specific computer architecture (e.g., Mac, PC, or other processor design). The object-oriented nature of JAVA, C++ and Smalltalk, makes them ideally suited for ABM, which can draw on the language's object-oriented structure to flexibly create an arbitrary number of individual (and heterogeneous) agents with numerous properties.

The computer science concept of *"abstraction"* means that modeling software shields modelers and users from many of the detailed constructs, such as pointers, class instantiation, memory management, which would have to be created manually in languages that operate at lower levels.

Several ABM platforms are designed to "abstract away" JAVA programming from users by translating user model designs from user-friendly interfaces (e.g., AnyLogic) or higher-level (i.e., easier to use) languages into JAVA behind the scenes. These high-level, structured programming tools make the creation, replication, and publication of simulation results easier and less prone to error (Ramanath and Gilbert 2004; Tobias and Hofmann 2004).

In this book, we are primarily relying on the NetLogo modeling environment (Wilensky 1999), a user-friendly ABM environment created by Dr. Uri Wilensky at Northwestern University's Center for Connected Learning and Computer-Based Modeling, which is located at Evanston, IL (since 2000).

Similar to the use of historical reference behavior at the system scale as a form of model verification, we can think of this concept as an application of structured scientific analysis to observed agent-level patterns, helping us to test alternative theories about how and why agents decide and act the way that they do. As a result, ABMs created in this manner are focused on understanding the patterns of collective action that emerge from large numbers of agents who follow observed rules of behavior. The complex social patterns that emerge at the macro-level are then viewed as the results of individual-level mechanisms for responding to external stimuli (Railsback and Grimm 2011).

Comparing Individual and Aggregate Modeling

As we discussed in Chapter 4, an important ethical responsibility of modelers involves first understanding a system and then evaluating the simulation methodology to apply; that is, the choice of modeling technique should be driven by its ability to properly represent and intervene in a problem, rather than problems being framed around the particular modeling technique known to modelers. With this in mind, we must observe how ABM parallels and augments approaches like SD, focusing on each technique's context-specific strengths and weaknesses in helping us to understand the system or problem of interest.

Like the SD method discussed in the last chapter, ABMs function as "virtual laboratories" where we can create controlled experiments to understand uncertainty and system organization. SD models treat system behavior in aggregate; variables summarize the actions of groups of individuals, enabling faster analysis, simulation, calibration, and model validation. As a result, SD models have spread in popularity, aided by the development of user-friendly, drag-and-drop software tools that facilitate model construction. However, there are numerous situations where models need to be able to explicitly represent the characteristics, behavior, states, and decisions of individual members of a population. Achorn (2004) argues that ABMs are useful for social science problems, where:

1. Agents are highly heterogeneous and problem solutions require individual observation and disaggregation from the group;
2. Agent interactions are discontinuous, complex, or highly nonlinear;
3. Uncertainty and complexity present challenges to useful study by other techniques;
4. Emergent behaviors are key outcomes seen in the systems being modeled;
5. Agents interact through complicated networks, either social or geographic; or
6. Agents can strategize or adapt over time, exhibiting dynamic or evolutionary behavior (Hofbauer and Sigmund 1998).

We can begin by considering four of these six factors to address situations where various facets of models are discrete, heterogeneous, informationally asymmetric, and spatially complex.

Discreteness and Heterogeneity

As the name implies, ABMs are typically well suited to domains where the natural unit of decomposition are individual, discrete entities. Specifically, ABMs can be useful for exploring *heterogeneous* (and hierarchical) phenomena, which may occur at multiple spatial or temporal scales. For example, ABMs may be advantageous where there are lots of different types of individuals interacting, and that heterogeneity could impact model outcomes. Explicit representation of heterogeneity, whether it be characterized by a typology of agents or by the "uniqueness" of individual agents (which vary continuously in a population; Brown and Robinson 2006), is a chief advantage of ABM, particularly when agent differences heavily influence adaptation and decision-making dynamics.

Information Asymmetry

One type of heterogeneity concerns the information that individuals can access. In contrast with aggregate models, where information is available system-wide, ABMs tend to define behaviors in terms of resources or information that are accessible to specific agents (Parunak et al. 1998). ABMs can be advantageous in situations where agents' internal behaviors and belief structures are intentionally kept from the rest of the system, which often occurs in conflict or competitive situations.

Spatial Complexity

Although aggregate models can represent space as discrete zones (e.g., Wils 1974), their abilities deteriorate quickly when we want to explore spatial interactions that occur across networks of individuals; for example, this applies to models of policies or diffusion where networks are interrupted (e.g., internet or electricity outages; disease quarantines; BenDor and Metcalf 2006) or dynamic (e.g., residents moving within a city; Metcalf and Paich 2005). In these cases, ABMs can be particularly useful as they can be used in a straightforward manner to understand dynamics within complex landscapes, including those that are uneven or "irregular" spaces (e.g., through a network of disconnected islands; BenDor and Kaza 2012).

The *"ecological fallacy"* occurs when we assume that the average of a population is representative of the individuals in that population or when we assume that any given individual will act like the average of their group (Robinson 1950; Keeling and Grenfell 2000). That is to say, the "average" person may be very rare in many situations.

In considering these modeling aspects—unit of analysis, heterogeneity, and spatial complexity—it becomes clear that choices to implement models at the agent level are inherently tied to "model fidelity" (a term we attribute to Nate Osgood [2009]); in some cases, even when an agent-centered perspective can lead to a more complex model, it can actually help to simplify the model creation process. This can occur for a variety of reasons.

For instance, if we have both data and a strong understanding of the system of concern at the individual level (e.g., individuals are either infected or not), the modeler can avoid many of the challenges associated with aggregate modeling. These challenges include issues with deriving causal relationships connecting aggregated quantities (e.g., average interaction and infection spread rates within a heterogeneous population). Deriving model parameters to describe aggregate relationships can lead to the loss of nuanced data about individual dynamics, thereby creating *ecological fallacy* problems. Finally, in participatory modeling situations, agent-driven model logic can lead to greater stakeholder acceptance if it is easier to understand, more transparent, and seen to be more representative of real-world behavior than similar aggregate models (Parunak, Savit, and Riolo 1998).

Decision Making and Agent Interactions in ABMs

Generally, agents in ABM models are represented entirely through their internal states (parameters specific to the agent; e.g., location, age, race, income, goals), their rules for modifying their states and their rules for behaving or interacting with other agents (decision making; e.g., agents that optimize their behavior over the short or long term, agents that plan).

Agent Decision Strategies

Specifying individual agent rules allows us to explore alternative representations of human behavior in conflicts, including situations where actors make decisions with limited resources, biased perceptions of reality, and incomplete or incorrect information. Moreover, exploring rules used by agents can allow us to explore the social dimensions of conflict, including aspects like imitation and competition, information exchange, cooperation, and other factors that heavily influence agent-level processes and macro-level emergent outcomes (Barnaud, Bousquet, and Trébuil 2008; Watkins et al. 2013).

A complete understanding and representation of agent-decision making behavior involves establishing heuristics that include the functions transforming agent interactions or stimulus into individual responses. Robinson et al. (2007) note that in computer science terms, these are commonly represented as a series of "IF... THEN" clauses, which can be informed by research, expert knowledge elicitation, or as we will discuss later in this chapter, by participant observations, role-playing games, and agent surveys.

The rules through which ABMs represent individual agent decisions and interactions can be extraordinarily complex or quite simple. Those who propose complex agents often contend that agents act on the basis of behavioral rules and objective functions (extending from sociological and cognitive psychological theories) that have evolved from their individual extended interactions and learning experiences in complex environments; that is, the structure of agent decision making is shaped uniquely by their environment (Billari et al. 2006).

Conversely, researchers like Robert Axelrod (1997, 2006) have argued that the most interesting modeling results occur when simple, agent-level rules produce complex patterns at the global level. The level of complexity that should be given to agents, therefore, may follow the dictum of the so-called "keep it simple, stupid" (KISS) principle, mirroring Antoine de Saint Exupéry's poignant reflection (1939, *Terre des Hommes*):

> It seems that perfection is reached not when there is nothing left to add, but when there is nothing left to take away.

ABMs across social science fields have produced stylized thinking about how to represent agent decisions, including mathematical representations of agent strategies that act as constrained optimization rules.[2] The choice of how to represent the behavior of human actors in an ABM has a strong influence on model results (Schlüter and Pahl-Wostl 2007). Decisions about behavior can include specific ways of representing rationality (Gintis 2000), uncertainty, and the embedded social complexity in which actors make decisions (Bryant 2003, 2010).

Billari et al. (2006) remind us of the specific nature of agent design, arguing that real-world agents attempt to reduce complexity according to their wants and needs (following seminal work by Herb Simon 1956), adjusting their social environment and relationships along the way. In fact, interactions among simple agents can create vast complexity, while interactions among complex agents can lead to simple emergent (global) behavior, despite their individual potential for incredible complexity. Figure 6.1 demonstrates Grimm et al.'s (2005) contention that there is a non-linear relationship between model complexity and model "payoff", which they define as a model's usefulness and "structural realism", or its ability to produce predictions that match real-world observations. They define a zone where models of intermediate complexity offer high payoffs and are guided by observation of multiple patterns (data) collected at different scales.

Agent Interactions

ABMs of social interactions start with rules and assumptions believed to replicate how individual agents behave, interact, and learn (Gilbert and Troitzsch 2005). Agent interaction can occur either through direct contact or communication or through mutually created changes in the surrounding environment (e.g., ecosystems or economy; Becu et al. 2008). Data recorded about these behaviors and emergent dynamics can then be treated as experiments, and analyzed inductively through qualitative or quantitative methods (Achorn 2004). This is a practice that Joshua Epstein (2006), a pioneer in agent-based techniques, calls "generative social sciences."

Among the many aspects to consider in structuring agent interactions, it is important to note that many different types of agents can be defined in a model, each with different

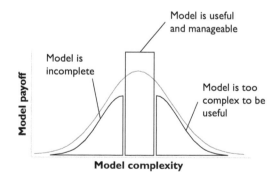

FIGURE 6.1
Trade-off between model complexity and model payoff. Adapted from Grimm et al. (2005).

---

[2] Originally introduced by Hirshleifer (1988, 1991), we offer a detailed discussion of this idea in the next chapter.

decision processes and behavioral frameworks. In an environmental illustration of this, Monticino et al. (2007) describe a model of land use in north-central Texas that includes four classes of agents, including landowners, homeowners, developers, and government agents. In this case, decisions made by one agent can influence other agent types; a landowner may decide to sell land, influencing future developer agent actions.

Spatial Interactions

Additionally, ABMs can realistically represent the spatial environments (i.e., landscapes, locations, spatial interconnections) where agent interactions occur (Parker et al. 2003; Heppenstall et al. 2011). Agents in these models require rules dictating how they move or interact across space as well as how location or spatial relationships to other agents affect their actions (Barnaud et al. 2013). This is important, as heterogeneity of agents can be directly due to an agent's location or their connectedness in social networks (Polhill, Gimona, and Aspinall 2011). For example, location can determine the quality of farmland and an agent's resulting farm-related income (see Chapter 10). Models that explicitly rely on spatial or network relationships can also integrate these factors into the way that agents seek to achieve their goals, including evolving their conflict strategies based on the location of other agents or changing their own location.[3]

Human Behavior and ABMs

The literature on ABM is now often segmented into the type of behavioral framework implemented, such as bounded rationality (Arthur 1994; Shakun 2001), utility maximization or optimization (Hirshleifer 1988, 1991; Reuveny, Maxwell, and Davis 2011), satisficing (Simon 1956; Scheffran 2000), character-centered motivations (Bryant 2004), or evolutionary processes (Hofbauer and Sigmund 1998; Reschke 2001). Although Robinson et al. (2007) divide behavioral models into decision making, learning, and adaptation models, it is now evident that the behavioral theories and mechanisms driving these models are becoming more diverse and sophisticated, blurring these divisions rapidly.

There has been a long history of efforts aimed at integrating theoretical models of human behavior into ABMs, starting with Robert Axelrod's (1984) exploration of the logic of cooperating agents in a repeated Prisoner's Dilemma game (discussed in detail in Chapter 3). Later work by Danielson (2002), Elliott and Kiel (2002), and many others has continued to explore the relationships between cooperation or competition and reciprocity or other behaviors, recognizing the need to empower agents with a wider range of behaviors. Part of this includes the use of more sophisticated theoretical frameworks to articulate refined model structures, which more accurately describe the cognitive dynamics of individual agents. Work in demography and population studies has also included efforts to integrate feedback structures that combine demographic behavioral theories at the individual level with larger-scale demographic outcomes (e.g., Billari et al. 2006).

---

[3] Additionally, Castella, Trung, and Boissau (2005) discuss the ways in which ABMs can integrate factors at hierarchical levels, including individual (e.g., household, farms) and landscape (e.g., village) level observations and dynamics.

ABM Disruptions to Economic Theory

In general, computational modeling has been used in the scientific process as a way of understanding and predicting facets of complex systems, usually within the boundaries of individual disciplines like hydrology or atmospheric sciences. Along these lines, economists have also long drawn on modeling techniques to understand social, economic, and ecological systems, especially those that entail coupled natural and human interactions (Prell et al. 2007; Link, Schneider, and Tol 2011). While many of these economic models have been based on fairly stringent assumptions such as perfect information, optimal behavior, and rational choice (Simon 1991), ABM is quickly becoming an important tool for analyzing alternative economic theories (Barnaud, Bousquet, and Trébuil 2008).

Much of the work occurring at the interface of modern social science and ABM has been driven by the movement of mathematicians, computer scientists, physicists, and other natural scientists, into areas traditionally studied in social sciences, such as urban growth, terrorism, and social dynamics (Cho 2009; Galam 2012). These researchers are gradually reconceptualizing economic dynamics by infusing economic theory with techniques from complexity theory and science, such as ABM. As a result, researchers are beginning to identify fundamental flaws in traditional, top-down economic models, including poor assumptions about "representative" agents, rigid decision rules about information and agent behavior, and market equilibrium constraints imposed on the emergent dynamics of the system.

For example, to create tractable and well-defined mathematical portrayals of "representative agents," economic models assume that the frequencies of individual characteristics vary under convenient and arbitrary statistical distributions (Cho 2009). These bell curves often have nothing to do with reality, fail to address unexpectedly extreme values ("fat tails"), and ignore correlations leading to economic booms and busts (Paich and Sterman 1993; Arquitt, Honggang, and Johnstone 2005).

As discussed in previous chapters, another criticism of economic modeling, generally, centers on the "rational actor paradigm," whereby traditional models assume that individuals know their preferences, have access to complete information about their environment, and have unlimited capacity to process information about their decisions and their potential consequences (Billari et al. 2006). Increasingly, some economists have objected to the rational actor paradigm for a variety of reasons, including unfounded assumptions about agent homogeneity, lack of agent interactions (e.g., Kirman 1992), oversimplification of the role of agent social networks (creating uncertainty, information deficits, and asymmetries; Grimm and Railsback 2005), and the unpredictable behavior of agents within this network.

> *Decision theory* is the field addressing the ranking and selection of the options that individuals are faced with, according to their preferences (Billari et al. 2006; Aknine 2012).

ABMs now allow economists to explore and relax classic assumptions by treating population groups as heterogeneous individuals that interact to create emergent economic dynamics that function less like monolithic economic forces, and more like the aggregate outcomes of many individual decisions (Billari et al. 2006). By defining heterogeneous agent characteristics and behavioral modes (including communication, interaction, and learning patterns), and calibrating models using experimental data on human subjects, ABMs can apply complex and robust aspects of *decision theory*, making "utility optimization" just one of many possible agent strategies or behaviors (Billari et al. 2006).

The evolving field of *agent-based computational economics* is focused on studying the economic dynamics associated with decentralized market economies, where autonomous, adaptive, and interacting agents create emergent macroeconomic patterns. By using ABMs to represent economic dynamics, the hope is that this emerging field will build integrated, nuanced economic theories of phenomena such as imperfect competition, trade network formation, inductive learning, and the coevolution of agent behaviors with economic institutions (Tesfatsion 2003).

Drama Theory

In one of the most sophisticated efforts to incorporate more advanced social behavior into ABMs, Jim Bryant (2004, 2015) introduces *drama theory*, a characterological depiction of conflict and cooperation, as a potentially powerful theoretical tool for exposing, simulating, and assessing the "characteristic patterns of confrontation management Bryant (2004, p. 87)." Drama theory depicts agent interactions as a set of "characters" that each hope to convince the other agents to adopt their advocated solutions by exerting influence. This influence comes in the form of "fallback actions" (e.g., continuing to pollute or harvest a resource at historical rates), which each agent may unilaterally adopt in an attempt to force others to comply, thereby creating credibility problems and questions about agent trustworthiness.

Bryant (2004) suggests that modelers could assess agents' collaborative styles and describe them in terms of how agents' characters would handle these dilemmas, thereby simulating how they would handle situations in ABMs of complex conflicts. For example, do agents tend to resign their position and concede to others? Do they become aggressive or hostile, digging in their heels and deepening their positions?

In assessing these traits, Bryant (2003) describes a series of six dilemmas that challenge parties as they attempt to establish a shared solution in the face of ongoing conflict or implement joint action during periods of cooperation. Four of these dilemmas undermine the agents' influence, whereby an agent (1) finds that their fallback action is detrimental to their own interests, (2) prefers solutions presented by other agents to their own fallback action, (3) prefers other agent's solutions to their own, and (4) sees that others prefer their fallback actions instead of solutions. Two other cooperation dilemmas emerge when an agent prefers their own course of action to an agreed solution, and when an agent believes that another agent may prefer to break an agreed solution.

The contention is that agents will seek to eliminate these dilemmas, which are uncomfortable and are reflected badly in the social arenas in which they operate. Therefore, ABM dynamics arise from the emergent behavior of interconnected agents creating and resolving these dramatic dilemmas. While we are not aware of specific examples of ABMs where agents' decisions are driven by drama-theoretic rules, it remains a novel and sophisticated way of integrating complex agent behavior that is supported by social–psychological theory.

It is worth noting that drama theory is an approach to structuring agent decision rules that is far more complex than the method we describe in the next chapter. However, the theoretical ability to portray the cognitive and collaborative styles of individual agents represents an important area for future work and in-depth application of our technique.

Validation and the Empirical Basis of ABMs

Throughout this chapter, we have discussed the many nuances, capabilities, and applications of ABM that have been inspired by real-world physical or social systems. Many models are able to closely represent complex facets of disputes (e.g., Becu et al. 2003), including multiscale, adaptive, noncontinuous, and dynamic interactions characterizing natural resource systems and conflicts (Holling 1978; Holling, Gunderson, and Ludwig 2002).

However, the enduring reality is that most models never move beyond a "proof-of-concept" level and are never empirically tested in the rigorous manner that professionals require for other modeling techniques (Schlüter and Pahl-Wostl 2007).[4] Our ability to generalize models from specific cases, and our ability to "scale up" the interaction processes of several agents to many agents, are two of the major remaining challenges in ABM (challenges that are often trade-offs between one another). These observations were made over a decade ago by Janssen and Ostrom (2006) and unfortunately, remain true today.

In response to this reality, some researchers and practitioners argue that ABMs have historically always been, and may always be, been better suited as exploratory tools than predictive mechanisms for representing human behavior (e.g., Bousquet et al. 1999; Billari et al. 2006). However, as confidence in ABM as a valid scientific technique has gradually increased, and relevant data have become available, there has been a growing push to begin informing ABMs using empirical techniques. The emergent nature of ABM dynamics only strengthens the need for rigorous analysis of simulations and how they contrast with reality.

Omissions of important social, physical, or other aspects of systems (or treating these factors as exogenous or static) and failures to represent the key feedbacks of real-world complex systems can lead to substantial model errors and stakeholder distrust in the model (Ramanath and Gilbert 2004). Without this integration, models will continue to fall short of predicting or understanding agent responses to environmental change (Pahl-Wostl 2006). It is the hope of many that systematic, interdisciplinary research into empirical ABM techniques will eventually overcome these challenges. As Edmonds and Hales (2003, p. 12.2) cautioned, "[i]f we are to be able to trust the simulations we use, we must independently replicate them. An unreplicated simulation is an untrustworthy simulation."

Methods for Empirically Informing ABMs

The rise of data science and "big data" has created faster growing demand for empirically parameterized ABMs (Wang et al. 2016). Since the early 2000s, research has explored mixed-method techniques for collecting data to parameterize ABM. These include the qualitative characteristics that describe heterogeneity and variation among agents, and the quantitative techniques to explain variations in the demographic, socio-economic, network connections, and landscape contexts of individual agents (Robinson et al. 2007). Of particular importance is the use of experimental design for learning about individual agent behavior and decision-making processes, collaborative styles (Luna-Reyes et al. 2006), preferences, and utility functions.

"Stylized facts" are empirical findings that are taken as being broadly true, which simplifies and removes the probabilistic nuance that traditionally accompanies statistically based analysis results. *"Face validity"* refers to the transparency or relevance of a model as it appears to participants. It is the extent to which a model is subjectively seen as representing the system it purports to represent.

---

[4] This statement is not without debate, as others argue that verification should be performed by experts (Bonabeau 2002) or by stakeholders and participants using models (e.g., see Barreteau, Bousquet, and Attonaty 2001; Castella, Trung, and Boissau 2005), rather than by quantitative methods.

There are several primary empirical methods for collecting data to inform and empirically test ABMs, including geographic information systems and remote sensing data, surveys (e.g., Brown and Robinson 2006; Monticino et al. 2007), expert knowledge elicitation, participant observation and ethnographic tools (e.g., Huigen et al. 2006), field or laboratory experiments, role-playing games (discussed in the next section), and "stylized facts" (Janssen and Ostrom 2006). Robinson et al. (2007, p. 47) note that these empirical data collection processes can help to answer important questions about the individual (micro) level behavior, including questions like

- What environmental or social factors influence actor decisions and how strong are they relative to each other?
- What are the primary types of agents? How many of each type are there? Are they different enough that we must define separate decision rules and goals (value functions) for each actor (e.g., Janssen and Ahn 2006)?
- How do agents interact and exchange information? What are the types, frequencies, and conditions of interaction? How does the nature of agent interactions (e.g., trade, family relationships, social interactions) affect emergent phenomena?
- How do agents make decisions? What decision models or cognitive processes do they use?
- When and how do actors learn or adjust their decision making? Do agents forecast future events? How do they remember the past? What information do they selectively accept or ignore?
- How can we characterize the "sequence and duration" of events and agent actions and interactions? When do agents update their information? That is, when do they "realize" that things have happened?

Robinson et al. (2007) address the role of data collection in identifying heterogeneity among agents, noting the advantages and drawbacks of several methods. Our ability to empirically discover information that allows us to segment different types of agents and determine how individuals vary within each type is completely constrained by the theories we use to guide our use of different data collection methods and experimental designs; in the end, identifying distinct agent types (segments or groups in the population) stands in a strong tension with viewing agent-level variation as a continuous function throughout the population (Robinson et al. 2007).

For example, survey data collection is guided by theoretical understanding of agent types; surveys eliciting information on the characteristics of individuals may characterize different types of agents based on researcher interpretation of previous data collected and understanding of the way a given culture or conflict functions. This interpretation may be biased or otherwise irrelevant for use in the real world.

Although laboratory and field experiments are quite adept at informing ABMs, including how agent learning processes are structured and why/when/how agents make repeated decisions (e.g., Evans, Sun, and Kelley 2006), these experiments require controlled settings to elicit information about agents' rule-based behavior that can help inform ABMs. This is a challenge that well-designed role-playing games can help alleviate. For example, Castillo and Saysel (2005) used experimental data collected via a gaming design to calibrate a behavioral model of collective action for five heterogeneous fishermen and crab hunters (agents) off Providence Island in the Colombian Caribbean Sea. The resulting simulations were able to replicate experimental data, explore extreme conditions, and explore assumptions about the incentives and behavioral characteristics of participants.

Another important barrier in deriving useful observations of social or conflict systems is that empirical observation may take extended periods of time. In many cases, data may be entirely unavailable to perform statistical analyses. Therefore, along the lines of the "social validation" processes we discuss in the next section on participatory modeling, it is critical that: (1) the model's agent structure and emergent behavior are plausible, given what we know about the conflict; (2) we understand the model's behavior; (3) the model helps us to understand our empirical data better than we did before; and (4) the model maintains "face validity" with other experts or stakeholders (Janssen and Ostrom 2006).

Participatory ABM

In the last two chapters, we spent a lot of time discussing participatory modeling, both in general terms and in its specific use in SD modeling interventions. Our discussion of ABM and its application to environmental conflict would be incomplete without looking at the nature of participatory ABMs, which tend to place more emphasis on the individual actors represented in the models than comparable SD interventions (Pahl-Wostl 2006).

As in other modeling approaches, our need to accurately represent human behavior in ABMs often requires input from the very people affected by the conflict ("stakeholders," see Chapter 4 for a more extensive definition). Stakeholders are intimately familiar with the complexity and detail of conflict systems; this means that their opinions, experiences, and understanding (perceptions) of a conflict can assist modelers in deriving agent-level behavioral modes as well as rules governing agent decision making and interactions.

In participatory simulations, which Ramanath and Gilbert (2004) call, "user-centered simulation," knowledge elicitation techniques capture subjective ideas and agent behavior, allowing the model to function as a medium for articulating each actor's goals, beliefs, expectations, and actions. ABM provides an ideal platform for use by stakeholders, as it allows for intuitive, elegant, and flexible descriptions of heterogeneous stakeholders, their behavior, and their environment without the need for complex mathematical details that describe the aggregate dynamics of the system (Barnaud, Bousquet, and Trébuil 2008).

Unlike many conflict-modeling interventions that seek to support high-level decision making and policy design, many ABM conflict interventions aim to influence the collective learning and action of local stakeholders (Barnaud, Bousquet, and Trébuil 2008). Rather than being entirely tied to complex economic optimization models, which are often very sophisticated, yet difficult to validate (Balmann 2000), models tend to be based entirely on observations of agent behavior expressed in the real world (or other forums like the role-playing games discussed later), which are influenced by profoundly important yet simple-to-represent factors like social networks and socio-economic differentiation (Barreteau and Bousquet 2000; Berger 2001).

Conflating "Successful" Participation with Intervention Context

Like many other conflict modeling techniques, outside of data collection ABMs have rarely involved stakeholders in the model construction process (Costanza and Ruth 1998; Barreteau, Bousquet, and Attonaty 2001; Boulanger and Bréchet 2005; Gurung, Bousquet, and Trébuil 2006). This lack of participation during the rest of the modeling process can undermine knowledge transfers between researchers, conflict stakeholders, and policy makers (Becu et al. 2008). Moreover, efforts to formalize, systematize, and communicate methods for developing useful ABM simulations have been lacking, particularly for participatory models that involve users in design and testing efforts (Ramanath and Gilbert 2004).

However, it is important to draw a major contrast with other participatory modeling techniques; participatory ABM applications have been disproportionately directed at the developing world, where stakeholders often have relatively low incomes. As a result, participation in these modeling interventions can be uniquely affected by a variety of factors, including stakeholders' immediate hope for empowerment, social norms (e.g., neighbor participation), and authoritarian pressure (if authorities are involved in the modeling exercise). For example, although farmers and other stakeholders in recent *companion modeling (ComMod)* applications (described later in this section) are inevitably quite busy, interventions can be designed to fit around their schedules, as evidenced by numerous, successful interventions, which extended over many days and/or weeks (e.g., Gurung, Bousquet, and Trébuil 2006).

Contrast this with *group model building* interventions (discussed in Chapter 4), where participation is also limited by the availability and interest of participants, which more often tend to be corporate or organizational leaders and elites. Therefore, the ability of group model building interventions to attract participation, buy-in, and future action may be limited, as participants may not see the direct applicability of the modeling effort and may have limited time and energy to devote to the process.

Role-Playing Games (RPGs) and Agent-Based Modeling (ABMs)

A variety of studies have implemented behavioral ABM simulations through RPGs (also described in Chapter 4). Agent-based RPGs act as a partial representation of agents' actual behavior, where, like the real world, players interact by playing roles, each of which follow certain rules of the RPG (Becu et al. 2008). The idea is that RPGs identify possible strategies for collective management as well as solutions to conflict, prior to making management or governance decisions. RPGs can be particularly valuable in developing countries, where authorities often see the populace as "backward peasants who make irrational use of resources (Castella, Trung, and Boissau 2005, p. 9)." Disadvantaged groups, such as farmers, are often given few opportunities to speak openly and publicly, thus limiting understanding of local context, and its influence on agent behavior and the adoption of certain practices.

RPGs are created assuming that stakeholders have difficulties understanding a computer model directly and would therefore have difficulty directly assessing model assumptions, suggesting scenarios, and interpreting results on their own. Instead, stakeholders are confronted with the model as articulated through a gaming mechanism, which can be particularly useful for involving stakeholders in simulations of conflicts where they are competing over scarce resources (Becu et al. 2008).

Typical environmental RPG designs introduce simple mechanisms (algorithms) to describe resource management practices specific to given development projects, communities, or environments (Castella, Trung, and Boissau 2005). We can quickly see how an agent-based view of these dynamics becomes essential, as these mechanisms are likely to vary across space (e.g., between villages) and across environments based on distribution of resources and extraction conditions (e.g., forestry practices).

Companion Modeling: A Platform for RPGs

The use of RPGs in ABMs has been pioneered through the *ComMod* platform developed at CIRAD[5], a French research center for agriculture, natural resources, and development (Barreteau, Bousquet, and Attonaty 2001; D'Aquino et al. 2002; Étienne 2013). ComMod interventions have been shown to be useful for producing knowledge about complex systems, while supporting stakeholder-led decision-making processes (Gurung, Bousquet, and Trébuil 2006; Barnaud, Bousquet, and Trébuil 2008) and easing communication between researchers and local stakeholders (Castella et al. 2007). When producing system understanding and knowledge, these RPGs attempt to improve stakeholder and researcher understanding of interactions related to conflicts, rather than directly creating conflict resolution strategies.

Barnaud, Bousquet, and Trébuil (2008) describe the ComMod process in the context of an intervention with conflicting groups of Thai farmers (Figure 6.2C). First, the

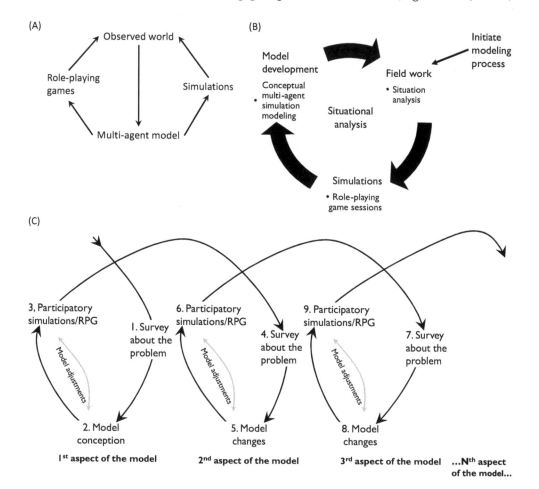

FIGURE 6.2

The ComMod process as portrayed by (A) Bousquet, Trébuil, and Hardy (2005) as a generalized ComMod approach; (B) Gurung, Bousquet, and Trébuil (2006) while mediating a water conflict in Bhutan; and (C) Barnaud, Bousquet, and Trébuil (2008) while working with farmers in Northern Thailand. Figures adapted from each study.

---

[5] CIRAD is the Centre de coopération internationale en recherche agronomique pour le développement.

research team developed a conceptual model capturing their assumptions regarding the behaviors and social and ecological dynamics. This early conceptual model was used to design the RPG and construct a temporary ("straw man") ABM for use in the game. After stakeholders were able to play the game, the researchers were able to update the underlying ABM to account for new decision rules and information learned during the course of game play. Stakeholder feedback prompted rapid adaptive change in the model. Following this, villagers suggested scenario simulations that helped to improve and validate the model.

ComMod has been used in diverse conflicts and settings, such as water management negotiations between communities in Bhutan (Gurung, Bousquet, and Trébuil 2006), as well as multiple modeling interventions in Thailand, where multiple factors, including greater state presence, closure of land frontiers, and the recent inability to relocate villages, have exacerbated longstanding conflicts among farmers (Becu et al. 2003; Barnaud, Bousquet, and Trébuil 2008). In one project in Northern Thailand (Mae Sa watershed, Chiang Mai province), existing ethnic tensions and unbalanced sociopolitical power had aggravated an upstream–downstream water allocation and management conflict between two villages. Simulations during this effort (Becu et al. 2008) were fairly accurate in predicting the emergence of a small group of wealthier ("entrepreneurial") farmers in the watershed. Combined with predictions of continued impoverishment of many in the basin, this work has led researchers into ongoing research around the role of credit in improving the conditions of poorly performing farmers.

In perhaps the most advanced ComMod implementation, which occurred in response to a conflict resulting from farmers deforesting landscapes in the mountains of northern Vietnam, Castella, Trung, and Boissau (2005) combined a narrative conceptual model, a spatially explicit ABM, an RPG, and a geographic information system. They found that scenarios simulated over longer periods than those used in the game allowed researchers to demonstrate the long-term implications of stakeholder choices in the game.

The ComMod process stands out for many reasons. In particular, the initial ABM created by researchers, which is originally guided by available empirical data collection techniques, can be corrected through the highly iterative and adaptive nature of the RPG. This allows modelers to identify agent types and heterogeneity that had not been realized previously. However, Robinson et al. (2007) also warns of the risk that significant, potential heterogeneity could be left out of ComMod models due to (1) the relatively small number of participants in many interventions, (2) the approximately one-to-one representation between modeled agents and real-world stakeholders, and (3) the need to simplify (or oversimplify) as a way of ensuring stakeholder involvement and model understanding.

Participatory ABMs and Social Validation

By engaging in RPGs, stakeholders help to "validate" the model's proposed environment, behavioral rules of agents, and the emergent properties from agent interactions during the game. Validation by stakeholders is defined as the process by which stakeholders suggest changes to aspects of the model that they did not believe matched reality (Barreteau, Bousquet, and Attonaty 2001 and Castella, Trung, and Boissau 2005). Models therefore become "valid" when consensus forms, and stakeholders accept the model as a "good" representation of stakeholders' and researchers' reality. It is worth noting that ABMs can be validated both in terms of their full system-level output as well as their individual-level behavior, which can be compared with observations of actual stakeholder agents (Parunak, Savit, and Riolo 1998).

In cases where the goal is to evaluate alternative conflict resolution strategies, system-oriented solutions that attempt to control for the emergent behavior of many conflict actors may be difficult to implement. Conversely, if the behavior of simulated agents closely resembles those of specific, corresponding stakeholders (or stakeholder groups, e.g., firms or interests), then validated ABMs may also be particularly effective at testing the impacts of agent behavioral changes on the system's emergent dynamics. This is what Parunak, Savit, and Riolo (1998) term, "translating [ABMs] back into practice," whereby recommendations for modified agent behavior are translated back into agents' real-life counterparts.

Gurung, Bousquet, and Trébuil (2006, p. 2) discuss the objectives of *social model validation*, noting that stakeholders are invited to play the game in the hopes that they will:

1. Understand the model, the simulations they are shown, and where the model differs with reality (Becu et al. 2008). Barreteau, Bousquet, and Attonaty (2001) call this process, "opening the black box of the model," thereby making rules explicit to participants and facilitating collective action;

2. Propose scenarios that should be assessed and discussed to better understand and validate the model. The agent-based model can then be used to more fully explore these scenarios that are collectively identified by stakeholders during game play;

3. Validate the model by examining the individual agent behaviors and interactions, and the properties of the system that emerge from agent interactions; and

4. Test the model by proposing modifications to agent behavior or interactions to find out if they would improve the accuracy of the model. This represents iterative feedback between stakeholders and researchers, thereby developing a shared representation of reality (Bousquet et al. 1999).

It is important to point out that social validation should not be confused with formal model validation procedures developed in the ABM or SD modeling communities (e.g., Barlas and Carpenter 1990). It is also worth considering that in most ABMs, agent behavioral modes are practically limited to the actions that modelers can infer from relationships found through rigorous statistical analysis. This logistical limitation evokes other highly constrained model frameworks, such as CA and Markov modeling (Karlin 2014), which both require specification of "transition probabilities" that dictate individual-level actions or change. While this limitation might initially appear to improve the rigor of ABMs, limiting the operations of ABMs to actions that are statistically supported can limit the effectiveness of ABMs as decision support systems. As a result, policy analysts and decision makers are deprived of the ability to evaluate the full spectrum of cause–effect relationships that may occur, either as theorized by stakeholders, or as may occur based on the totality of social and ecological knowledge (Scoones 1999).

Robinson et al. (2007) argue that validating our representations of agent decision-making structures against their real-world counterparts is nearly impossible. As a result, we must instead validate our models based on how well agent *actions* (i.e., the outcomes of their simulated decision-making processes) correspond with the real-world actions of the stakeholders represented in the model. That is, although we may theorize the reasoning behind behavioral shifts, most techniques for data collection appear to be more useful for identifying the thresholds or events that influence agents to change their behavior or take different actions.

Summary

Throughout this book, we have explored conflict as a dynamic and complex form of human interaction, which emerges from individual agents who modify their actions and behaviors in response to the emergent dynamics in conflict systems. In this chapter, we have focused on exploring the emerging science and technology of modeling based on this agent-based perspective, wherein representations of agent *behavior* can be shown to organically create emergent system *structure*. As we explored in the last chapter, this *system structure* creates *system behavior*, which in our case, represents the life history of an environmental dispute.

We have explored the history and core facets of ABM, as well as applications of the technique to conflicts. We have also explored the links connecting ABM and complexity science, and drawn contrast between the approaches and philosophies of ABM and SD modeling. We discussed ABM development platforms and software environments, issues concerning how agents interact, make decisions, and account for human behavior. Finally, we discussed issues around validation, the empirical basis of ABMs, and the 20 year history of participatory ABM research.

Unfortunately, one of the shortcomings with current participatory agent-based simulations is that they tend to view all conflict resolution processes as originating at the local level, with "bottom-up" negotiations creating institutions for intervening in conflict. However, as significant evidence has demonstrated, innovations that lead to change in real-world disputes, whether they are technical or social, often do not systematically emerge from stakeholder negotiations at the local scale (Castella, Trung, and Boissau 2005). Instead, many conflicts are shifted by policies imposed from above ("top-down"), such as financial incentives or improved infrastructure, creating change that was previously out of reach to stakeholders. Frameworks for modeling conflicts at the agent level must be flexible enough to incorporate the realities of localized actions and stakeholder negotiation processes, as well as conflict interventions by governments or other powerful institutions.

Having reviewed several of the more sophisticated applications of ABM to environmental conflict, we will present a framework in the next chapter that builds on previous efforts to incorporate changing agent preferences and actions. In doing so, we will introduce an ABM paradigm that allows us to represent negotiated ("bottom-up") solutions, alongside the top-down policies as they are typically implemented in the real world. Our hope is to create an application of ABM that can be a tool for unifying and structuring the ways that we consider the social, economic, and environmental factors affecting conflict.

Questions for Consideration

1. *What is an agent?* Billari et al. (2006, pp. 3–4) define an agent as any entity (organizational or not) that is "…able to act according to its own set of rules and objectives." In fact, the term "agent" in ABMs implies *agency* or the capacity of an entity to act

in the world (Hitlin and Elder 2007). *What is a more complete definition of agency? What factors do you think that a good definition should include? What aspects of you and your environment give you agency?* See Emirbayer and Mische (1998) and Fuchs (2001) for a detailed discussion and critique of the concept of *agency.*

2. In Chapter 1, we drew on the definition of environmental conflict established by the Swiss Peace Foundation and ETH Zurich Center for Security Studies (Libiszewski 1992) as "…traditional conflicts induced by *environmental degradation.* Environmental conflicts are characterized by the principal importance of degradation in one or more of the following fields: overuse of renewable resources, overstrain of the environment's sink capacity (pollution), or impoverishment of the 'space of living.'" *How might our thinking around environmental conflict be affected by the way that ABMs represent interactions of conflicting parties? How could we reframe this definition of conflict through the lens of an ABM?*

3. In Thomas Schelling's (1971) model of racial segregation, homogenous agents could only tolerate a proportion of their neighbors being different from them, resulting in repeated migration and eventually leading to neighborhood and city-level racial segregation. *What proportion of nonsimilar neighbors would agents need to tolerate in order to prevent segregation? Why?* Hint: See Clark (1991) for an empirical evaluation of Schelling's findings. You can also experiment with Schelling's original model (and extended versions) at http://netlogoweb.org (search the models library for "segregation").

4. ABM is often discussed alongside cellular automata (CA) modeling, a spatially explicit technique for representing individual actions and agent behavior. *Where are CA models still being used today?* See Wolfram (1983, 2002) and White and Engelen (1997) for more information on CA modeling and explore http://netlogoweb.org for many CA examples.

5. *What are some examples of emergent effects that result from local agent actions and interactions?* See Rudel (2011) for an examination of this topic.

6. Castella, Trung, and Boissau (2005) refer to the concept of "social validation," or the process by which stakeholders suggest changes to the model and consensus forms that the model is a "good" representation of reality. However, Becu et al. (2003) argue that the term "validation" may not be adequate to describe model behavior that cannot be tested experimentally. Instead, they suggest that stakeholder feedback helps to "authenticate" models through firsthand accounts of model legitimacy. *Do you believe that social validation is a legitimate way of validating a model? Why or why not? In what circumstances could you be wrong?*

7. Evaluating the success of participatory ABM interventions is difficult. Efforts to formalize, systematize, and communicate methods for developing useful and successful social simulations have been lacking, particularly for participatory models that involve users in design and testing efforts. Like other participatory techniques, many ABM interventions have been documented in anecdotal ways, leaving the field fragmented. *How should we document ABM interventions? How should we measure "success" of an intervention?* Ramanath and Gilbert (2004) is a good resource to begin your exploration.

8. Some economists have objected to the "rational actor paradigm" for many reasons, including poor assumptions about agent homogeneity and lack of agent interactions (e.g., Kirman 1992). They also criticize the oversimplification of

agents' social networks and the roles networks have in creating uncertainty, unpredictable behavior, and information deficits and asymmetries (Grimm and Railsback 2005). Even if all individuals maximized their utility, which has been challenged repeatedly (Simon 1991; Arthur 1994), some scholars have attacked the idea that heterogeneous agents could be accurately modeled from a single individual, whose choices and behavior represent the hypothetical mean of the population (Kirman 1992). *How does this critique fit with our earlier discussion of the "ecological fallacy?" Based on this criticism, give an example of an inaccurate prediction made by the rational actor paradigm? How could we fix that prediction? What assumptions would we need to change?*

9. Janssen and Ahn (2006) discuss the use of discrete agent *typologies* (i.e., agents who are risk averse, somewhat risk adverse, or risk tolerant) and the use of *continuous distributions* of agent parameters (e.g., agents with risk aversion/tolerance on a 0–100 scale). *What are the advantages and disadvantages of each approach? Create a conceptual agent for a topic that interests you. Give an example of each approach—typology and distribution—for a facet of that agent. Which approach would be easier to communicate from the model?*

10. The concept of resilience originated in ecology (Holling 1973) and is now used widely in diverse fields, including urban and regional planning (Beatley 2009), economics (Briguglio et al. 2009), and political science (Chu 1999). *What does resilience mean in your field? What sorts of controversies revolve around that definition? For more in-depth definitions of resilience in ecology, see Holling 1973, 2009).*

11. Although Parunak, Savit, and Riolo (1998) long ago called for comparative studies of SD models and ABMs, especially the models leading to qualitatively different behaviors (e.g., Wilson 1998), little work has answered their call. They argued that ABM and SD model differences are likely due to SD's use of "averaged" behavior, homogeneity assumptions, and the inabilities of SD models to add nonlinearities at the individual level. *Give two examples of differences that we might see between an ABM and an SD model of the same problem or system. Why do these differences emerge? Would there be any way to resolve any differences between the models?*

12. Throughout the modeling process, researchers can actually emerge as stakeholders themselves due to their substantial interaction with other stakeholders and their extensive involvement in the conflict system. This reality can create ethical dilemmas that could disrupt the balance of power between conflict stakeholders. ComMod researchers have discussed this issue in their group's charter as one of the many facets of the modeling process over which researchers have little control, but bear great responsibility (http://cormas.cirad.fr/ComMod/en/charter/; Becu et al. 2008). *Read the ComMod group's charter. What other ethical issues do modelers face? What responsibilities should modelers bear for stakeholders' interests? What responsibilities should they bear for their own interests?*

Additional Resources

Interested in learning more about ABM? In this book, we do not completely review the theory and techniques of ABM, which can often be software-specific. However, there are many resources out there for those wanting to learn to build ABMs. The Santa Fe Institute's Complexity Explorer is a fantastic resource for learning more about complexity science (https://www.complexityexplorer.org/courses/). The site also hosts numerous courses on complex systems topics, including courses in ABM, Dynamic Game Theory, NetLogo, and Dynamical Systems and Chaos.

Resources for Learning General ABM Concepts

Grimm, Volker, and Steven F. Railsback. 2005. *Individual-Based Modeling and Ecology.* Princeton, NJ: Princeton University Press.

Gilbert, Nigel. 2007. *Agent-Based Models.* London, England: Sage Publications.

There are many software platforms that have been developed over the last 30 years for creating ABMs. See Gilbert and Bankes (2002), Railsback, Lytinen, and Jackson (2006), and Nikolai and Madey (2009) for comparisons of the many well-established available platforms. While we use NetLogo throughout this book, we refer interested readers with no previous NetLogo experience to Stigberg's (2012) in-depth introduction to the NetLogo modeling environment, as well as Railsback and Grimm's (2011) and Wilensky and Rand's (2015) comprehensive expositions and guides to the NetLogo language and use.

We recommend that users wishing to gain familiarity with NetLogo and agent-based simulations begin by experimenting with the large and highly instructive library of sample models that is included with the software, which can be downloaded (https://ccl.northwestern.edu/netlogo/) or used in a web browser (http://www.netlogoweb.org/). Users can tinker with sample models at their leisure or follow Netlogo's established web tutorials to immediately construct their first model.

Resources for Learning NetLogo Modeling Software

Railsback, Steven F., and Volker Grimm. 2011. *Agent-Based and Individual-Based Modeling: A Practical Introduction.* Princeton, NJ: Princeton University Press.

Wilensky, Uri, and William Rand. 2015. *An Introduction to Agent-Based Modeling: Modeling Natural, Social, and Engineered Complex Systems with NetLogo.* Cambridge, MA: MIT Press.

Resources for Learning AnyLogic Modeling Software

Borshchev, Andrei. 2013. *The Big Book of Simulation Modeling: Multimethod Modeling with AnyLogic 6.* Chicago, IL: AnyLogic North America.

Grigoryev, Ilya. 2015. *AnyLogic 7 in Three Days: A Quick Course in Simulation Modeling* (2nd Edition). Charleston, SC: CreateSpace Independent Publishing Platform.

Interested in ABM applied to specific areas? ABM has become a popular technique in a wide range of fields, including economics (Schredelseker and Hauser 2008), political science (Nelson, Cain, and Yang 2015), sociology (Halpin 1999; Macy and Willer 2002), psychology (McCarter, Budescu, and Scheffran 2011), energy (Abdollahian, Yang, and Nelson 2013), geography (Rand et al. 2005), demography (Billari et al. 2006), urban planning

(White, Engelen, and Uljee 2015), environmental sciences (Grimm 1999; Grimm et al. 2005), and ecology (BenDor and Metcalf 2006; BenDor et al. 2009).

There are many resources across these fields that are noteworthy.

Social Sciences

Durlauf, Steven N., and H. Peyton Young. 2004. *Social Dynamics*. Cambridge, MA: MIT Press.

Campbell, Heather E., Yushim Kim, and Adam M. Eckerd. 2015. *Rethinking Environmental Justice in Sustainable Cities: Insights from Agent-Based Modeling*. New York: Routledge.

Epstein, Joshua M. 2006. *Generative Social Science: Studies in Agent-Based Computational Modeling*. Princeton, NJ: Princeton University Press.

Billari, Francesco C., Thomas Fent, Alexia Prskawetz, and Jürgen Scheffran. 2006. *Agent-Based Computational Modelling: Applications in Demography, Social, Economic and Environmental Sciences*. Berlin, Germany: Physica-Verlag HD.

Dam, Koen H. van, Igor Nikolic, and Zofia Lukszo. 2012. *Agent-Based Modelling of Socio-Technical Systems*. Berlin, Germany: Springer Science & Business Media.

Conflict

Müller, Heinz J., and Rose Dieng. 2012. *Computational Conflicts: Conflict Modeling for Distributed Intelligent Systems*. Berlin, Germany: Springer Science & Business Media.

Epstein, Joshua M. 2002. "Modeling Civil Violence: An Agent-Based Computational Approach." *Proceedings of the National Academy of Sciences of the United States of America* 99 (Suppl 3): 7243–50.

Fang, Liping, Keith W. Hipel, and D. Marc Kilgour. 1993. *Interactive Decision Making: The Graph Model for Conflict Resolution*. New York: John Wiley & Sons.

Bousquet, François. 2005. *Companion Modeling and Multi-Agent Systems for Integrated Natural Resource Management in Asia*. Manila, Philippines: International Rice Research Institute.

Business or Management

North, Michael J., and Charles M. Macal. 2007. *Managing Business Complexity: Discovering Strategic Solutions with Agent-Based Modeling and Simulation*. Oxford, England: Oxford University Press.

Complex Systems and Adaptation

Perez, Pascal, and David F. Batten. 2006. *Complex Science for a Complex World: Exploring Human Ecosystems with Agents*. Canberra: ANU E Press.

May, Robert McCredie. 2001. *Stability and Complexity in Model Ecosystems*. Princeton, NJ: Princeton University Press.

Alonso, Eduardo, Daniel Kudenko, and Dimitar Kazakov. 2003. *Adaptive Agents and Multi-Agent Systems: Adaptation and Multi-Agent Learning.* Berlin, Germany: Springer Science & Business Media.

Interested in the interaction of SD and ABM? The trade-offs between agent-based and aggregate modeling approaches are the subject of an on-going discussion in the SD and ABM literatures Rahmandad and Sterman (2008). Some of the earliest published SD models characterized interactions among individual firms (i.e., the seminal 1961 work, *Industrial Dynamics* by Jay Forrester), and many recent SD models have used approaches that explicitly represent individuals or draw on formulations to combine individual-based and aggregate approaches. See Osgood (2009) for numerous applications to diverse areas, including strategy and marketing, environmental management and ecology, health policy, urban planning, and transportation management.

Osgood (2009) goes on to note that behavior within individual-based models can be represented in many different fashions, including the use of:

1. classic differential equation methods (along with "subscripting," representing the states of individuals, which is typical in SD modeling software; see Appendix I in Andrew Ford's (1999) *Modeling the Environment*);
2. discrete objects and rules (as is typical in classic ABM); or
3. hybrid approaches, such as hybrid automata (Lynch, Segala, and Vaandrager 2003) or "grid-cell" modeling, in which SD models are copied into individual grid cells, essentially creating complex CA models (Maxwell and Costanza 1997).

Framing the "ABM or SD" issue in this manner, the choice of modeling technique can be structured around the problem at hand and the data that is available. For example, if we have data and a strong understanding of the system of concern at the individual level, the modeler can avoid many of the challenges associated with aggregate modeling. When BenDor, Metcalf, and Paich (2011) sought to model whether or not specific contaminated land parcels are chosen for remediation and re-development, they only had aggregate data on the land area of contaminated sites. As a result, they could not use ABM. However, if they had a different data set, one that had information on individual sites, they could have created an ABM to represent the likelihood that sites were redeveloped at the individual level.

Interested in ABM and spatial analysis? As we discussed in this chapter, ABMs can be particularly useful when we want to understand spatial movement across a landscape, such as the spread of disease, organisms, or innovations through an uneven or irregular space (BenDor and Metcalf 2006). Parunak, Savit, and Riolo (1998) generalize this discussion, arguing that ABMs help us to conceptually separate physical space from interaction space; that is, rather than relying on physical space to facilitate interactions, ABMs allow us to represent any sort of arbitrary network or spatial topologies that we want (BenDor and Kaza 2012). For example, location can determine the quality of farmland and an agent's resulting farm-related income (see Chapter 10). Models that explicitly rely on spatial or network relationships can also integrate these factors into the way that agents seek to achieve their goals, including evolving their conflict strategies based on the location of other agents, or changing their own location.

Interested in how agents can hide information in ABM? In aggregate models, information is available system-wide; if information exists, everyone has that information.

However, ABMs tend to define behaviors in terms of resources or information that are accessible to agents (Parunak, Savit, and Riolo 1998). As in real life, agents' internal behaviors and belief structures may be intentionally kept from the rest of the system. While we discussed this briefly from a model realism perspective, the ability to keep information private can also be advantageous in modeling interventions. In some cases, stakeholders' decision-making processes and goals are proprietary information and cannot be shared with other stakeholders. To keep an agent's decision rules hidden, agents can precompile these rules, making it so that not even modelers know how agents make decisions.

References

Abdollahian, Mark, Zining Yang, and Hal Nelson. 2013. "Techno-Social Energy Infrastructure Siting: Sustainable Energy Modeling Programming (SEMPro)." *Journal of Artificial Societies and Social Simulation* 16 (3): 6.

Achorn, E. 2004. "Integrating Agent-Based Models with Quantitative and Qualitative Research Methods." In *AARE 2004 International Education Research Conference*, edited by Peter Jeffrey, Australian Association for Research in Education: Deakin, Australia.

Aknine, Samir. 2012. "A Multi-Agent Model for Overlapping Negotiations." *Group Decision and Negotiation* 21: 747–90.

Allen, Timothy, and Thomas W. Hoekstra. 1993. *Toward a Unified Ecology.* New York: Columbia University Press.

Anderies, J. M., B. Walker, and A. Kinzig. 2006. "Fifteen Weddings and a Funeral: Case Studies and Resilience-Based Management." *Ecology and Society* 11 (1): 21.

Arquitt, Steve, Xu Honggang, and Ron Johnstone. 2005. "A System Dynamics Analysis of Boom and Bust in the Shrimp Aquaculture Industry." *System Dynamics Review* 21 (4): 305–24.

Arthur, B. W. 1994. "Inductive Reasoning and Bounded Rationality." *American Economic Review* 84 (2): 406–11.

Axelrod, Robert. 1984. *The Evolution of Cooperation.* New York: Basic Books.

Axelrod, Robert. 1997. *The Complexity of Cooperation: Agent-Based Models of Competition and Collaboration.* Princeton, NJ: Princeton University Press.

Axelrod, Robert. 2006. "Advancing the Art of Simulation in the Social Sciences." In *Handbook of Research on Nature Inspired Computing for Economy and Management*, edited by Jean-Philippe Rennard, 90–100. Hersey, PA: Idea Group.

Balmann, A. 2000. "Modeling Land Use with Multi-Agent Systems: Perspectives for the Analysis of Agricultural Policies." In *Microbehavior and Macroresults: Proceedings of the Tenth Biennial Conference of the International Institute of Fisheries Economics and Trade* (July 10–14, 2000), edited by R. S. Johnston. Corvallis, OR.

Barlas, Y., and S. Carpenter. 1990. "Philosophical Roots of Model Validation." *System Dynamics Review* 6 (2): 148–66.

Barnaud, Cécile, François Bousquet, and Guy Trébuil. 2008. "Multi-Agent Simulations to Explore Rules for Rural Credit in a Highland Farming Community of Northern Thailand." *Ecological Economics* 66 (4): 615–27.

Barnaud, Cécile, Christophe Le Page, Pongchai Dumrongrojwatthana, and Guy Trébuil. 2013. "Spatial Representations Are Not Neutral: Lessons from a Participatory Agent-Based Modelling Process in a Land-Use Conflict." *Environmental Modelling & Software*, Thematic Issue on Spatial Agent-Based Models for Socio-Ecological Systems, 45 (July): 150–59.

Barreteau, O., and François Bousquet. 2000. "SHADOC: A Multi-Agent Model to Tackle Viability of Irrigated Systems." *Annals of Operation Research* 14: 139–62.

Barreteau, O., François Bousquet, and J. M. Attonaty. 2001. "Role-Playing Games for Opening the Black Box of Multi-Agent Systems: Method and Lessons of Its Application to Senegal River Valley Irrigated Systems." *Journal of Artificial Societies and Social Simulation* 4 (2): 12.

Batten, David F. 2000. *Discovering Artificial Economics: How Agents Learn and Economies Evolve.* Boulder, CO: Westview Press.

Beall, Allyson, and Andrew Ford. 2007. "Participatory Modeling for Adaptive Management: Reports from the Field II." In *25th International Conference of the System Dynamics Society.* Albany, NY: International System Dynamics Society. https://labs.wsu.edu/collaborativemodeling/documents/2016/01/reports-from-the-field-isds-conference-proceedings.pdf/

Beatley, Timothy. 2009. *Planning for Coastal Resilience: Best Practices for Calamitous Times.* Washington, DC: Island Press.

Becu, Nicolas, Andreas Neef, Pepijn Schreinemachers, and Chapika Sangkapitux. 2008. "Participatory Computer Simulation to Support Collective Decision-Making: Potential and Limits of Stakeholder Involvement." *Land Use Policy* 25 (4): 498–509.

Becu, Nicolas, P. Perez, A. Walker, O. Barreteau, and C. Le Page. 2003. "Agent Based Simulation of a Small Catchment Water Management in Northern Thailand: Description of the CATCHSCAPE Model." *Ecological Modelling* 170: 319–31.

BenDor, Todd. 2012. "The System Dynamics of U.S. Automobile Fuel Economy." *Sustainability* 4: 1013–42.

BenDor, Todd, and Nikhil Kaza. 2012. "A Theory of Spatial System Archetypes." *System Dynamics Review* 28 (2): 109–30.

BenDor, Todd, and Sara Metcalf. 2006. "The Spatial Dynamics of Invasive Species Spread." *System Dynamics Review* 22 (1): 27–50.

BenDor, Todd, Sara Metcalf, and Mark Paich. 2011. "The Dynamics of Brownfield Redevelopment." *Sustainability* 3 (6): 914–36.

BenDor, Todd, James Westervelt, J. P. Aurambout, and William Meyer. 2009. "Simulating Population Variation and Movement within Fragmented Landscapes: An Application to the Gopher Tortoise (Gopherus Polyphemus)." *Ecological Modelling* 220 (6): 867–78.

Berger, T. 2001. "Agent-Based Spatial Models Applied to Agriculture: A Simulation Tool for Technology Diffusion, Resource Use Changes and Policy Analysis." *Agricultural Economics* 25 (2–3): 245–60.

Berger, Thomas, and Christian Troost. 2014. "Agent-Based Modelling of Climate Adaptation and Mitigation Options in Agriculture." *Journal of Agricultural Economics* 65 (2): 323–48.

Bhavnani, R., D. Backer, and R. Riolo 2006. *Agent-Based Computational Modelling: Applications in Demography, Social, Economic and Environmental Sciences.* Berlin, Germany: Physica-Verlag HD.

Bhavnani, R., D. Backer, and R. Riolo. 2008. "Simulating Closed Regimes with Agent Based Models." *Complexity* 14 (1): 36–44.

Bonabeau, Eric. 2002. "Agent-Based Modeling: Methods and Techniques for Simulating Human Systems." *Proceedings of the National Academy of Sciences of the United States of America* 99 (Suppl 3): 7280–87.

Borshchev, Andrei. 2013. The Big Book of Simulation Modeling: Multimethod Modeling with AnyLogic 6 . Chicago, IL: AnyLogic North America.

Boulanger, P.-M., and T. Bréchet. 2005. "Models for Policy-Making in Sustainable Development: The State of the Art and Perspectives for Research." *Ecological Economics* 55: 337–50.

Bousquet, François, O. Barreteau, C. Le Page, C. Mullon, and J. Weber. 1999. "An Environmental Modelling Approach. The Use of Multi-Agent Simulations." In *Advances in Environmental and Ecological Modelling,* edited by F. Blasco and A. Weill, 113–22. Amsterdam, Netherland: Elsevier.

Bousquet, François, and C. Le Page. 2004. "Multi-Agent Simulations and Ecosystem Management: A Review." *Ecological Modelling* 176 (3–4): 313–32.

Bousquet, François, G. Trébuil, and B. Hardy, eds. 2005. *Companion Modeling and Multi-Agent Systems for Integrated Natural Resource Management in Asia.* Manila, Philippines: International Rice Research Institute.

Briguglio, Lino, Gordon Cordina, Nadia Farrugia, and Stephanie Vella. 2009. "Economic Vulnerability and Resilience: Concepts and Measurements." *Oxford Development Studies* 37 (3): 229–47.

Brown, Dan. G., and Derek T. Robinson. 2006. "Effects of Heterogeneity in Preferences on an Agent Based Model of Urban Sprawl." *Ecology and Society* 11 (1): 46.

Bryant, James William. 2003. *The Six Dilemmas of Collaboration: Inter-Organisational Relationships as Drama*. Chichester, England: Wiley.

Bryant, James William. 2004. "Drama Theory as the Behavioural Rationale in Agent-Based Models." *In Analysing Conflict and its Resolution*, 87–90. Oxford, England: The Institute of Mathematics and its Applications.

Bryant, James William. 2010. "The Role of Drama Theory in Negotiation." In *Handbook of Group Decision and Negotiation*, edited by D. Marc Kilgour, Colin Eden, Tung Bui, Guy Olivier Faure, Gregory Kersten, and Peyman Faratin, Vol. 4, 223–45. Advances in Group Decision and Negotiation. Amsterdam, Netherlands: Springer.

Bryant, James William. 2015. *Acting Strategically Using Drama Theory*. Boca Raton, FL: CRC Press.

Byrne, D. S. 1998. *Complexity Theory in the Social Sciences: An Introduction*. New York: Routledge.

Castella, Jean-Christophe, Suan Pheng Kam, Dang Dinh Quang, Peter H. Verburg, and Chu Thai Hoanh. 2007. "Combining Top-down and Bottom-up Modelling Approaches of Land Use/Cover Change to Support Public Policies: Application to Sustainable Management of Natural Resources in Northern Vietnam." *Land Use Policy* 24 (3): 531–45.

Castella, Jean-Christophe, Tran Ngoc Trung, and Stanislas Boissau. 2005. "Participatory Simulation of Land-Use Changes in the Northern Mountains of Vietnam: The Combined Use of an Agent-Based Model, a Role-Playing Game, and a Geographic Information System." *Ecology and Society* 10 (1): 27.

Castillo, Daniel, and Ali Korom Baysel. 2005. "Simulation of Common Pool Resource Field Experiments: A Behavioral Model of Collective Action." *Ecological Economics* 55 (3): 420–36.

Cho, Adrian. 2009. "Ourselves and Our Interactions: The Ultimate Physics Problem?" *Science* 325 (5939): 406–8.

Chu, Yun-han. 1999. "Surviving the East Asian Financial Storm: The Political Foundation of Taiwan's Economic Resilience." In *The Politics of the Asian Economic Crisis*, edited by T. J. Pempel, 184–202. Ithaca, NY: Cornell University Press.

Cioffi-Revilla, Claudio, and Nicholas Gotts. 2003. "Comparative Analysis of Agent-Based Social Simulations: GeoSim and FEARLUS Models." *Journal of Artificial Societies and Social Simulation* 6 (4).

Clark, William A. V. 1991. "Residential Preferences and Neighborhood Racial Segregation: A Test of the Schelling Segregation Model." *Demography* 28 (1): 1–19.

Coleman, James S., and Thomas J. Fararo, eds. 1992. *Rational Choice Theory: Advocacy and Critique*. London, England: Sage.

Costanza, Robert, and Matthias Ruth. 1998. "Using Dynamic Modeling to Scope Environmental Problems and Build Consensus." *Environmental Management* 22 (2): 183–95.

Danielson, P. 2002. "Competition among Co-Operators: Altruism and Reciprocity." *Proceedings of the National Academy of Sciences* 99 (Suppl. 3): 7237–42.

D'Aquino, Patrick, Olivier Barreteau, Michel Etienne, Stanislas Boissau, Sigrid Aubert, François Bousquet, Christophe Le Page, and William S. Daré. 2002. "The Role Playing Games in an ABM Participatory Modeling Process: Outcomes from Five Different Experiments Carried out in the Last Five Years." In *Integrated Assessment and Decision Support, 1st Meeting iEMSs*, edited by A. E. Rizzoli and A. J. Jakeman, 275–80. Lugano, Switzerland: International Environmental Modelling and Software Society.

Davidsson, Paul. 2000. "Multi Agent Based Simulation: Beyond Social Simulation." In *International Workshop on Multi-Agent Systems and Agent-Based Simulation*, 97–107. Berlin: Springer.

DeAngelis, D. L., and L. J. Gross. 1992. *Individual-Based Models and Approaches in Ecology: Populations, Communities, and Ecosystems*. New York: Chapman & Hall.

Edmonds, Bruce, and David Hales. 2003. "Replication, Replication and Replication: Some Hard Lessons from Model Alignment." *Journal of Artificial Societies and Social Simulation* 6 (4): 11.

Elliott, E., and L. D. Kiel. 2002. "Exploring Cooperation and Competition Using Agent-Based Modeling." *Proceedings of the National Academy of Sciences* 99 (Suppl. 3): 7193–94.

Emirbayer, Mustafa, and Ann Mische. 1998. "What Is Agency?" *American Journal of Sociology* 103 (4): 962–1023.

Epstein, Joshua M. 2002. "Modeling Civil Violence: An Agent-Based Computational Approach." *Proceedings of the National Academy of Sciences of the United States of America* 99 (Suppl 3): 7243–50.

Epstein, Joshua M. 2006. *Generative Social Science: Studies in Agent-Based Computational Modeling.* Princeton, NJ: Princeton University Press.

Epstein, Joshua M., and R. L. Axtell. 1996. *Growing Artificial Societies: Social Science from the Bottom Up.* Cambridge, MA: MIT Press.

Étienne, Michel. 2013. *Companion Modelling: A Participatory Approach to Support Sustainable Development.* Berlin, Germany: Springer Science & Business Media.

Etienne, Michel, Christophe Le Page, and Mathilde Cohen. 2003. "A Step-by-Step Approach to Building Land Management Scenarios Based on Multiple Viewpoints on Multi-Agent System Simulations." *Journal of Artificial Societies and Social Simulation* 6 (2): 2.

Evans, T. P., W. Sun, and H. Kelley. 2006. "Spatially Explicit Experiments for the Exploration of Land Use Decision-Making Dynamics." *International Journal of Geographical Information Science* 20 (9): 1013–37.

Findley, Michael G. 2008. "Agents and Conflict: Adaptation and the Dynamics of War." *Complex* 14 (1): 22–35.

Ford, Andrew. 1999. *Modeling the Environment: An Introduction to System Dynamics Modeling of Environmental Systems.* Washington, DC: Island Press.

Forrester, Jay W. 1961. *Industrial Dynamics.* Cambridge, MA: MIT Press.

Fuchs, Stephan. 2001. "Beyond Agency." *Sociological Theory* 19 (1): 24–40.

Galam, S. 2012. *Sociophysics: A Physicist's Modeling of Psycho-Political Phenomena.* Berlin, Germany: Springer.

Gilbert, Nigel, and Steven Bankes. 2002. "Platforms and Methods for Agent-Based Modeling." *Proceedings of the National Academy of Sciences of the United States of America* 99 (Suppl 3): 7197–98.

Gilbert, Nigel, and Klaus Troitzsch. 2005. *Simulation for the Social Scientist.* New York: Open University Press.

Gintis, H. 2000. *Game Theory Evolving: A Problem Centered Introduction to Modeling Strategic Interaction.* Princeton, NJ: Princeton University Press.

Grimm, Volker. 1999. "Ten Years of Individual-Based Modelling in Ecology: What Have We Learned and What Could We Learn in the Future?" *Ecological Modelling* 115 (2): 129–48.

Grimm, Volker, and Steven F. Railsback. 2005. *Individual-Based Modeling and Ecology.* Princeton, NJ: Princeton University Press.

Grimm, Volker, Eloy Revilla, Uta Berger, Florian Jeltsch, Wolf M. Mooij, Steven F. Railsback, Hans-Hermann Thulke, Jacob Weiner, Thorsten Wiegand, and Donald L. DeAngelis. 2005. "Pattern-Oriented Modeling of Agent-Based Complex Systems: Lessons from Ecology." *Science* 310 (5750): 987–91.

Gurung, Tayan Raj, François Bousquet, and Guy Trébuil. 2006. "Companion Modeling, Conflict Resolution, and Institution Building: Sharing Irrigation Water in the Lingmuteychu Watershed, Bhutan." *Ecology and Society* 11 (2): 36.

Halpin, B. 1999. "Simulation in Society." *American Behavioral Scientist* 42 (10): 1488–1508.

Heppenstall, Alison J., Andrew T. Crooks, Linda M. See, and Michael Batty. 2011. *Agent-Based Models of Geographical Systems.* Berlin, Germany: Springer Science & Business Media.

Hilborn, Robert C. 2004. "Sea Gulls, Butterflies, and Grasshoppers: A Brief History of the Butterfly Effect in Nonlinear Dynamics." *American Journal of Physics* 72 (4): 425–27.

Hirshleifer, Jack. 1988. "The Analytics of Continuing Conflict." *Synthese* 76: 201–33.

Hirshleifer, Jack. 1991. "The Paradox of Power." *Economics and Politics* 3: 177–200.

Hitlin, Steven, and Glen H. Elder Jr. 2007. "Time, Self, and the Curiously Abstract Concept of Agency." *Sociological Theory* 25 (2): 170–91.

Hofbauer, J., and K. Sigmund. 1998. *Evolutionary Games and Population Dynamics.* Cambridge, England: Cambridge University Press.

Holland, J. H. 1998. *Emergence: From Chaos to Order.* Reading, MA: Helix Books.

Holling, C. S. 1973. "Resilience and Stability of Ecological Systems." *Annual Review of Ecology and Systematics* 4: 1–23.

Holling, C. S. 1978. *Adaptive Environmental Assessment and Management.* London, England: John Wiley and Sons.

Holling, C. S. 2009. "Engineering Resilience versus Ecological Resilience." In *Foundations of Ecological Resilience,* edited by L. H. Gunderson, C. R. Allen, and C. S. Holling, 25–62. Washington, DC: Island Press.

Holling, C. S., L. H. Gunderson, and D. Ludwig. 2002. "In Quest of a Theory of Adaptive Change." In *Panarchy: Understanding Transformations in Systems of Humans and Nature,* edited by L. H. Gunderson and C. S. Holling, 3–24, Washington, DC: Island Press.

Huigen, M., K. Overmars, and W. De Groot. 2006. "Multi-Actor Modeling of Settling Decisions and Behavior in San Mariano Watershed, the Philippines; a First Application with the MameLuke Framework." *Ecology and Society* 11 (2): 33.

Itami, Robert M. 1994. "Simulating Spatial Dynamics: Cellular Automata Theory." *Landscape and Urban Planning* 30 (1–2): 27–47.

Janssen, Marco A., and T. K. Ahn. 2006. "Learning, Signaling and Social Preferences in Public Good Games." *Ecology and Society* 11 (2): 21.

Janssen, Marco A., and Elinor Ostrom. 2006. "Empirically Based, Agent-Based Models." *Ecology and Society* 11 (2): 37.

Karlin, Samuel 2014. *A First Course in Stochastic Processes.* New York: Academic press,

Keeling, M. J., and B. T. Grenfell. 2000. "Individual-Based Perspectives on R0." *Journal of Theoretical Biology* 203 (1): 51–61.

Keylock, Christopher J., and Danny Dorling. 2004. "What Kind of Quantitative Methods for What Kind of Geography?" *Area* 36 (4): 358–66.

Kirman, A. P. 1992. "Whom or What Does the Representative Individual Represent?" *Journal of Economic Perspectives* 6 (2): 117–36.

Lewin, R. 1992. *Complexity: Life at the Edge of Chaos.* New York: Macmillan.

Libiszewski, Stephan. 1992. *What Is an Environmental Conflict? (Occasional Paper No. 1).* Zurich, Switzerland: Environment and Conflicts Project, Swiss Peace Foundation, Berne, and Center for Security Studies and Conflict Research.

Link, Peter Michael, Uwe Schneider, and Richard Tol. 2011. "Economic Impacts of Changes in Fish Population Dynamics: The Role of the Fishermen's Harvesting Strategies." *Environmental Modeling and Assessment* 16 (4): 413–29.

Luna-Reyes, Luis Felipe, Ignacio J. Martinez-Moyano, Theresa A. Pardo, Anthony M. Cresswell, David F. Andersen, and George P. Richardson. 2006. "Anatomy of a Group Model-Building Intervention: Building Dynamic Theory from Case Study Research." *System Dynamics Review* 22 (4): 291–320.

Lynch, N., R. Segala, and F. Vaandrager. 2003. "Hybrid I/O Automata." *Information and Computation* 185 (1): 105–57.

Macy, M. W., and R. Willer. 2002. "From Factors to Actors: Computational Sociology and Agent-Based Modeling." *Annual Review of Sociology* 28: 143–66.

Maxwell, Thomas, and Robert Costanza. 1997. "A Language for Modular Spatio-Temporal Simulation." *Ecological Modelling* 105: 105–13.

McCarter, M., D. Budescu, and Jürgen Scheffran. 2011. "The Give-or-Take-Some Dilemma: An Empirical Investigation of a Hybrid Social Dilemma." *Organizational Behavior & Human Decision Processes* 116: 83–95.

Metcalf, Sara, and Mark Paich. 2005. "Spatial Dynamics of Social Network Evolution." In *Proceedings of the 23rd International Conference of the System Dynamics Society.* Boston, MA: International System Dynamics Society.

Minar, N., R. Burkhart, C. Langton, and M. Askenazi. 1996. *The Swarm Simulation System: A Toolkit for Building Multi-Agent Simulations (Working Paper 96-06-042)*. Santa Fe, NM: Santa Fe Institute.

Monticino, Michael, Miguel Acevedo, Baird Callicott, Travis Cogdill, and Christopher Lindquist. 2007. "Coupled Human and Natural Systems: A Multi-Agent-Based Approach." *Environmental Modelling & Software* 22 (5): 656–63.

Mora, Thierry, and William Bialek. 2011. "Are Biological Systems Poised at Criticality?" *Journal of Statistical Physics* 144: 268–302.

Moreno, Niandry, Fang Wang, and Danielle J. Marceau. 2009. "Implementation of a Dynamic Neighborhood in a Land-Use Vector-Based Cellular Automata Model." *Computers, Environment and Urban Systems* 33 (1): 44–54.

Moss, S., and B. Edmonds. 2005. "Towards Good Social Science." *Journal of Artificial Societies and Social Simulation* 8 (4): 13.

Nelson, H. T., N. L. Cain, and Z. Yang. 2015. "All Politics Is Spatial: Integrating an Agent-Based Model with Spatially Explicit Landscape Data." In *Rethinking Environmental Justice in Sustainable Cities: Insights from Agent-Based Modeling*, 168–89. New York: Routledge.

Neumann, J. von, and Arthur W. Burks. 1966. *Theory of Self-Reproducing Automata*. Urbana, IL: University of Illinois Press.

Nikolai, Cynthia, and Gregory Madey. 2009. "Tools of the Trade: A Survey of Various Agent Based Modeling Platforms." *Journal of Artificial Societies and Social Simulation* 12 (2): 2.

North, M. J., T. R. Howe, N. T. Collier, and J. R. Vos. 2007. "A Declarative Model Assembly Infrastructure for Verification and Validation." In *Advancing Social Simulation: The First World Congress*, edited by S. Takahashi, D. L. Sallach, and J. Rouchier. Heidelberg, Germany: Springer.

Osgood, Nathaniel. 2009. "Lightening the Performance Burden of Individual-Based Models through Dimensional Analysis and Scale Modeling." *System Dynamics Review* 25 (2): 101–34.

Pahl-Wostl, Claudia. 2006. "The Importance of Social Learning in Restoring the Multifunctionality of Rivers and Floodplains." *Ecology and Society* 11 (1): 10.

Paich, Mark, and John D. Sterman. 1993. "Boom, Bust, and Failures to Learn in Experimental Markets." *Management Science* 39 (12): 1439–58.

Parker, D. C., S. M. Manson, Marco A. Janssen, M. J. Hoffmann, and P. Deadman. 2003. "Multi-Agent Systems for the Simulation of Land-Use and Land-Cover Change: A Review." *Annals of the Association of American Geographers* 93 (2): 314–37.

Parunak, H. Van Dyke, Robert Savit, and Rick Riolo. 1998. "Agent-Based Modeling vs. Equation-Based Modeling: A Case Study and Users' Guide." In *Multi-Agent Systems and Agent-Based Simulation*, edited by Jaime Sichman, Rosaria Conte, and Nigel Gilbert, Vol. 1534: 277–83, Berlin/Heidelberg, Germany: Springer.

Perrings, C. 2006. "Resilience and Sustainable Development." *Environment and Development Economics* 11: 417–27.

Polhill, J. Gary, Alessandro Gimona, and Richard J. Aspinall. 2011. "Agent-Based Modelling of Land Use Effects on Ecosystem Processes and Services." *Journal of Land Use Science* 6 (2–3): 75–81.

Prell, Christina, Klaus Hubacek, Mark S. Reed, Claire Quinn, Nanlin Jin, Joe Holden, Tim Burt, et al. 2007. "If You Have a Hammer Everything Looks like a Nail: Traditional versus Participatory Model Building." *Interdisciplinary Science Reviews* 32 (3): 263–82.

Radzicki, Michael J. 1989. "Reference Modes and the Optimal Shape Parameter." *System Dynamics Review* 5 (2): 192–98.

Railsback, Steven F., and Volker Grimm. 2011. *Agent-Based and Individual-Based Modeling: A Practical Introduction*. Princeton, NJ: Princeton University Press.

Railsback, Steven F., Steven L. Lytinen, and Stephen K. Jackson. 2006. "Agent-Based Simulation Platforms: Review and Development Recommendations." *Simulation* 82 (9): 609–23.

Ramanath, Ana Maria, and Nigel Gilbert. 2004. "The Design of Participatory Agent-Based Social Simulations." *Journal of Artificial Societies and Social Simulation* 7 (4): 1.

Rand, William, Dan Brown, Rick Riolo, and D. Robinson. 2005. "Toward a Graphical ABM Toolkit with GIS Integration." In *Proceedings of the Agent 2005 Conference on Generative Social Processes, Models, and Mechanisms*. Chicago, IL.

Reiter, D. 2003. "Exploring the Bargaining Model of War." *Perspectives on Politics* 1 (1): 27–43.

Reschke, C. H. 2001. "Evolutionary Perspectives on Simulations of Social Systems." *Journal of Artificial Societies and Social Simulation* 4 (4): 8.

Reuveny, Rafael, John W. Maxwell, and Jefferson Davis. 2011. "On Conflict over Natural Resources." *Ecological Economics* 70 (4): 698–712.

Richardson, G. P. 1991. *Feedback Thought in Social Science and Systems Theory*. Philadelphia, PA: University of Pennsylvania Press.

Robinson, W. S. 1950. "Ecological Correlations and the Behavior of Individuals." *American Sociological Review* 15 (3): 351–57.

Robinson, Derek T., Daniel G. Brown, Dawn C. Parker, Pepijn Schreinemachers, Marco A. Janssen, Marco Huigen, Heidi Wittmer, et al. 2007. "Comparison of Empirical Methods for Building Agent-Based Models in Land Use Science." *Journal of Land Use Science* 2 (1): 31–55.

Rudel, Thomas. 2011. "Local Actions, Global Effects? Understanding the Circumstances in Which Locally Beneficial Environmental Actions Cumulate to Have Global Effects." *Ecology and Society* 16 (2): 19.

Scheffran, Jürgen. 2000. "The Dynamic Interaction between Economy and Ecology: Cooperation, Stability and Sustainability for a Dynamic-Game Model of Resource Conflicts." *Mathematics and Computers in Simulation* 53: 371–80.

Schelling, Thomas C. 1971. "Dynamic Models of Segregation." *Journal of Mathematical Sociology* 1: 143–86.

Schlüter, Maja, and Claudia Pahl Wostl. 2007. "Mechanisms of Resilience in Common-Pool Resource Management Systems: An Agent-Based Model of Water Use in a River Basin." *Ecology and Society* 12 (2): 4

Schredelseker, Klaus, and Florian Hauser. 2008. *Complexity and Artificial Markets*. Berlin-Heidelberg, Germany: Springer.

Scoones, I. 1999. "New Ecology and the Social Sciences: What Prospects for a Fruitful Engagement?" *Annual Review of Anthropology* 28 (1): 479–507.

Shakun, Melvin F. 2001. "Unbounded Rationality." *Group Decision and Negotiation* 10 (2): 97–118.

Simon, Herbert A. 1956. "Rational Choice and the Structure of the Environment." *Psychological Review* 63 (2): 129–38.

Simon, Herbert A. 1991. "Bounded Rationality and Organizational Learning." *Organization Science* 2 (1): 125–34.

Slantchev, B. 2003. "The Principle of Convergence in Wartime Negotiations." *American Political Science Review* 97: 621–32.

Smaldino, Paul E., Jimmy Calanchini, and Cynthia L. Pickett. 2015. "Theory Development with Agent-Based Models." *Organizational Psychology Review* 5 (4): 300–17.

Sterman, John D. 2000. *Business Dynamics: Systems Thinking and Modeling for a Complex World*. New York: Irwin/McGraw-Hill.

Stigberg, David. 2012. "An Introduction to the NetLogo Modeling Environment." In *Ecologist-Developed Spatially-Explicit Dynamic Landscape Models*, edited by James D. Westervelt, Gordon L. Cohen, and Bruce Hannon, 27–41. Modeling Dynamic Systems. Boston, MA: Springer US.

Stillman, Richard A., Kevin A. Wood, and John D. Goss-Custard. 2016. "Deriving Simple Predictions from Complex Models to Support Environmental Decision-Making." *Ecological Modelling* 326: 134–41.

Tango, Peter J., and Richard A. Batiuk. 2013. "Deriving Chesapeake Bay Water Quality Standards." *JAWRA Journal of the American Water Resources Association* 49 (5): 1007–24.

Taylor, Glenn, Michael Quist, and Allen Hicken. 2009. "Acquiring Agent-Based Models of Conflict from Event Data." In *Proceedings of the 21st International Joint Conference on Artificial Intelligence*, 318–23. Pasadena, CA: Morgan Kaufmann Publishers Inc.

Tesfatsion, Leigh. 2003. "Agent-Based Computational Economics: Modeling Economies as Complex Adaptive Systems." *Information Sciences* 149 (4): 262–68.

Tobias, R., and C. Hofmann. 2004. "Evaluation of Free Java-Libraries for Social-Scientific Agent Based Simulation." *Journal of Artificial Societies and Social Simulation* 7 (7): 6.

Vitro, Kristen A., Miranda E. Welsh, Todd K. BenDor, and Aaron Moody. 2017. "Ecological Theory Explains Why Diverse Island Economies Are More Stable." *Complex Systems* 26 (2): 135–56.

Voinov, Alexey, and François Bousquet. 2010. "Modelling with Stakeholders." *Environmental Modelling & Software* 25 (11): 1268–81.

Wagner, R. H. 2000. "Bargaining and War." *American Journal of Political Science* 44: 469–84.

Wang, Shiyong, Jiafu Wan, Daqiang Zhang, Di Li, and Chunhua Zhang. 2016. "Towards Smart Factory for Industry 4.0: A Self-Organized Multi-Agent System with Big Data Based Feedback and Coordination." *Computer Networks*, Industrial Technologies and Applications for the Internet of Things, 101 (June): 158–68.

Watkins, Cristy, Dean Massey, Jeremy Brooks, Kristen Ross, and Moira Zellner. 2013. "Understanding the Mechanisms of Collective Decision Making in Ecological Restoration: An Agent-Based Model of Actors and Organizations." *Ecology and Society* 18 (2): 32.

Wheeler, Scott. 2005. *On the Suitability of NetLogo for the Modelling of Civilian Assistance and Guerrilla Warfare (DSTO-TN-0623)*. Edinburg, South Australia: Australian Defence Science and Technology Organization, Systems Sciences Laboratory.

White, Roger, and Guy Engelen. 1997. "Cellular Automata as the Basis of Integrated Dynamic Regional Modelling." *Environment and Planning B* 24: 235–46.

White, Roger, Guy Engelen, and Inge Uljee. 2015. *Modeling Cities and Regions as Complex Systems: From Theory to Planning Applications*. Cambridge, MA: MIT Press.

Wilensky, Uri. 1999. "NetLogo." http://ccl.northwestern.edu/netlogo/. Evanston, IL: Northwestern University Center for Connected Learning.

Wilensky, Uri, and William Rand. 2015. *An Introduction to Agent-Based Modeling: Modeling Natural, Social, and Engineered Complex Systems with NetLogo*. Cambridge, MA: MIT Press.

Wils, Wilbert. 1974. "Metropolitan Population Growth, Land Area, and the Urban Dynamics Model." In *Readings in Urban Dynamics*, edited by Nathaniel J. Mass, 103–14. Cambridge, MA: Wright-Allen Press.

Wilson, W. G. 1998. "Resolving Discrepancies between Deterministic Population Models and Individual-Based Simulations." *American Naturalist* 151 (2): 116–34.

Wolfram, Stephen. 1983. "Statistical Mechanics of Cellular Automata." *Reviews of Modern Physics* 55: 601–44.

Wolfram, Stephen. 2002. *A New Kind of Science*. General Science. Champaign, IL: Wolfram Media.

Wu, C. T. 2009. *An Introduction to Object-Oriented Programming with Java*. New York: McGraw-Hill Education.

7

Modeling Conflict and Cooperation as Agent Action and Interaction

I think that the unity we can seek lies really in two things. One is that the knowledge which comes to us at such a terrifyingly, inhumanly rapid rate has some order in it. We are allowed to forget a great deal, as well as to learn. This order is never adequate. The mass of un-understood things, which cannot be summarized, or wholly ordered, always grows greater; but a great deal does get understood. The second is simply this: we can have each other to dinner. We ourselves, and with each other by our converse, can create, not an architecture of global scope, but an immense, intricate network of intimacy, illumination, and understanding. Everything cannot be connected with everything in the world we live in. Everything can be connected with anything.

J. Robert Oppenheimer,
The Growth of Science and the Structure of Culture (1958)

Introduction

We have spent the last several chapters demonstrating how mathematical modeling and computer simulation can contribute to a deeper understanding of conflict resolution, cooperation, and peace building (van den Belt 2004). Models can also help us understand and formalize decision making and interagent interactions during conflict and cooperation. In this chapter, we introduce an approach for modeling environmental conflicts by representing the interaction of shared environmental conditions—including the state or quality of resources—and the individual goals, resources, and decisions of agents involved in conflict.

To do this, we will create a conceptual and theoretical framework that unifies many computer modeling approaches for the purpose of better understanding, representing, exploring, and resolving environmental conflict. These methods—including agent-based modeling (ABM), evolutionary game theory, system dynamics (SD) modeling, and network analysis—are unified under the umbrella of *complex systems* analysis (discussed in Chapters 5 and 6), which provides a useful paradigm for approaching conflict analysis.

This chapter will draw on all of these methods to introduce the VIABLE framework (*Values and Investments for Agent-Based interaction and Learning in Environmental systems*) in qualitative and conceptual terms, leaving much of the mathematical nuance toward the end of the chapter and specific applications of environmental conflict presented in Part III of this text.

We will begin this chapter by discussing the basic concepts around how we will represent human decision making, including the concept of action pathways and the use of capital, agent interactions, and their effects on rational and adaptive decision making. Next, we will argue for the use of multiple complex systems techniques in conjunction with modeling environmental conflict. Following this, we will introduce *viability theory*, the mathematical basis of our VIABLE modeling approach and explore how we can reconceptualize environmental conflict states using this theory. Next, we introduce the VIABLE framework, first conceptually, and then mathematically. We will start this discussion in the context of the decisions of a single agent, and expand it to explore multiple agents making decisions in conjunction with each other. Finally, we will use the VIABLE framework to mathematically explore the conditions for conflict and cooperation, offering a conceptual example that will lead us to more concrete examples in the following chapters.

Agent Decision Making

Capability and Capital

The extent to which environmental disputes are able to damage a society or lead to constructive social change critically depends on the way they are handled by the agents involved. Actions can only be taken if the *"capability"* to act exists. This is commonly understood as the ability to change a course of action (Nussbaum and Sen 1993), which includes any means of giving people the ability to change their natural and social environment.

If actions are directed at producing something of value, capability is often associated with the "capital" used to create goods and services (Walsh 1935; Bourdieu 1983; Castle 2002). In economic theory, the most common dimensions of capital are natural capital (usable resources of an ecosystem), physical capital (assets made by humans for production), and financial capital (monetary wealth). A good example for this is seen in aggregated optimal growth models in economics (Koopmans 1967), where a "production" function is designed to represent the flow of economic output, which depends on capital, labor, and technology. Increasingly, human production factors are included, such as human capital (workers' skills and abilities), social capital (such as social networks; Gutierrez, Hilborn, and Defeo 2011), political capital (instruments and institutions in political decision making), and cultural capital (knowledge, skills, education status, and personal advantages).

- Capital concerns the first aspect of our approach: agents must be able to respond to a changing environment by adjusting their capabilities and investments to pursue values and avoid risks.

Rationality and Utility

Significant effort has gone into creating decision tools that can help to rank and select among possible courses of action (e.g., Cegan et al. 2017). As traditionally characterized by economic theory, "rational" decision makers choose the most preferred or optimal option. This is usually taken as an option that maximizes an agent's individual "utility," which is described in rigorous mathematical terms called a "utility function" (sometimes also called an "objective function"; see Bhavnani, Backer, and Riolo 2008).

The rational approach to characterizing decisions—particularly those decisions made in a conflict setting—assumes a decision maker who has complete knowledge of their situation, and who selects an optimal, time-discounted decision path. In this context, simulation models typically consider a single rational decision maker optimizing a global "welfare" function (essentially "optimizing" the entire system around certain objectives).

Unfortunately, this approach is unrealistic in situations undergoing long-term environmental change, as the world is shaped by numerous agents who act according to their own interests, capabilities, and rules.

This means that many factors and interactions are highly uncertain and beyond the decision maker's direct control. Furthermore, as we explored in the last chapter, the complex socio-economic interactions among multiple agents undermine predictability of the system and its dynamics. Alternatively, other techniques, such as certain types of cost–benefit analysis, can seek decision pathways that balance high benefit, low cost, and minimal risk (Feuillette et al. 2016).

- Rationality and decision making concern the second aspect of our approach: agents must be able explicitly select among a range of action pathways according to either formal or informal (even subconscious) rules, preferences, and criteria.

Adaptation in the Presence of Others

As we have discussed throughout this book, changes in environmental systems induce a wide range of individual and collective responses, including conflict. Each agent involved in a conflict can select among potential strategies for handling the situation, which incorporate their capabilities (resources) and willingness (e.g., risk tolerance) to take action, as well as their values and goals (including risks avoided and expected benefits), which are all shaped by personal and social experiences. The vulnerability of agents to the decisions of others prompts adaptation mechanisms, the structure of which depends on the historic effects agents have had on each other, as well as the manner by which agents are able to respond to each other according to their decision-making apparatuses. For example, adaptive responses to environmental change are shaped by the social context in which agents live, including their livelihoods, social networks, institutions, and governance.

Adaptation increases the complexity of action–reaction dynamics, leading to more complex interactions between agents and the natural environment and between the agents in their social environment (including security risks and conflicts). Conflict stakeholders face the risks of environmental change not as isolated individuals, but rather as members of social groups that can help protect individuals against risks. Furthermore, they can mutually enhance their capabilities through adaptation mechanisms that lead to reduced vulnerability (Mearns and Norton 2009).

In a modeling context, we can argue that by observing the outcomes of different agent actions, and adjusting each agent's invested resources in the next time period, agents can adapt and learn. Social evolution processes allow for the adjustment of interaction parameters as a result of learning among agents. Agents often have the ability to select and continue their more stable interactions, while the unstable ones are considered less "successful" according to fitness criteria in societal competition.

Unifying Complex Systems Approaches to Studying Conflict

Several approaches have been developed in complexity and decision sciences to assess the interplay between multiagent responses to environmental change and the social interactions that result from these responses (Scheffran 2006b). ABM, for example, has been shown to be particularly useful in using methods from statistical physics and nonlinear dynamics to replicate patterns of collective action—such as *self-organization*—that emerge from large numbers of agents that behave in relatively homogeneous ways. Network analysis can also help us by describing the connectivity between agents and providing structural analysis of the social interaction in conflicts (Flint et al. 2009; Maoz 2010). For example, network diffusion models of disease spread can also be used to study the spread of interagent behavioral patterns. This may help to describe the proliferation of violence and conflict as well as the diffusion of environmentally beneficial technical innovations and social practices, such as climate change mitigation and adaptation (Carmin, Anguelovski, and Roberts 2012; Berke 2014).

Self-organization is the tendency for a complex system to create an *emergent* global order or replicable patterns. In this process, disorganized *individual decisions* create *globally organized behavior,* like herding buffalo or swarming bees (see Kelly 1994; Schweitzer 1997; Weidlich 2006; Helbing 2010).

In situations where agents are not homogenous and are attempting to achieve diverse goals, conflicts may occur when the agents' preferred actions are incompatible. In these situations, we can turn to *game theory* (discussed in Chapter 3), which provides an optimizing framework for analyzing interdependent decision making and negotiations among players with respect to environmental conflicts. In a *dynamic game,* multiple agents (usually only a handful to remain mathematically tractable) mutually adapt their targets, values, and actions to those of other players to change the outcome in their own favor (Scheffran 2002; Reuveny, Maxwell, and Davis 2011).

Each of these approaches to studying conflict as a complex system has advantages and weaknesses. For example, while solutions derived from game-theoretic analysis are useful in situations with only a few agents (or only a few goals), they are often difficult to apply in settings with large numbers of agents, multiple decision criteria, and agents operating under bounded rationality. Likewise, ABM faces difficulty in selecting rules that adequately describe real-world agent decisions (Janssen and Ostrom 2006).

However, if used in conjunction, each of the failings of these techniques can benefit from the strengths of the others. ABM is an approach that is ideally situated to examine many agents and complex decision rules, while dynamic game theory can define agent decision rules through utility optimization and other approaches. Combining dynamic game theory with ABM allows us to represent agents' goals and decision rules as they evolve together through time, providing the basis for developing new approaches for conflict decision support (Billari et al. 2006).

One of these approaches was articulated by Reuveny, Maxwell, and Davis (2011), who implemented a dynamic, game-theoretic simulation in which agents maximize their efficiency in achieving their goals or enhancing their positions. Their decision maker-centered approach focused on charting the capabilities of agents, determining appropriate courses of action and investments. However, by largely focusing on optimization of individual decisions, this approach does not explicitly address the impacts of the decisions of other agents. This is an area that is often the focus of participatory modeling approaches, which emphasize conflict structure and the impacts of group behavior.

It is clear that we must consider both agent and group behavior, incorporating the goals and utilities of those making decisions, as well as the goals of others in the group embroiled in conflict. If we consider this problem as a challenge of decision making in the context of a lot of other decision makers and uncertainty, then we can turn to innovative techniques for setting goals and boundaries at both agent and group levels. We frame these boundaries and their role in containing an evolving conflict system using an innovative mathematical approach, known as *viability theory*, through which we will develop an integrative framework for modeling human action and interaction.

Viability Theory

In 1971, biochemist and Nobel Laureate Jacques Monod introduced the idea that many natural and social systems evolve in deterministic and stochastic ways, but are bound by constraints that force them to remain "viable" in their environment. Mathematician Jean-Pierre Aubin later formalized these ideas into the mathematics subfield called *viability theory*, which describes a framework for determining ways of keeping a dynamic system within constraints, defined by either objective limits or value-based judgments (Aubin 1991). The "viability" concept, and the general application of mathematical analysis toward maintaining a dynamic system within constraints, evokes many complex systems concepts, such as *stability, confinement, homeostasis, resilience,* and *adaptation.*

Viability theory's main analytical approach involves determining a system's *viability kernel,* which describes the subset of the dynamic state space of a system that is *viable*; this means that the system's future dynamics (evolution path) will remain confined within rigorously defined constraints at all times (Aubin and Saint-Pierre 2007; Aubin, Bayen, and Saint-Pierre 2011). In some simple systems, the viability kernel can be determined analytically, but most often it is done numerically or graphically.

Viability theory also includes a set of mathematical methods describing how to confine system evolution to defined boundaries, calculating control mechanisms when boundaries are approached or exceeded. For example, we can define a tolerable window of global temperature change that is mathematically estimated by greenhouse gas concentrations (Petschel-Held et al. 1999; Scheffran 2008a) and necessary emission reductions as a function of projected trajectories to avoid exceeding critical thresholds of CO_2 concentration and global average temperature change, such as the limit of 1.5–2°C as defined in the Paris Treaty of 2015 (Scheffran 2016).

Rather than go into the intricacies of mathematical viability theory, we will instead explore the application of viability theory in multi-agent situations. However, we direct interested readers to in-depth discussions of the mathematical details of viability theory in Aubin and Saint-Pierre (2007) and Aubin, Bayen, and Saint-Pierre (2011). We should warn readers that these texts can be quite mathematically demanding and are written for a professional mathematical audience. However, we can take the qualitative ideas behind the mathematics and translate them into boundary conditions for our multi-agent framework, as we later make explicit in the applications in Part III of this text.

Viability Theory and Resilience

One important and common question for resource-based conflicts concerns whether the intensity and rate of environmental change exceeds the viable limits of natural and social systems, and if so, which control mechanisms can and should be applied. The challenge is to identify admissible "guard rails" for action, within which conflict negotiations can maneuver. In doing this, we also have to take into consideration stakeholder weaknesses, opportunities, adaptive capacities, and critical thresholds, such as system behavior that leads directly to environmental or economic collapse. Examples of these collapses can be seen in extreme water or resource shortages, as well as species extinctions (Ceballos and Ehrlich 2002; Thomas et al. 2004).

Viability theory, at least in its application to environmental matters, deserves an important place within the broader discussion of system *resilience* (on this relationship see Scheffran 2016). In the last chapter, we discussed and defined resilience as the ability of a system to cope with disturbance and adapt under stress to maintain structure and function. It is worth mentioning again that many different fields have developed approaches to thinking about the stability properties of different systems, including ecosystems (Holling 1973; Walker and Salt 2012) and cities (Perrings 2006; Briguglio et al. 2009; Holling 2009). There is now a growing acknowledgment that resilience is largely determined by the close coupling of natural and social systems. Here, the interactions between biophysical and social processes determine the capacity of the system to adaptively respond to stress and changes in the environment (Schlüter and Pahl-Wostl 2007).

Viability Theory and Multi-Agent Conflict

When closely familiar with a system, we can often describe how agents implement actions based on locally updated information and decision rules, which can be used by each agent responding to the changing state of a system; this is known as a theory of *adaptive control* (e.g., Astrom and Wittenmark 1994; Dercole, Prieu, and Rinaldi 2010). Adaptive control enables agents to adjust to the actions of other agents, as well as the changing environment, in order to reach their goals or targets. Agents decide and act on the basis of incomplete knowledge, and are restrained to a spatial and temporal window of information; in this sense, we can think of agents as acting to reach a goal that they can "see" (i.e., a locally "optimal" state based on their restrained information) or a goal that is feasible if all other agents and possibilities are taken into account (i.e., a globally optimal state). We can also describe situations where agents can learn to evolve their behavior between locally and globally optimal strategies.

Adaptive control approaches guide strategies that constrain and adjust actions so that the system and individual agents remain viable over the long term (Figures 7.1 and 7.2). The set of strategies and outcomes yielding the state of being viable is known as a *viability domain* (or *viability kernel*), which we might articulate, for example, as a specific range of carbon concentrations or global average temperatures that prevent run-away climate change, or as a range of financial returns that guarantee continued solvency of a company. Likewise, adaptive controls might consist of decision rules determining when to invest in reductions of environmental pollution or the use of natural resources to avoid critical thresholds, like runaway climate change (Oubraham and Zaccour 2018).

To estimate whether current trends can be perceived as tolerable or require changes in the current course of action, it is imperative to have up-to-date information about the difference between the current state of the system and the critical thresholds that lead to system collapse (Scheffran 2008a). In order to accurately model how agents consider the consequences of their actions through time, we will have to adjust for the considerable inertia and time lags that are inherent to natural and social systems; this means taking into consideration the rates of change of the system and the behavior of individual agents.

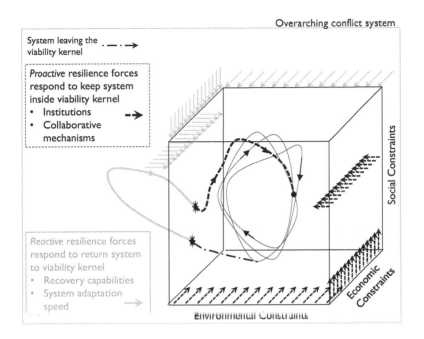

FIGURE 7.1

Using a box, we can visualize the evolutionary trajectory of an environmental *system* within viability constraints. *Proactive* and *reactive* forces can increase the resilience of the system to external shocks (e.g., droughts or floods) and internal chaotic dynamics (e.g., runaway feedback, arms races, retaliation). Proactive efforts to increase resilience often include institutions that increase communication and collaboration (i.e., plans, regulations, business associations). Reactive resilience efforts could include forces and institutions that help to realign the system trajectory after the system has moved beyond viability constraints (e.g., disaster relief funds, emergency plans, monitoring efforts). Figure 7.2 depicts the viability of individual agents.

A Conceptual Introduction to the VIABLE Framework

The VIABLE modeling approach applies viability theory to generate a formalized understanding of the stability and complexity of *conflict*, which we characterize as a dynamic interaction process involving agents who are driven into non-viable states. As a result, they fail to reduce their conflict potential to tolerable levels. We characterize *conflict potential* using the following factors:

- *Values* and *goals* describe what agents want, depending on their interests, needs and motivations, objectives and targets, and in the context of the risks and dangers they are willing to take. Agents will evaluate the impacts of actions and define targets for future actions based on how they affect their value (V) or goal (target value; V^*)

- *Action paths (k)* describe ways that agents strategize and act to better their value (V) and achieve their goals (V^*). Agents can select multiple paths from a range of decision options and potential strategies (e.g., choices of technologies, modes of behavior).

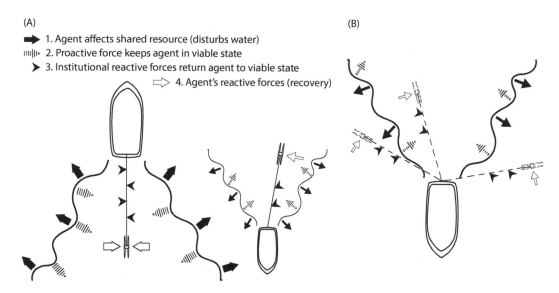

FIGURE 7.2

Water skiers on a lake are an apt metaphor for agent interactions in a viability framework. (A) As agents move across the water, they send disturbances (waves) outward that affect other skiers (1). These disturbances of the water—which is akin to a shared resource affected by conflict—can be counteracted by different resilience forces on the part of the agent. First, the agent is capable of creating proactive resilience forces, which disrupt the waves from other skiers to keep the agent within a viable state; for example, agents can invest in larger boats that create larger wakes (2), which more forcefully keep the skier from leaving the viable space. (B) In cases where a skier is thrown outside of their wake (i.e., the shared resource disturbances put them outside of their viability kernel), the agent relies on both their tether and their own strength to pull them back into a viable space. The tether (3) represents institutional forces (e.g., disaster management, emergency loan programs) that help an agent recover, while the agent's own skill or ability (4) to return to the wake depends on how the agent has prepared for such an occurrence (e.g., factors such as savings, ingenuity, alternative strategies, financial management, or access to additional, private resources). Figure design assisted by Ellen Emeric.

- *Investments* (C) concern the instruments and efforts that agents can use to change the state of the environment in pursuing their values and goals. We can also think of these as "capabilities," including capital, resources, efforts, or financial investments, which can be split among multiple action paths (k) used by the agent to better the agent's value (V) to achieve its goals (V^*). The agent uses rules to determine the fraction of investments (r) they should place into each action path (k).

Along any of these dimensions, agents may draw a line (sometimes implicit) between what they are willing to tolerate and what they find intolerable (viability constraints). When an observed state of the environment is intolerable for a given set of agents, they naturally seek to use their capabilities to change this state to their benefit. Conflict is provoked when these responses cause intolerable outcomes for other agents; for example, while expanded logging increases revenues and prevents layoffs at a lumber mill, it increases erosion, damaging the water resources of other agents downstream.

The complex, emergent dynamics from agent actions and interactions may either increase or decrease the potential for conflict. In the former case, these dynamics produce an unstable and escalating conflict. If this remains unresolved, an escalation continues until some of the conflicting agents reach their resource limits (e.g., go bankrupt), give up (i.e., seek other types of value), or cease to exist (i.e., are eliminated from the conflict).

Alternatively, if agents succeed in reducing conflict potential, they improve the chances for conflict resolution and a more stable, long-term interaction. Learning, cooperation, and negotiation are important mechanisms to allow agents to adapt their goals, resources, and actions to those of other agents—working together rather than against each other.

Unresolved conflicts can prevent the successful goal attainment for some, or even all, agents. Even though agents might behave in ways that are individually rational, their collective interaction can lead to collectively detrimental behavior and produce negative outcomes. A classic example is the tragedy of the commons, where individual use of a shared resource leads to a collective overexploitation and even destruction of this resource (Hardin 1968).

Another example is the so-called "security dilemma" (e.g., Jervis 1978a, b), where individual attempts to protect against perceived threats by other agents results in countermeasures (such as military capability) that provoke more threatening responses. For instance, nations spend their military budget (investment) into increasing armament (action) to improve their own security (value), which at the same time undermines the security of other nations who increase their own armament to compensate for this value loss, leading to an escalating downward spiral of insecurity and conflict (an "arms race"). This interplay can eventually lead to an escalating action–reaction pattern (Wallace and Wilson 1978; Schoffran 1989) that is collectively irrational to all agents.

Conflict prevention (and avoidance of conflict escalation) aims to mediate between contradictory positions and strengthen cooperation. Modeling these interdependencies examines the conditions under which conflicting interactions emerge, as well as the possibility that agents jointly influence the system toward a more mutually beneficial direction.

Four Relationships Determining Conflict and Cooperation

In Chapter 1, we introduced the concept that conflict can be an investment strategy on the part of agents that are vying for limited resources. We can now reframe and expand this discussion to view conflict explicitly as the product of agent adaptation and interaction as they seek disparate goals. Conflict often emerges from incompatible actions, values, behavioral rules, and priorities of agents who are unable to reduce their differences to tolerable levels. Whether conflict or cooperation prevails depends on the responses of each agent, as represented by formalized decision rules, investment priorities, and action paths, as well as by each agent's potential for learning and adaptation to their conflict environment.

The mutual impacts (linkages) between agents can be described for the following four situations:

1. If two agents are disconnected, then they are independent and have no impact on each other's values. The agents may be influencing different environmental systems, which are not connected, or the impacts of an agent's investments do not significantly affect the other agent. In this *neutral relationship*, an agent's actions are not directly influenced by other agents, and are thus kept constant, as long as each agent's own goals (values) do not change. For example, a deer hunter and a computer company could be seen to have a *neutral relationship* as their values are not influenced by the other agents' value-driven actions.

2. In a *hostile* or *competitive relationship*, the investment of Agent A in a course of action needs to be increased to compensate for the efforts of another Agent B. Agent A experiences this as a loss, as its investment is used to combat its value loss due to the actions of Agent B and cannot be used for other value-generating

purposes. This can even result in the failure of agent A to attain its goal. The actions taken in response to conflict problems may undermine each agent's values, create tension, and provoke responses that generate further losses. An inherently unstable conflict escalates actions and interactions by the conflicting parties, which aggravates conflict tension and intensity. If unresolved, conflicts can consume enormous amounts of resources, pushing the conflict parties toward extreme actions, including the use of violence or war, until the capability to act by some agents is exhausted or destroyed (e.g., bankruptcy, surrender, organizational collapse).

3. In a *friendly* or *cooperative relationship*, both agents benefit from each other and may reduce their own efforts accordingly to realize their goal more easily or even expand their goal for the same investment. Conflicts can be diminished when agents pursue cooperative actions that provide value to all agents (win–win), without being completely optimal in achieving the goals of individual agents. The goals of agents engaged in a cooperative framework can be still individual goals and need not necessarily be commonly held.

4. In a *mixed case*, one agent cooperates and the other does not, leading to unilateral exploitation of one agent by the other agent, a situation that may not be stable in the long run as the exploited agent may also switch to noncooperation (provided this is not excluded by some sort of punishment that prevents a switch).

Even in the competitive case (2), a state of mutual satisfaction is often possible, but the question is whether the investment rates required for agents to maintain that state can be sustained viably over the long term. If an agreement cannot be achieved, even with the maximum possible effort on the part of all agents, an unresolved conflict will continue unless agents acquire more or different capabilities or change their own goals.

Alternatively, one or both agents can change their behavior by switching to other action paths that make their actions more efficient and less threatening to other agents. Conflict resolution processes can help to reduce the conflict tension and stabilize the interaction by involving agents in learning and adjusting their actions until interagent agreement is reached (Kriesberg 2009). We therefore refer to *cooperation* as the process by which agents adjust their goals and actions to achieve mutual benefits (Giuliani and Castelletti 2013). This transition from conflict to cooperation requires adaptation toward common positions and mutually beneficial actions that stabilize the interaction. Whether this transition is successful depends on the institutional and governance capacity of societies to prevent or manage conflicts.

Single Agent VIABLE Modeling

We can first consider the actions taken by a single agent, which have impacts on the natural and social environment, provoking environmental feedback responses that later affect the agent. These feedbacks depend on (1) the type of action chosen by the agent, (2) the feedback response of the environment to the agent's actions, and (eventually) (3) the feedback response emerging from the social interaction among the other agents, who may view the action as beneficial or damaging (we will consider agent interactions in the next section). To formalize these feedbacks, we first need to understand them in detail.

Interaction between agent values and capabilities: If an agent increases investment in a course of action ($\Delta C > 0$), and this enhances the value to the agent ($\Delta V > 0$), then we can think of this investment as having a positive self-impact on this agent. However, if a course of action results in a value loss ($\Delta V < 0$), a further application of investment ($\Delta C > 0$) along this course will further aggravate the situation, against the agent's belief that the added investment will lead to a value gain. Such an "irrational" behavior may lead to a downward spiral if the action path remains unchanged (i.e., "throwing good money after bad"; investments that go nowhere). Instead, agents can learn and evaluate the effectiveness of their investment, change their actions to improve the situation, and compensate for the losses they experience. Alternatively, agents can change their value goal V^* to avoid negative side effects of their investments.

Impact of agent actions on natural resources: it is important to be explicit about how we are characterizing the "natural system" with respect to conflicting agents. Significant and longstanding work has attempted to shift the oversimplified dichotomy of "natural space" vs. "human space," (e.g., cities vs. wilderness) and toward an understanding that agents exist and act *within* environmental systems (i.e., cities *are* environmental systems, people live in wilderness areas; McHarg 1969). Here, we must disentangle the (1) *conflict system* as characterized by the overarching ecosystem in which humans live, interacting with their surroundings and with each other (often in conflict) and the (2) *resource system* that forms the basis of the conflict, whether that be food, land, water, or another element of the larger system. In this book, we view conflicts as occurring *within* the former, overarching system, and *about* the latter environmental system or resource system, which we symbolize as x.

That being said, the traditional course of urban and agricultural development is resource-intensive, where agents' goals require consumption or use of natural resources. The environmental Kuznets hypothesis states that beyond a certain level of per capita income, per-capita environmental degradation begins to decline (Kuznets 1955; Stern 2017). However, the basis for this assertion is controversial, and rooted in context-dependent (economy- and culture-dependent) relationships between agent activities and natural resource sustainability (see discussion in Dasgupta et al. 2002).

Impact of agent actions on inter-agent (or societal) stability: investments that are made in pursuit of individual goals have a vast range of consequences for society. One possibility is that societies are more stable if more people are satisfied regarding their values. If this is true, then significant threats to agents' goals (values) could undermine societal or system stability and create conflict. However, this also means that value losses may be offset by adjustments made to the courses of actions that agents take, including interventions from institutions and governance mechanisms, or choice and reformulation of values to create new preference orders.

For example, flood losses can create huge social problems and potential conflict in damaged communities. However, in the United States, for instance, government interventions lower this conflict through insurance compensation mechanisms that restore the status (and value) of affected agents (e.g., Shively 2017). Similarly, agents involved in automobile collisions could run into intense conflict due to sudden value loss. Here, the public has intervened by establishing institutions and relationships; governments can require drivers to carry private insurance (as in the USA and Europe), which offset value losses and reduce conflict. Again, context-dependent analysis is needed to identify how certain actions and investments can either intensify conflicts or promote long-term agent cooperation.

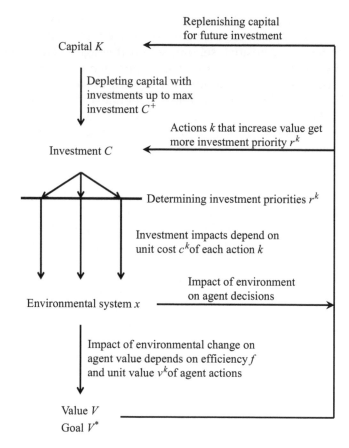

FIGURE 7.3

Adaptive feedback cycle linking a single agent's resources, its investment decisions in different action paths, the impacts of those actions on environmental systems, and the resulting change in the agent's value and proximity to its goals.

Figure 7.3 shows the primary elements of this adaptive control approach for a single agent, who starts with some total level of capital stock K, which is applied (invested) as a stream of assets or capabilities (C) to given action paths (k), selected based on relative investment priorities (r^k), which change an environmental system (x). The observed impacts are evaluated based on the agent's values or goals (V), which are a function of the benefits, costs, and risks of the actions taken. Over time, an agent's actions are adjusted through repeated feedback and learning cycles according to decision and adaptation rules in response to environmental changes, and the assessment of expected benefits, costs, and risks associated with each action path. These actions are adjusted according to an agent's ability to (1) evaluate the natural and social environment around them and (2) implement their decision rules.

We will define a series of variables throughout this chapter and use them extensively in the following application chapters. We provide a quick reference list of variable explanations in Box 7.1.

BOX 7.1 VARIABLE DEFINITIONS

All are given for each agent $i = 1,...,$ n unless otherwise specified

V = value achieved by an agent that indicates something positive or valuable, such as utility or benefit.

V^* = a value goal or target value that an agent aims to achieve (variables with a * superscript indicate a target or goal).

C = rate of total agent investment.

C^+ = the highest rate of investment available for an agent.

C^* = an agent's target investment rate.

K = capital stock available to agent (accumulated over time and depleted by investment C).

x = state of environmental system (can stand in for a vector of many environmental state variables).

σ = unit cost of environmental change (investment C per unit change of x for each agent).

k = action path for acquiring value ($k = 1,..., m$)

r^k = fraction of investment (priority; ranges from 0 to 1) committed to action path a^k.

$\tilde{C}$ = response curves representing the dynamic adaptation of investments made by agents in the face of investment changes by other agents.

$\hat{C}$ = the equilibrium point of response curves, which corresponds to the investment behavior whereby all agents are able to meet target values. This equilibrium may emerge as a steady, inducing conflict or cooperation.

v = the unit value or marginal value, found by taking the derivative of the agent's value function V, with respect to the priority of an action path r, $v_r = \partial V / \partial r$.

f = the efficiency of producing value output for a given unit of investment (also can be thought of as a benefit–cost ratio; v/c).

f_{ij} = the interaction efficiency describing the effect of investment by each agent i on the value of each other agent j. These mutual efficiencies depend on the allocation of investments and action paths taken by both Agent i and other agents j ($i, j=1,..., n$).

α = adaptation rate of an agent to change priority r^k for investing in an action path k.

β = adaptation rate an agent changes investment C to a target level C^*.

κ = the constant adaptation factor within the logistic growth function that affects the speed at which adaptation rates α or β move from 0 to 1 as r or C increase (given as κ^r or κ^C, in the adaptation rule for r and C, respectively).

Multi-Agent VIABLE Modeling

Multi-agent settings are increasingly relevant when multiple regions, countries, businesses, or citizens are affected by environmental change and respond through individual or collective actions that lead to social interaction. At the global level, the primary decision makers are governments of nations or groupings among them. At the local level, individual citizens and consumers are key players who affect (or are affected by) environmental change. Local and global decision making processes are connected through several layers of aggregation (from billions of citizens to a few diplomats representing their countries), with each layer having its own characteristic decision procedures for setting targets and implementing them as real actions (Scheffran 2008b).

We can adapt the single-agent framework in Figure 7.3 to describe the interaction between multiple agents i, each with their values and goals (V_i), capabilities and investments (C_i), action paths (k), and action priorities and rules (r_i^k). The action paths selected by each agent affect the environmental system (x), which then impacts other agents' values and capabilities (Figure 7.4).

Inter-agent impacts are almost entirely dependent on the action paths selected and thus on the respective priorities and rules of agent behavior. The agents are also interconnected directly through communication processes for exchanging information, which can also be treated as actions (the decision to share information or not; e.g., Kaza and Hopkins 2009). We can now express conflict-related agent behavior, actions, inter-agent social interactions, and resource dynamics, including behavior endogenous to the resource (e.g., fish stock

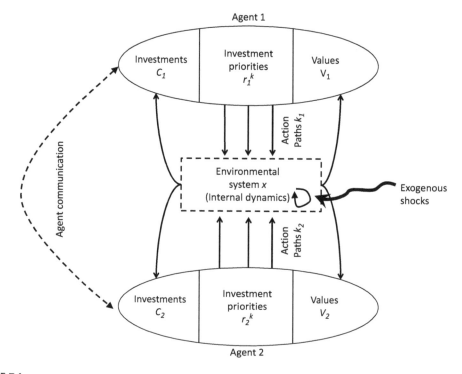

FIGURE 7.4

Interaction between two agents in a shared environmental system. Agents may directly interact (e.g., agent-to-agent communication), or they may indirectly interact through the environmental system.

reproduction rate; forest regrowth) and exogenous environmental shocks (e.g., drought and diseases), within the VIABLE framework.

Mathematical Modeling Using the VIABLE Framework

To analyze environmental conflict using viability theory, we will start by discussing the case of a single-agent feedback cycle, which we mathematically formalize in this framework through a phase-transition model. This framework will describe how and why changes in the goal and preference structures at the individual agent level can rapidly have an impact on the macro-scale conflict system. We will then extend this concept into a multi-agent model framework.[1]

Modeling the Individual Conflict Agent

An essential precondition for analyzing and modeling conflict is a strong understanding of the actions of individual agents. Actions occur in a system environment, characterized by relevant system variables that define a current system state $x(t)$ at a given time t. We can generally assert that agents evaluate and choose preferred system states based on their own internal preference structure. Under certain conditions, the preference order can be represented by a *value function* $V(x)$ and target values V^*.

As discussed earlier, to achieve these targets, the agent decides on investing its *capability* C (these can also be interpreted as invested resources, efforts, or "costs" for the agent) into a particular *action* k—which is a function of the system state and investment costs—to change the system state by Δx and generate value V, which is then compared to the target value V^*. Value that is added $V > 0$ can be consumed, directly reinvested, or accumulated to form productive capital K. Through this process, an agent can control its total efforts applied to system change, as well as its direction given by its priority r^k for allocating investment into action path k.

In the following, we define *unit variables* as the ratios between effect and cause. For instance, by investing a flow of investment $C(t)$ over time, we can assert that an agent changes the system state by $\Delta x(t)$, which leads to the unit cost $c(t) = C(t)/\Delta x(t)$. Inversely, the unit value $v(t) = V(t)/\Delta x(t)$ is the impact on value $V(t)$ of a change in $\Delta x(t)$. From this definition, we can determine the effects $\Delta x(t) = C(t)/c(t)$ and $V(t) = v(t)\Delta x(t)$. Cause and effect are only related linearly if $c(t)$ and $v(t)$ are constant, which is often not the case (and may be state or time dependent).

Investment results in a value gain $V(t) = v(t)\Delta x(t) = f(t)C(t)$, where $v(t)$ is the value impact per unit of system change (Δx), and $f(t) = v(t)/c(t)$ is the efficiency of producing value output for a given unit of investment (benefit–cost ratio). If, at a given time t, an agent aims to achieve a particular value $V^*(t)$, the required effort to bridge the gap toward the desired state is given as

---

[1] The roots of the model framework go back to Scheffran (1989) and have appeared under different forms and acronyms, including "SCX," "VCX," and "VCAPS", see in particular Scheffran (2000, 2001) and Scheffran and Hannon (2007). Early applications were focused on issues of arms races, proliferation, and missile defense (see Jathe and Scheffran 1995; Jathe, Krabs, and Scheffran 1997; Scheffran 2003). Other applications can be found in application chapters in Part III of this book.

$$C^*(t) = \frac{V^*(t)}{f(t)} \tag{7.1}$$

Thus, we can represent an agent's *value* targets (goals) as investment targets (again, thinking of investment as "cost" or *effort* on the part of the agent to change the system for their benefit).

Using these assumptions, we can now better understand how agents can adapt their investment rates. If at a given time t, the actual investment flow of an agent $C(t)$ differs from what it wants or needs to invest, given by its target effort $C^*(t)$, the gap can be bridged by a process to adapt effort according to the decision rule $\Delta C(t) = \beta D(C)$, where β is the adaptation rate at which an agent can alter its investment behavior, and $D(C)$ is some decision rule for adapting the investment to pursue a particular action path. The most basic decision rule would simply close the gap between our current investment rate and our target investment rate, yielding $D(C) = C^*(t) - C(t)$. Drawing on Equation 7.1, this means that

$$\Delta C(t) = \beta D(C) = \beta \left[C^*(t) - C(t) \right] = \frac{\beta}{f(t)} \left[V^*(t) - V(t) \right] \tag{7.2}$$

In pursuing its target investment rate $C^*(t)$, an agent has only a limited amount of resources for investment under its control, given by an upper limit C^+. In the case where $C^+ < C^*$, the agent's available investment rate will be insufficient to achieve its goal. In this case, an agent can try to improve the efficiency f of its investments (again, efficiency is the ratio of value output for a given unit of investment), which implies that the same value V can be achieved at lower cost C.

Improving the effectiveness of investments can be accomplished by switching from an action $k = 1$ to an action $k = 2$ with a higher efficiency $f_2 > f_1$ than the first one. Perhaps the most important factor of this framework is the representation of agent *learning*, which captures agents' ability to adapt the allocation of investment C to achieve their target value V_i^*. Learning involves finding a more preferable action path, and allocating an increasing fraction (priority) r of the investment $C(t)$ to this new option, and a lower share of the total investment, $1 - r$, to the previous action (assuming only two action paths). Then, the overall efficiency of the mixed action is $f(r) = (1-r)_1 + rf_2$, which hopefully will be greater than f_1 itself.

If agents behaved optimally, they would allocate all their resources to the new, more efficient path immediately. However, the ability of real world agents to adjust their actions is limited, almost ubiquitously, by common issues such as institutional bureaucracy and delays in shifting technology or strategies. We can take this into account in our modeling framework by introducing a value-dependent decision rule for the action priorities, $\Delta r(t) = \alpha D(V, r)$, similar to the decision rule we introduced for adjusting investment levels, $D(C)$. For example, if r^* is the priority for *Action Path 1* that best serves an agent's interest (i.e., increases value, thereby making the decision rule value-dependent), then a decision rule $D(r)$ that would quickly optimize shifts in action paths to arrive at r^* would be $\Delta r = \alpha(r^* - r)$. That is, the rule would change the fraction of total investment (priority) going to *Action Path 1* to close the gap between its current value r and its target value r^*, depending on adaptation rate α. For $r^* = 1$, all investment is spent on *Action Path 1*, with no investment made towards Action Path 2.

Among the many decision rules possible, we can consider the case where an agent pursues action paths based on the value gains that they expect. Stated another way, agents change their priority r^k for a given action path based on the marginal value that they expect to yield from that new action. This can be mathematically represented as an agent moving along a *value-priority gradient*, by taking the derivative of the agent's value function V, with respect to an action path's priority r^k, $v_r^k = \partial V / \partial r^k$, indicating the direction of pathways toward growing value (moving uphill). We will explore this option more in the context of fishery conflicts in Chapter 8.

The speed of adaptation to strategic or investment changes is represented by adaptation rates α and β, respectively. These rates can either be fixed, a function of the system state, or a result of other factors. See the Appendix 7.1 for a further exploration of these rates.

"What are *difference equations?*" *Differential* equations are used in situations where we can create analytically tractable (i.e., solvable) equations that can be simulated through continuous time, yielding changes in states, such as dV, which occur over an infinitesimally small temporal change dt (Parunak et al. 1998; Sterman 2000). In cases where we cannot create closed-form solutions with analytically tractable equations, we are forced to calculate discrete changes in C, V, or r (e.g., dC, dV, or dr) over a small (discrete) unit change in time, which we still call dt.

Drawing on all of the previous concepts that we have explored in this chapter, these illustrations of value-driven adaptive decision rules together form a set of *difference equations* for a single agent[2]:

$$\Delta C(t) = \beta \left[C^*(t) - C(t) \right] = \beta / f \left[V^*(t) - V(t) \right]$$

$$\Delta V(t) = v(t)\Delta x(t) = f(t)\Delta C(t) \tag{7.3}$$

$$\Delta r(t) = \alpha \left[(r^*(t) - r(t) \right]$$

These three equations explain the dynamics—changes in investments C, and priorities r^k associated with a given path k—associated with a single agent seeking target goals V^*. This is a generalized way of representing the efforts made by single agents toward achieving their goals. Using this framework, we can therefore think about conflict as the occasion when multiple agents interact, and those interactions disturb the abilities of certain agents to meet their goals.

As a side note, we should mention that adaptation rates are not necessarily constant and could respond to state variables. In particular, the choice $\beta = \kappa^C C(t)[C^+(t) - C(t)]$ would represent the situation that the adaptation rate becomes zero at upper investment threshold $C(t) = C^+(t)$ and lower investment threshold $C(t) = 0$, thus preventing departure from the investment boundaries. Here κ^C is the modified adaptation rate for investment. Overall, this then gives us a dynamic equation (Scheffran 1989; Scheffran and Hannon 2007):

$$\Delta C(t) = \kappa^C C(t) \left[C^+(t) - C(t) \right] \left[C^*(t) - C(t) \right] \tag{7.4}$$

Another modification assumes that the dynamics of priority are driven by the partial derivative v_r ($v_r = \partial V / \partial r$) within upper and lower boundaries $r = 1$ and $r = 0$:

---

[2] Additional information on the derivation of these equations in given in Appendix 7.2.

$$\Delta r(t) = \kappa^r v_r(t) \, r(t) \left[1 - r(t) \right] \tag{7.5}$$

Multi-Agent Interaction, Stability, and Conflict

Thus far, we have described the dynamic system for single agents who pursue their goals by investing their resources across (up to) two specified paths, which thereby changes the environment. If n agents act to consume or otherwise modify the same environmental system, then emergent dynamics can evolve from the complex interactions between agents' strategies. During the interaction process, each agent A_i $(i = 1, ..., n)$ invests its efforts C_i into action paths that affect system state(s) x (which can represent a vector of many different system states). These system states can represent resource levels or quality, or any other aspect of a conflict system. Each agent then evaluates the outcome of these action paths according to its own values V_i, deriving new actions for the next time step.

The only condition for the investment allocations r_i^k, the relative portions of investments in different action paths, is that they sum to 100% $\left(\sum_{k=1}^m r_i^k = 1 \right)$. Agents can adapt the amount C_i and allocation r_i^k of investments to impact the state x of the system (environment), leading to an induced change in the system states $\Delta x_i^k = r_i^k / c_i^k \, C_i$ as brought on by each of the agents i (c_i^k is the unit cost of investment of agent i to change the system state x; in terms of units, c_i^k is cost per unit change of x; r_i^k has no unit).

Rigorously Defining Interactions

Now that we have many agents finding value by altering the ecosystem (e.g., catching fish or impacting resources), we need to consider the effects of agents' *interactions* on their values and on the ecosystem. Take the situation where the value V_i of each agent increases in proportion to the investments of all agents C_j with $j = 1, ..., n$:

$$V_i = \sum_j \left\{ f_{ij} \, C_j \right\} + V_i^U \tag{7.6}[3]$$

This means that in addition to *externally* driven value changes V_i^U, an agent's total value is affected by the investments of all agents, weighted by the interaction efficiencies f_{ij} that describe the mutual couplings between a pair of agents, including self-effects f_{ii}. Previously, we defined f as an agent's efficiency of producing value output for a given unit of investment (benefit–cost ratio; $f = v/c$). In this case, each *interaction efficiency* f_{ij} between agent i and j, f_{ij}, depends on the mixed allocation of investments C_j of Agent j where r_j^k represents the investment priorities that Agent j allocates for each action path k.[4] This means that the overall effect of an agent's investment depends on how they allocate it to the different possible pathways. This is akin to determining which products you can buy within a given budget that create a higher or lower total value for you and for other agents.

---

[3] The interaction efficiencies f_{ij} include the sum over action pathways k, which is discussed more in Appendix 7.3.
[4] A complete definition of f_{ij} is given in Appendix 7.3.

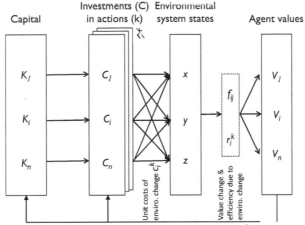

FIGURE 7.5
A representation of feedbacks present during interactions among multiple agents in the VIABLE modeling framework. This visualization builds on the singe-agent feedback cycle in Figure 7.3, and adds the potential for multiple environmental system states x, y, z (e.g., water quality, habitat, recreational resources).

Although changes in value also depend on any externally driven value V_i^U changes (e.g., a natural disaster or exogenous shock to agents' goals), for simplicity, we leave these out of our further analyses. In expanding our framework to multiple agents, we can now observe that the target conditions $V_i = V_i^*$ for one agent depend on the actions and resources of all agents (including the agent i itself). Given these interagent interactions, Figure 7.5 depicts the conceptual relationship between values (goals), capital, investments, actions paths, and resulting environmental system states.

Investment Targets and Equilibria

We now want to understand how much investments $\tilde{C}_i$ each agent would need to achieve their target values V_i^*, depending on the investments of all other agents and their respective efficiencies, f_{ij}. Resolving the respective relationship between target values and investments of multiple agents in Equation 7.6 leads to target investment $\tilde{C}_i$, which can be referred to as an *investment response function*. These define the required investment of each agent to meet its own value target in the face of other agents' investments. These response functions depend on the investments and value targets of all other agents, as well as the interaction efficiencies between agents. In Appendix 7.3, we discuss the functional form of $\tilde{C}_i$.

To achieve their target values V_i^*, each agent needs to adapt its actual investment $C_i(t)$ in a given time period t toward the response curves $\tilde{C}_i(t)$. If all agents follow these adaptation mechanisms, this leads to a dynamic interaction of investments among multiple agents. These reaction curves geometrically form multidimensional surfaces in more complex systems with more state variables.

The intersection of the response curves in the vector space of investments $C = (C_1, ..., C_n)$ of all n agents is represented by the equilibrium vector $\hat{C} = F^{-1}V^*$. This corresponds to the balance of investments or satisfaction point where all agents are able to meet their target investments $\hat{C}_i$ (together with all the other agents, these points form the vector $\hat{C}_i = (\hat{C}_i, ..., \hat{C}_n)$). $F = (f_{ij})$ represents the interaction matrix of all mutual efficiencies, coupling the investments and the values for each pair of agents $i, j = 1, ..., n$.

Building on our previous discussion, we can now continue to explore different forms of relationships between agents. For example, if the action priorities r_i^k create *non-cooperative* relationships, the equilibrium investment rates $\hat{C}_1$ and $\hat{C}_2$ for the two agents increase (i.e., the efforts needed to achieve value targets V_i^* increase; Figure 7.6A); conversely, in a cooperative relationship, equilibrium investments are lower. Both of these relationships are

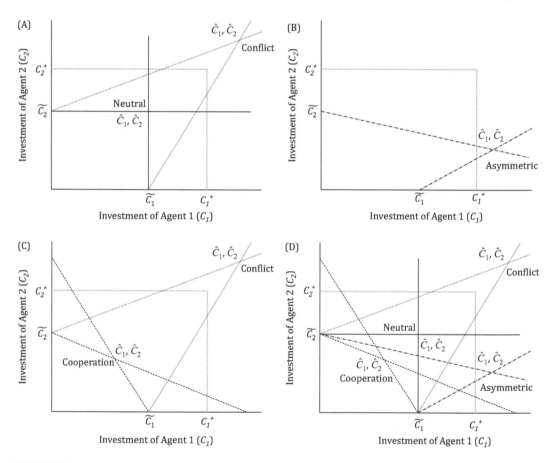

FIGURE 7.6

Hypothetical investment response curves $\tilde{C}_i$ and intersection points $\hat{C}_i$ denoting neutral (solid lines), cooperative (dashed lines), conflicting (dotted lines), and asymmetric relationships (dot-dashed lines) between Agent 1 and Agent 2. C_i^+ represents the upper limit of investments for each agent. Panel (A) compares response curves $\tilde{C}_i$ in neutral (i.e., equilibrium $\hat{C}_i$) and conflict situations. Panel (B) shows these curves in an asymmetric situation, where Agent 2's investments force Agent 1 to exceed their maximum possible investment rate C_1^+. Panel (C) compares a cooperative situation with a conflict situation; note that increasing levels of investment by each agent lowers the other agent's required investment to meet target values V_i^*. Panel (D) depicts all of these scenarios together.

shown as *symmetric* cases (Figure 7.6C). Likewise, in *asymmetric* cases (Figure 7.6B), one player may profit from the actions of another player, who in return, may lose from the interaction (i.e., invest until they run out of resources and cannot continue).

To alleviate the problems that arise in the conflicting or asymmetric cases, agents can instead try to improve the choices of their own priorities r_i^k, which can be expressed as a dynamic game problem. A conflict situation can be prevented as long as the adverse impacts of the investments of other agents that affect an environmental system x can be compensated for by an agent's own investments, allowing them to achieve their goal within their budget limits ($\hat{C}_i < C_i^+$).

If the equilibrium investment point $\hat{C}$—the point where a group of agents can all satisfy their goals—is outside of the investment abilities for at least one agent (i.e., above their maximum investment resources C_i^+), then the conflict potential remains unresolved. If this agent does not want to give up its goals, it may also be possible to increase efficiency or acquire capital resources elsewhere (e.g., through support from other agents who join into an alliance or coalition; Axelrod et al. 1995). More advanced forms of this can include efforts of multiple agents to form coalitions by pooling some of their resources, and redistributing gains and losses to individual agents—we will discuss this example in the chapter on fishery competition and cooperation (Chapter 8). These "transfer payments," as we can think of them, may contribute to keep conflict under control as long as the resource gap can be bridged, but they may also further escalate a conflict if they damage the interests of others.

A conflict becomes unresolvable and tends to escalate if the response curves $\tilde{C}$ are parallel, or their slopes do not permit a joint goal to be set within the admissible investment domain (where $\hat{C}_i < C_i^+$). In this case, no compensation can help and the investment adjustment process fails; the interaction between the agents becomes unstable, escalatory, and eventually exceeds limits (leading to conflict). Under these conditions, agents increasingly generate harm to each other and their reactions in return lead to an escalation of growing investments and declining values, until some agents have used up all their capital (cannot invest further) or vanish if their value losses become existential threats.

Mathematically Reframing Conditions for Conflict and Cooperation

There are several available measures that can be used to resolve or end conflicts. First, an external entity (e.g., a government or other third-party institution) could intervene to entirely halt the supply of resources or "dry out" the conflict (e.g., fishing or development moratoriums). Alternatively, agents with similar interests could support each other to bolster their current practices (e.g., join an alliance). Finally, agents could attack other "adversarial" agents, attempting to reduce their abilities to achieve their goals or force them to modify their goals.

However, we argue that an alternative to these classic, and often failing, options is to change behavior by adjusting agent goals and directing action paths to create mutual benefits among the conflicting agents. While the classic solutions can be applied asymmetrically against the will of adversarial agents (e.g., by a blockade or use of force), determining action paths that create shared benefits between agents requires mutual adjustment, consent, and negotiation. In many ways, this is an analytical framing of the mutual gains approach, whereby actions are under-taken during multiparty negotiations, including preparation, value creation, value distribution, and "follow-through" (i.e., execution of negotiated agreements), to maintain collaborative efforts that produce benefits for all parties. This approach was developed by Cruikshank and Susskind (1989) and Susskind and Landry (1991) and discussed exten-sively in Chapter 1.

We translate these qualitative assessments of conflict into mathematical conditions in Appendix 7.2. Of particular interest are stabil-ity conditions at a given investment equilibrium. Since we are simulating a dynamic conflict system, we can apply an established approach from stabil-ity theory (Sanchez 1979) that draws on the *Jacobi matrix*, a concept from vector calculus that we can use to describe the interactions between agents (Arrowsmith and Place 1992). Given that we have a set of dynamic equations for each agent's value

A *Jacobi matrix* is made up of all the first-order partial derivatives of a series of dynamic equations, taken with respect to every independent variable. In our case, the Jacobi Matrix F serves as an interaction matrix between agents where the interconnecting elements $f_{ij} = v_i^k / c_j^k$ represent the agents' value changes due to other agents' investment changes. We will normally represent f_{ij} as incremental unit change using difference equations, but in analyti-cally tractable cases (where we could consider time to be continuous), we could use partial differential equa-tions that represent infinitesimally small differential change.

function, then the elements f_{ij} of the matrix (the "Jacobians") describe the rates of change of each agent's value function V_i due to changes in the investments of each agent C_j (Figure 7.7). So, if we had five agents, then the Jacobi matrix would yield 25 entries, representing the rates of value change that would result from changes in the investments of all five agents (including self-induced effects f_{ii}).

In mathematical terms, "competitive" interactions occur when Agents 1 and 2 experi-ence positive "self-impacts" of their investments C_i on their own values V_i, (meaning that *sensitivities* $f_{11} > 0$ and $f_{22} > 0$), while they inflict negative mutual impacts on each other's values ($f_{12} < 0$ and $f_{21} < 0$). A switch to cooperation means that the mutual impacts become positive ($f_{12} > 0$ and $f_{21} > 0$), thus switching from an adverse to a symbiotic relationship. Even under conditions of competition, it may be possible to move the mutual satisfaction point $\hat{C}$ (see Figure 7.6) into the admissible range of investment by increasing the effi-ciency of an agent's investment on their own value (i.e., increasing the self-impact f_{ii}). See Appendix 7.4 for a more detailed exploration of stability within agent interactions.

To make this more realistic, we can take into account that, since the interaction coeffi-cients $f_{ij}(r_j^k)$ between the agents depend on the allocation priorities r_j^k of all agents, each agent can influence the dynamic system by changing their own priorities r_i^k to push toward

$$C_1 \to C_1^* \, C_i \to C_i^* \, C_n \to C_n^*$$

$$F(C) = \begin{bmatrix} f_{11} & \cdots & f_{1n} \\ \vdots & \ddots & \vdots \\ f_n & \cdots & f_{nn} \end{bmatrix} \begin{matrix} \to V_1 \to V_1^* \\ \to V_i \to V_i^* \\ \to V_n \to V_n^* \end{matrix}$$

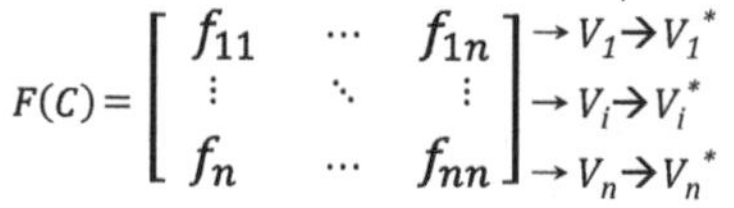

FIGURE 7.7

The multi-agent interaction matrix with investment inputs C_j and value outputs V_i connected through interac-tion efficiencies f_{ij}. It is each agent's hope that as their investments move toward their targets C_i^*, the interaction matrix moves agent values V_i toward their target values V_i^*.

BOX 7.2 EXAMPLE CONFLICT SIMULATION

To understand the basic operation of the VIABLE framework, we take a hypothetical example of a conflict between two agents and use the tools that we have developed to explore agent interactions and conflict stability. This example, which will help us classify certain types of behavior, is based on the assumption of the special case of constant f_{ij}. Even in this simplified case, we can see the complexity produced by this dynamic system (Scheffran and Hannon 2007).

Agents 1 and *2* are interested in using their investments C_i to each control a variable Y_i to move toward a given target Y_i^* of this variable; thus, both aim at closing the gap $Y_i(t) - Y_i^*$. In addition, they take not only this actual distance to the target into consideration but also the actual rate of change or trend $\Delta Y_i(t)$ in each time step, weighted with a response time τ_i. Both of these factors can be combined in the value function $V_i(t) = Y_i(t) - Y_i^* + \tau_i \, \Delta Y_i(t)$, where $\Delta Y_i = \sum_j f_{ij}(r_j)C_j$ is affected by the investments of all agents (see Equation 7.6). From this equation, we can see that agents would meet their targets at $V_i^* = 0$. According to Equation 7.4, for $V_i(t) < 0$, the investment response is positive ($\Delta C_i > 0$), for $V_i(t) > 0$ the investment response is negative ($\Delta C_i < 0$), and for $V_i(t) = 0$ the investment response is zero ($\Delta C_i = 0$). In the last (third) case, we have $\Delta Y_i(t) - (1/\tau_i)(Y_i^* - Y_i(t))$; thus, the rate of change $\Delta Y_i(t)$ is proportionate to the gap between Y_i^* and $Y_i(t)$, which is asymptotically closing over time. The actual state is approaching (homing in on) the target with a response time τ_i that defines the weight of change.

Take the initial parameters of the agent to be similar in nearly every way, where initial investments are $C_i(0) = \$30$ million and upper investment limits $C_i^+ = \$60$ million, and value targets are normalized to zero ($V_i^* = 0$). The interaction efficiencies $f_{ii} = 0.01$, $f_{ij} = -0.01$ are fixed (which also implies a single fixed action path) and the investment response parameters $\tau_i = 0.005$ are equal for both agents $\tau = \tau_i$. However, the agents' value functions are initialized asymmetrically as $V_1(0) = -0.3$ and $V_2(0) = 0.3$, which corresponds to a dynamic *zero-sum game*, where the value gained by one agent diminishes the value of the other "antagonist" agent by the same amount.

When both agents have an initial response time $\tau = 0$ (i.e., agents are only satisfied by an immediate decline of the gap without any delay), the agents show increasing oscillations around their target values $V_i^* = 0$ toward increasing investments (see Case 1 in Figure 7.8).

When we gradually increase the response time τ, this escalating oscillation is eventually damped, leading to stable oscillation when $\tau = 1$ (Case 2), and eventually to converge toward the target values when $\tau = 10$ (Case 3).

If we continue to ratchet up the response time, thus giving more weight to the rate of change ΔY_i compared to the actual gap $Y_i - Y_i^*$, we see fluctuations start to occur around the target value at approximately $\tau = 23$. Afterwards, these fluctuations continue to grow with increasing τ, until various patterns begin to occur, including those that are nearly periodic and break down after $t = 40$ ($\tau = 35$; Case 4), are seemingly chaotic ($\tau = 45$; Case 5) or separate positive and negative value domains of both agents ($\tau = 50$; Case 6), and again are semiperiodic, but with a different period ($\tau = 60$; Case 7). The unpredictable (chaotic) behavior seen in the investment patterns is the result of overshooting behavior on the part of each agent as they attempt to close

(Continued)

BOX 7.2 (*Continued*)   EXAMPLE CONFLICT SIMULATION

the gap between their current values V_i and their target values V_i^*, but increasingly respond to the rate of change rather than the actual gap. These dynamics resemble the bifurcation pattern for the logistic equation in chaos theory (Rasband 1990), with τ being the parameter that represents overshooting behavior (Jathe and Scheffran 1995).

In most social and natural systems, nonlinearities can trigger a sequence of instabilities (e.g., Cases 5–7 in Figure 7.8) surrounding thresholds and "tipping point" elements of social systems. Along the action pathways taken by agents, seemingly "minor" events could provoke major qualitative changes of the conflict system, something that is characteristic for chaotic structures. A self-reinforcing chain reaction could also increase the potential risk of runaway antisocial behavior (retaliation) that could put the whole system at risk.

We can now experiment by modifying the interaction efficiencies slightly (Figure 7.9); in Case A we will first leave the positive self-impacts untouched at $f_{ii} = 0.01$, while further diminishing the already negative mutual impacts to $f_{ij} = -0.011$. Stronger negative mutual impacts create interactions that induce increasing investments and declining values for both agents, leading to an unstable and harmful conflict (with a negative stability index $S_{ij} < 0$; see Appendix 7.4).

In Case B, we increase the positive self-impacts to $f_{ii} = 0.011$, leaving the negative mutual impacts at their previous value $f_{ij} = -0.01$. Here, interactions are mutually beneficial and stable ($S_{ij} > 0$), reducing investment costs and converging toward agents' target values (Figure 7.9). Investments go to zero since the value gap is positive; that is, agents exceed their value targets.

From these experiments, we learn that conflict resolution can require movements from damaging to beneficial interactions, requiring behavioral changes that are articulated through different action path priorities r^k. Response time τ plays a significant role for the transition to chaos but not as an indicator for instability.

their respective goals. We can determine if agents' actions maintain a stable or unstable investment balance represented by the equilibrium $\hat{C}$, or if they destabilize the equilibria as a result of their adaptation responses.

Summary

In this chapter, we have presented the VIABLE framework, an integrated conceptual approach to modeling environmental conflict and cooperation. This approach highlights the importance of agent responses and interactions as key factors in analyzing the impact of environmental change on security, stability, and conflict. This framework also facilitates both conceptual and mathematical representations of the interaction of multiple agents, who each pursue their objectives by allocating their capabilities to various action paths.

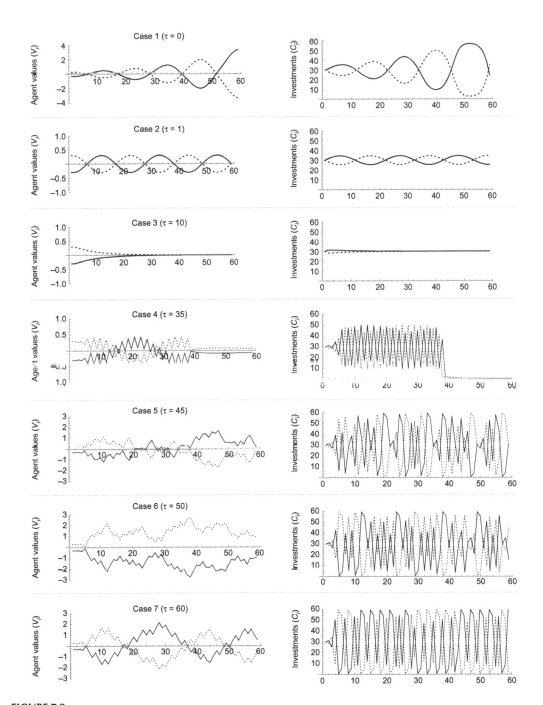

FIGURE 7.8

Transition from predictability (determinism) to chaos as response parameter τ increases in a zero-sum dynamic game between antagonistic Agent 1 (solid line) and Agent 2 (dotted line). Agent values $V_i(t)$ are shown in the left column, and investments $C_i(t)$ are on the right.

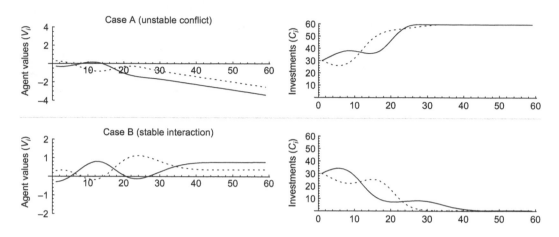

FIGURE 7.9

Impacts on agent values V_i and investments C_i in an unstable, escalating conflict (Case A) and a stabilized conflict (Case B) between Agent 1 (solid line) and Agent 2 (dotted line).

Unlike many other modeling approaches, we are focused on the feedback-laden nature of how agents enhance their "value" (i.e., achieve their goals), evaluate their access to (and ability to use) resources, observe system states, and update their decisions in light of other agents' information and actions. Together, these factors affect the ability of agents to maintain their own viability, as well as the emergent viability of natural systems. Within this framework, we argue that it is possible to study the complexity and instability of multi-agent constellations and the transition to cooperative states for conflicts in a wide range of fields. In repeated learning cycles, agents can adjust their actions and resources to those of other agents, thus including shifts from conflict to cooperation.

Conflict is a dynamic and complex form of human interaction, often emerging from incompatible actions, values, and goals and consuming a considerable amount of resources (Scheffran and Hannon 2007). Conflicts may express and contribute to system instability and chaotic escalation between adversaries, causing a breakdown of social and natural systems. Whether the feedback effects from environmental change on agents and their interactions are positive or negative depends on the preferences and strategies of agents. These feedback effects can also vary dramatically due to even the smallest changes, an expression of the nondeterministic behavior of agent interactions and their decision-making choices and abilities.

Rapid or drastic environmental change can challenge the stability of social systems, requiring new agent decision rules and responses. For example, societies that are particularly vulnerable to environmental and climatic change (e.g., in the Maldives islands, Bangladesh, East Africa) need to develop and build their capacity for adaptation and resilience. They also must recognize the speed and intensity of agent or institutional responses in comparison to the speed and intensity of environmental change. Provided that agents collectively develop effective and creative coping strategies, environmental change may also offer opportunities for constructive societal change and institutional development.

Environmental management and sustainable development strategies often fail due to their inability to adequately take into consideration the effects of important social interactions such as conflicts, dialogs, negotiations, coalition formation, and institution building. Several key issues are important in this context. First, technological change plays a key role in agent–environment interactions. As we will see throughout the application chapters in Part III of this book, heterogeneity in available technologies can vastly alter the inner workings of conflicts.

Second, conflicting agents can adjust their actions to cooperatively stabilize their interactions and—sometimes—to form stable coalitions (Scheffran 2006a). These adjustments can include the use of innovative social mechanisms, including governance and institutional structures that are more effective for facilitating conflict resolution. These innovative mechanisms can build on participative approaches to sustainable development and peace building.

This leads us to a final thought about agent interactions: adaptive strategies can be designed to restore livelihood and maintain resilience for people living in regions with changing environmental conditions. Some of these strategies could include, for example, the use of existing natural resources more efficiently, producing or using new types of natural resources that reduce environmental stress, providing a sustainable energy supply, and improving disaster management (e.g., Berke et al. 1996).

A Summary of the VIABLE Process

We can briefly summarize the VIABLE process across nine primary steps that we will use during our applications in Chapters 8–10:

1. Describe the issues of concern, identify key agents involved in this issue, identify the manner of interactions among agents.
2. Specify the relevant system-wide and agent-specific variables and viability constraints.
3. Determine available action paths a^k and evaluate how agents choose among and prioritize r^k these pathways.
4. Determine investments C^k associated with action paths k and how agents make investments.
5. Using information from Steps [1–4], create agent-value functions V_i.
6. Calculate marginal values v^k when changing investments among different action paths (i.e., partial derivatives $v_r^k = \partial V/\partial r^k$).
7. Establish decision rules that agents will use to change between action pathways.
8. Calculate interaction efficiencies f_{ij} (value gain for a given unit of investment) between agents, including the full interaction ("Jacobi") matrix, which represents each agent's investment effects on the value of all other agents ($f_{ij} = v_i^k/c_i^k$).
9. Analyze constraints, interaction stability, and viability corridors.
   These nine steps are the extent that we delve in the VIABLE technique in this book. To push further, readers should consider four additional steps.
10. Collect data and define the problems and issues through participatory processes and expert knowledge elicitation. These data will help enrich the skeleton model structure from Steps [1–9]. If stakeholders are *only* included in Steps [1–9] for data collection purposes, it may limit stakeholder buy-in and social learning (see discussion in Chapter 4).
11. Work with stakeholders to calibrate and validate the model.
12. Simulate a variety of *useful* and *relevant* scenarios, as well as visualizations of the model and its output.
13. Iterate through Steps 10–12 and evaluate model intervention outcomes.

Institutions and Conflict

In pursuing their individual interests, interacting agents trigger a cascading sequence of events, whereby actions taken by one agent can incite more intense responses by other agents. To address these and other risks, it is crucial that agents involved in environmental conflicts cooperate in managing the transition from individual antagonism to collective action. Through negotiations, agents can adapt to each other's actions and seek common solutions that lead to mutual benefits, reduced costs, or diminishing risks (Cruikshank and Susskind 1989; Susskind and Landry 1991). Moreover, to avoid shortsighted agent tendencies to game each other (e.g., the prisoners' dilemma discussed in Chapter 3) and evade cooperation, institutions are needed to establish binding and verifiable agreements.

Environmental conflicts often create collective action problems that require joint efforts to avoid environmental deterioration and make sure that the cumulative damage inflicted by all agents will not exceed critical limits. In the next three chapters, we explore the challenges inherent to finding institutional mechanisms to guarantee that each agent's individual investments are properly assigned to each agent, and that their compliance is ensured, thereby avoiding commons problems and other free-riding issues (e.g., fisheries declines, water resource conflicts; Delli Priscoli and Wolf 2009). To address these underlying social and institutional dilemmas, new rules, norms, and innovations need to develop. In the next chapter, we will discuss conditions under which joint action can sustain system resources if individuals coordinate their actions to yield a viable and fair share of the resources. The type of agent interactions that prevail will depend on the tools for (and rules of), communication available before, during, and after the interactions, as well as the institutional settings that are jointly created (Ostrom 1990; Ostrom et al. 1999).

There is a lot of work that still needs to be done to explore and expand the VIABLE framework for use with real-world conflicts. Aspects such as sensitivity testing, uncertainty analysis, calibration, and validity testing are all avenues needing further research. In the next several chapters, we will introduce several applications of this framework to give readers a sense of its potential and where they can make their own extensions and applications.

Appendix 7.1: Adaptation Rates α and β

The speed of adaptation to strategic and investment changes is represented by adaptation rates α and β, respectively. For example, one natural choice to describe the strategic rate of change α may be a logistic growth function, which is commonly used in evolutionary game theory (Hofbauer and Sigmund 1998), where $\alpha = \kappa^r r(1-r)$ and κ^r is a constant response factor that affects the rapidity of moving from 0 to 1 as r increases. This function dampens the rate of change near the boundary values (since α is a rate, these boundaries are 0 and 1).

Accordingly, we can consider investment adaptation to also be a logistic growth function, but this time as a function of the current investment rate $C(t)$, which is limited by the maximum investment rate $C^+(t)$, yielding $\beta(t) = \kappa^C C(t)[C^+(t) - C(t)]$, where κ^C is a constant adaptation factor affecting how quickly value targets drive the adaptation of investment $C(t)$ between minimum and maximum.

Appendix 7.2: Derivation of Single Agent Governing Equations

Since investment $C(t)$ creates value $V(t) = v(t)\Delta x(t) = f(t)C(t)$, we can determine the target investment as

$$C^*(t) = \frac{V^*(t)}{f(t)} \tag{7.1}$$

to realize a certain target value $V^*(t)$ at a given time t. The difference between target investment $C^*(t)$ and actual investment rate $C(t)$ is a driver to change investment:

$$\Delta C(t) = \beta D(C) = \beta \left[C^*(t) - C(t) \right] = \left(\frac{\beta}{f(t)} \right) \left[V^*(t) - V(t) \right] \tag{7.2}$$

Using the logistic function of β derived in Appendix 7.1, we obtain

$$\Delta C(t) = \kappa^C C(t) \left[C^+(t) - C(t) \right] \left[C^*(t) - C(t) \right]$$
$$= \left(\frac{\kappa^C}{f(t)} \right) C(t) \left[C^+(t) - C(t) \right] \left[V^*(t) - V(t) \right] \tag{7.4}$$

Furthermore, preference changes $\Delta r(t) = \kappa^r v_r r(t)[1 - r(t)]$ represent a situation where preference change is driven by marginal value v_r and slows down near the boundary conditions $r = 1$ and $r = 0$. Together, these equations form a set of difference equations describing the actions of a single agent.

Appendix 7.3: Response Curves $\tilde{C}_i$ and Multi-Agent Interaction Efficiency Matrix f_{ij}

Assuming that investment fraction $r_j^k C_j$ induces a change of $\Delta x_j^k = r_j^k C_j / c_j^k$ along pathway k, this results in a value change $V_{ij}^k = v_i^k \Delta x_j^k$ of agent i by agent j via path k. Total value change of agent i is the sum of the individual contributions of all agents:

$$V_i = \sum_j \sum_k V_{ij}^k = \sum_j \sum_k v_i^k r_j^k C_j / c_j^k = \sum_j f_{ij} C_j \tag{7.7}$$

where $f_{ij} = \sum_j f_{ij}^k r_j^k$ and $f_{ij}^k = v_i^k / c_j^k$ is the interaction efficiency along path k. This means that these efficiencies f_{ij} are the sum of the efficiencies over all of the paths k, as weighted by agents' priorities r_i^k. Setting $V_i = V_i^*$ in Equation 7.7 (calculating conditions when agents have reached their goal values), we can now resolve this equation for target investment

$$\tilde{C}_i = \left\{ V_i^* - \sum_{j \neq i} f_{ij}(r)C_j \right\} / f_{ii}$$

(7.8)

which is the investment response curve $\tilde{C}_i$ of agent i.

Appendix 7.4: Stability within Agent Interactions

The general stability condition in system dynamics is that the effects of disturbances decay, keeping a system within a stable set of states (usually an equilibrium or fixed point). Mathematically, this implies that all eigenvalues of the Jacobi matrix are negative (negative eigenvalues yield an exponential decay of a disturbance [stability], positive eigenvalues exponential increase of a disturbance [instability]). For two agents with $f_{ii} > 0$ and $f_{ij} < 0$, the stability condition boils down to the essential requirement:

$$S_{ij} = f_{ii}f_{jj} - f_{ij}f_{ji} > 0$$

(7.9)

This definition of stability reflects the intuitive idea that the product of the self-imposed (positive) impacts f_{ii} and f_{jj} must exceed the product of the (negative) mutual impacts f_{ij}, and f_{ji} from other agents to create conditions that agents altogether are able to avoid or contain mutually damaging interactions leading to a downward spiral. This condition holds as long as additional agents do not interfere, or as long as their impact on the two-agent interaction can be ignored. Therefore, for an unstable conflict between two agents, in the presence of the instability condition $[f_{11} f_{22} - f_{12} f_{21}] < 0$, a stable agreement point does not exist and the conflict will escalate until agents either exhaust their ability to invest ($C_i > C_i^+$) or agents change their goals.

As the number of agents involved expands, the complexity of the interaction matrix increases, creating the potential for more instability, which can hinder formation of collaborative coalitions. For a larger number of agents with $f_{ii} > 0$ and $f_{ij} < 0$ (where $i \neq j$), a sufficient stability condition is $\sum_j f_{ij} = \left[f_{ii} + \sum_{j \neq i} f_{ij} \right] > 0$ for each $i = 1,\dots, n$. This has the intuitive meaning that the efficiency of Agent 1's actions in creating value for itself is stronger than the combined efficiency of all the other agents acting against Agent 1's value. If several agents try to increase efforts to act against Agent 1, it is still able to overcome this combined adverse action by increasing its own effort to achieve its goal (because it acts very efficiently).

For more than two agents ($n > 2$), the stability condition of the more complex $n \times n$ interaction matrix can be determined based on eigenvalue analysis techniques that find the characteristic roots (where they are set to zero) of the Jacobi matrix equation (discussed for example, in Atkinson 1972 or other texts). A more general stability condition is given in Scheffran (2001) and options for control are studied in Jathe, Krabs, and Scheffran (1997) and Krabs, Pickl, and Scheffran (2000).

Questions for Consideration

1. *Find out more information on one of the following examples and determine what led to increases in hostilities.*

   Was there an equivalent to self-induced f_{ii} or mutual f_{ij} impact efficiencies? Could you imagine an analog in these conflicts? What happens when agents get caught in the vicious cycle of investment and value loss? Unresolved conflicts can prevent the successful goal attainment for some, or even all, agents. For example, Reuveny and Maxwell (2001) reference numerous instances where resource conflicts have done great societal damage, including:

   - Deforestation and land degradation that lead to social strife and population displacements in the Philippines (post-1945; Hawes 1990);

   - The 1969 war caused by land scarcity in El Salvador, ensuing migration to Honduras, and the resulting competition over land between natives and migrants (Durham 1979); and

   - Aggravations of the ethnic conflict between India and Bangladesh due to cropland and food scarcities since the 1970s (Ashok 1996).

   - Other well-known cases of conflict intensification include the escalation spiral leading to World War I (Richardson 1960; Flint et al. 2009; Chi et al. 2014), and the nuclear arms race between the former superpowers in the Cold War that continuously increased the number of nuclear weapons and offensive first strike capabilities (Scheffran 1989; Scheffran and Hannon 2007).

2. In the two-agent example given in this chapter, the interaction efficiencies were first fixed at $f_{ii} = 0.01$, $f_{ij} = -0.01$, leading to a zero-sum game, continued conflict, and value loss for both agents. Next, the interaction efficiencies were modified, keeping positive self-impacts at $f_{ii} = 0.01$, while strengthening the negative mutual impacts to $f_{ij} = -0.011$. This created interactions that induced increasing costs and declining values for both agents, leading to an unstable and harmful conflict. Finally, we increased the positive self-impacts to $f_{ii} = 0.011$ and left the negative mutual impacts at their previous value $f_{ij} = -0.01$. In this case, interactions were mutually beneficial and stable, reducing investment costs and converging toward agents' target values. *What do you think would happen if self-impacts f_{ii} were set to zero and mutual impacts f_{ij} were set to positive values? What would happen if self-impacts f_{ii} were instead set to negative values and mutual impacts f_{ij} were set to zero? Why do you think the system creates these behaviors?*

Additional Resources

Interested in learning more about viability theory? We direct interested readers to in-depth discussions of the mathematical details of viability theory in Aubin, Bayen, and Saint-Pierre (2011), Aubin and Saint-Pierre (2007), and Krawczyk and Pharo (2013). Additionally, Aubin, Bernado, and Saint-Pierre (2005) present viability theory specifically in connection with climate change issues.

Interested in applications of the VIABLE framework? The modeling framework that we introduce has been applied in various fields, including the arms race between nuclear-armed missiles and missile defenses, fishery conflicts, energy, and climate change. In addition to the applications presented in the later chapters of this text, we refer readers to detailed applications in arms races, proliferation, and missile defense (Scheffran 1989, 1994; Jathe and Scheffran 1995; Jathe, Krabs, and Scheffran 1997; Scheffran 2003); energy and climate change (Scheffran and Jathe 1996; Scheffran 1999; Scheffran and Pickl 2000; Ipsen, Rösch, and Scheffran 2001; Scheffran 2008a); environmental conflict (Scheffran 1996, 1999); fishery conflict (Scheffran 2000; BenDor, Scheffran, and Hannon 2009); bioenergy (Scheffran and BenDor 2009); water conflict (Link et al. 2012); coalition formation (Scheffran 2006a); and cooperation in trading (Scheffran 2013). Earlier versions of this model framework have appeared under different forms and varying acronyms, including "SCX," "VCX," and "VCAPS." These iterations extend back to Scheffran (1989), and include Scheffran (2002), and Scheffran and Hannon (2007).

Interested in exploring conflicts with only a few agents? While solutions derived from game-theoretic analysis are adequate in situations with only a few agents, they are difficult to apply in settings with large numbers of agents, multiple decision criteria, and agents operating under bounded rationality. However, in cases where there are only a few agents with disparate goals, techniques like multicriteria decision analysis (see Chapter 3) can be simpler methods of deriving solutions that are said to be "Pareto optimal" (i.e., that yield the set of combined actions allowing the maximum joint improvement for all agents). Stated differently, in a Pareto optimal state, any further improvements for some agents will end up harming other agents (Fudenberg and Tirole 1983). We should note that Pareto optimality does not infer any information about the overall well-being of the stakeholders involved in a conflict and does not necessarily result in any type of socially desirable distribution of actions or resources (Barr 2004).

Just what is cooperation? We have defined "cooperation" as the process by which agents adjust their goals and actions to achieve mutual benefits. This transition from conflict to cooperation requires adaptation toward common positions and mutually beneficial actions that stabilize the interaction. Whether this transition is successful depends on the governance capacity of societies to prevent or manage conflicts.

Chapter 1 discusses this issue in detail in the context of "the mutual gains approach." See Cruikshank and Susskind (1989) and Susskind and Landry (1991) for more detail on the use of the "mutual gains approach" in negotiation. We can also note that in many areas of society, the public has intervened by establishing institutions and governance structures for conflict prevention and maintenance. In one example that we do not often think of as a conflict-related measure, governments can require drivers to carry private insurance, which offsets value losses from crashes, thereby preventing or reducing conflict. Other examples include the emergence of innovative institutional designs for environmental market mechanisms, which reduce conflict in enforcing environmental protection policies (e.g., Harlow 1974; BenDor and Doyle 2010).

We refer readers interested in this topic to Elinor Ostrom's (1990, 2009) groundbreaking *Institutional Analysis and Development* framework, which was developed to help with the analysis of *collective action problems* that require social structures, positions, and rules. Mancur Olson (1971 [1965]) first developed the concept that individuals in any group pursue *collective action*, as they undertake a joint effort to achieve a common objective. *Collective action problems* are situations in which many individuals would all benefit from an action, but the action's associated cost makes it implausible that any individual will undertake it alone.

Interested in issues around chaos or chaotic dynamics? During our overview of the history of conflict modeling in Chapter 3, we discussed the way in which Saperstein (1984, 1986) drew on the concept of *chaos* as a model for arms races and war outbreak. He demonstrated that even simple, deterministic arms race simulations may lead to the breakdown of predictability. The problem of chaotic dynamics in arms race models was further investigated by Grossmann and Mayer-Kress (1989) using nonlinear difference equations, which were based on Richardson's (1960) arms race model, but with discrete time and a dampened growth of expenditure at upper and lower limits.

A note on investment response functions and unit values. By understanding the relationship between target values and investments of multiple agents, we can determine *investment response functions* $\tilde{C}_i$, which define the investment targets of each agent needed to meet its own value target. These investment responses are functions of investment and value targets of all other agents, as well as the interaction efficiencies. In economic theory, these responses have been represented using a "tâtonnement" process (Bowley 1924), in which an iterative adjustment procedure takes the form of a "Walrasian" auction (Walras 1900 [2014]; Laves and Williams 2007). Here, many agents simultaneously calculate their demand for a good and submit prices to an auctioneer simultaneously. Leon Walras suggested that equilibrium would be achieved through a process of "tâtonnement" (a French term for "groping"), a form of "hill climbing" convergence.

Investments result in a value gain $V(t)=v(t)\Delta x(t)=f(t)C(t)$, where $v(t)$ is the value impact per unit of system change (Δx), and $f=v/c$ is the efficiency of producing value output for a given unit of investment (benefit–cost ratio). This value is philosophically similar to Reuveny, Maxwell, and Davis's (2011) "extraction efficiency" value, which is attached to each conflicting agent and follows a functional form expressed by Brander and Taylor (1998) and Maxwell and Reuveny (2005). Additionally, Bhavnani, Backer, and Riolo (2008) discuss the common use of utility functions in conflict resolution models.

A note on our hypothetical example model. Agents 1 and 2 are interested in using their investments C_i to each control a variable Y_i and move it toward a given target Y_i^*; thus, both aim at closing the gap $Y_i(t)-Y_i^*$. In other versions of this demonstration model, the variable Y_i represented a security function that was targeted to be positive (Scheffran 1989). In a more generalized version, Y_i has represented other variables as well, such as CO_2 emissions from agents aiming to meet an emission target (Scheffran and Hannon 2007). In the viability framework, this corresponds to the task to stay within the viability domain and take corrective actions to not exceed system boundaries. The value function would thus be an indicator of the viability of the dynamic system.

An essential precondition for analyzing and modeling conflicts—even simple ones—is a strong understanding of the actions of individual agents. We refer readers to the fascinating discussion of individual agent behavior as a determinant of system structure and dynamics in Grimm et al. (2005) and Railsback and Grimm (2011). We also discuss this in detail in Chapter 6.

References

Arrowsmith, D. K., and C. M. Place. 1992. *Dynamical Systems*. London, England: Chapman & Hall.

Ashok, Swain. 1996. "Displacing the Conflict: Environmental Destruction in Bangladesh and Ethnic Conflict in India." *Journal of Peace Research* 33 (2): 189–204.

Astrom, Karl Johan, and Bjorn Wittenmark. 1994. *Adaptive Control* (2nd Edition). Boston, MA: Addison-Wesley Longman Publishing.

Atkinson, F. V. 1972. *Multiparameter Eigenvalue Problems*. Mathematics in Science and Engineering. New York: Academic Press.

Aubin, Jean-Pierre. 1991. *Viability Theory*. Berlin, Germany: Birkhäuser.

Aubin, Jean-Pierre, Alexandre M. Bayen, and Patrick Saint-Pierre. 2011. *Viability Theory: New Horizons* (2nd Edition). Berlin, Germany: Springer-Verlag.

Aubin, Jean-Pierre, Telma Bernado, and Patrick Saint-Pierre. 2005. "A Viability Approach to Global Climate Change Issues." In *The Coupling of Climate and Economic Dynamics*, edited by Alain Haurie and Laurent Viguier, Vol. 22, 113–43. Advances in Global Change Research. Amsterdam, Netherlands: Springer.

Aubin, Jean-Pierre, and Patrick Saint-Pierre. 2007. "An Introduction to Viability Theory and Management of Renewable Resources." In *Advanced Methods for Decision Making and Risk Management in Sustainability Science*, edited by Jürgen P. Kropp and Jürgen Scheffran, 56–95. New York: Nova Science Publishers.

Axelrod, Robert, Will Mitchell, Robert E. Thomas, D. Scott Bennett, and Erhard Bruderer. 1995. "Coalition Formation in Standard-Setting Alliances." *Management Science* 41 (9): 1493–508.

Barr, N. 2004. *Economics of the Welfare State*. Oxford, England: Oxford University Press.

Belt, Marjan van den. 2004. *Mediated Modeling: A System Dynamics Approach To Environmental Consensus Building*. Washington, DC: Island Press.

BenDor, Todd, and M. W. Doyle. 2010. "Planning for Ecosystem Service Markets." *Journal of the American Planning Association* 76 (1): 59–72.

BenDor, Todd, Jürgen Scheffran, and Bruce Hannon. 2009. "Ecological and Economic Sustainability in Fishery Management: A Multi-Agent Model for Understanding Competition and Cooperation." *Ecological Economics* 68 (4): 1061–73.

Berke, P., Dale J. Roenigk, Edward J. Kaiser, and Raymond Burby. 1996. "Enhancing Plan Quality: Evaluating the Role of State Planning Mandates for Natural Hazard Mitigation." *Journal of Environmental Planning and Management* 39 (1): 79–96.

Berke, Philip. 2014. "Rising to the Challenge: Planning in the Age of Climate Change." In *Adapting to Climate Change: Lessons from Natural Hazards Planning*, edited by Bruce Glavovic and Gavin Smith, 171–92. London: Springer.

Bhavnani, R., D. Backer, and R. Riolo. 2008. "Simulating Closed Regimes with Agent Based Models." *Complexity* 14 (1): 36–44.

Billari, Francesco C., Thomas Fent, Alexia Prskawetz, and Jürgen Scheffran. 2006. "Agent-Based Computational Modelling: An Introduction." In *Agent-Based Computational Modelling: Applications in Demography, Social, Economic and Environmental Sciences*, edited by Francesco C. Billari, Thomas Fent, Alexia Prskawetz, and Jürgen Scheffran, 1–16. New York: Physica-Verlag HD.

Bourdieu, Pierre. 1983. "The Field of Cultural Production, or: The Economic World Reversed." *Poetics* 12 (4–5): 311–56.

Bowley, A. L. 1924. *The Mathematical Groundwork of Economics*. Oxford, England: Oxford University Press.

Brander, J. A., and M. S. Taylor. 1998. "The Simple Economics of Easter Island: A Ricardo-Malthus Model of Renewable Resource Use." *American Economic Review* 88: 119–38.

Briguglio, Lino, Gordon Cordina, Nadia Farrugia, and Stephanie Vella. 2009. "Economic Vulnerability and Resilience: Concepts and Measurements." *Oxford Development Studies* 37 (3): 229–47.

Carmin, JoAnn, Isabelle Anguelovski, and Debra Roberts. 2012. "Urban Climate Adaptation in the Global South: Planning in an Emerging Policy Domain." *Journal of Planning Education and Research* 32 (1): 18–32.

Castle, Emery N. 2002. "Social Capital: An Interdisciplinary Concept." *Rural Sociology* 67 (3): 331–49.

Ceballos, Gerardo, and Paul R. Ehrlich. 2002. "Mammal Population Losses and the Extinction Crisis." *Science* 296 (5569): 904–7.

Cegan, Jeffrey C., Ashley M. Filion, Jeffrey M. Keisler, and Igor Linkov. 2017. "Trends and Applications of Multi-Criteria Decision Analysis in Environmental Sciences: Literature Review." *Environment Systems and Decisions* 37 (2): 123–33.

Chi, S. H., C. Flint, P. Diehl, J. Vasquez, Jürgen Scheffran, S. M. Radil, and T. J. Rider. 2014. "The Spatial Diffusion of War: The Case of World War I." *Journal of the Korean Geographical Society* 49 (1): 57–76.

Cruikshank, Jeffrey, and Lawrence Susskind. 1989. *Breaking The Impasse: Consensual Approaches To Resolving Public Disputes*. New York: Basic Books.

Dasgupta, Susmita, Benoit Laplante, Hua Wang, and David Wheeler. 2002. "Confronting the Environmental Kuznets Curve." *Journal of Economic Perspectives* 16 (1): 147–68.

Delli Priscoli, J., and A. T. Wolf. 2009. *Managing and Transforming Water Conflicts*. Cambridge, England: Cambridge University Press.

Dercole, Fabio, Charlotte Prieu, and Sergio Rinaldi. 2010. "Technological Change and Fisheries Sustainability: The Point of View of Adaptive Dynamics." *Ecological Modelling* 221 (3): 379–87.

Durham, William. 1979. *Scarcity and Survival in Central America: The Ecological Origins of the Soccer War*. Palo Alto, CA: Stanford University Press.

Eaves, James, and Jeffrey Williams. 2007. "Walrasian Tatonnement Auctions on the Tokyo Grain Exchange." *Review of Financial Studies* 20 (4): 1183–218.

Feuillette, Sarah, Harold Levrel, Blandine Boeuf, Stéphanie Blanquart, Olivier Gorin, Guillaume Monaco, Bruno Penisson, et al. 2016. "The Use of Cost–benefit Analysis in Environmental Policies: Some Issues Raised by the Water Framework Directive Implementation in France." *Environmental Science & Policy* 57 (March): 79–85.

Flint, Colin, Paul Diehl, Jürgen Scheffran, John Vasquez, and Sang-hyun Chi. 2009. "Conceptualizing ConflictSpace: Toward a Geography of Relational Power and Embeddedness in the Analysis of Interstate Conflict." *Annals of the Association of American Geographers* 99 (5): 827–35.

Fudenberg, D., and J. Tirole. 1983. *Game Theory*. Cambridge, MA: MIT Press.

Giuliani, M., and A. Castelletti. 2013. "Assessing the Value of Cooperation and Information Exchange in Large Water Resources Systems by Agent-Based Optimization." *Water Resources Research* 49 (7): 3912–26.

Grimm, Volker, Eloy Revilla, Uta Berger, Florian Jeltsch, Wolf M. Mooij, Steven F. Railsback, Hans-Hermann Thulke, et al. 2005. "Pattern-Oriented Modeling of Agent-Based Complex Systems: Lessons from Ecology." *Science* 310 (5750): 987–91.

Grossmann, S., and G. Mayer-Kress. 1989. "Chaos in the International Arms-Race." *Nature* 337: 701–4.

Gutierrez, Nicolas L., Ray Hilborn, and Omar Defeo. 2011. "Leadership, Social Capital and Incentives Promote Successful Fisheries." *Nature* 470 (7334): 386–89.

Hardin, G. 1968. "The Tragedy of the Commons." *Science* 162: 1243–47.

Harlow, Robert L. 1974. "Conflict Reduction in Environmental Policy." *Journal of Conflict Resolution* 18 (3): 536–36.

Hawes, Gary. 1990. "Theories of Peasant Revolution: A Critique and Contribution from the Philippines." *World Politics* 42 (2): 261–98.

Helbing, Dirk. 2010. *Quantitative Sociodynamics: Stochastic Methods and Models of Social Interaction Processes*. New York: Springer Science & Business Media.

Hofbauer, J., and K. Sigmund. 1998. *Evolutionary Games and Population Dynamics*. Cambridge, England: Cambridge University Press.

Holling, C. S. 1973. "Resilience and Stability of Ecological Systems." *Annual Review of Ecology and Systematics* 4: 1–23.

Holling, C. S. 2009. "Engineering Resilience versus Ecological Resilience." In *Foundations of Ecological Resilience*, edited by L. H. Gunderson, C. R. Allen, and C. S. Holling, 25–62. Washington, DC: Island Press.

Ipsen, Dirk, Roland Rösch, and Jürgen Scheffran. 2001. "Cooperation in Global Climate Policy: Potentialities and Limitations." *Energy Policy* 29 (4): 315–26.

Janssen, Marco A., and Elinor Ostrom. 2006. "Empirically Based, Agent-Based Models." *Ecology and Society* 11 (2): 37.

Jathe, M., W. Krabs, and Jürgen Scheffran. 1997. "Control and Game Theoretical Treatment of a Cost-Security Model for Disarmament." *Mathematical Methods in the Applied Sciences* 20: 653–66.

Jathe, M., and Jürgen Scheffran. 1995. "Modelling International Security and Stability in a Complex World." In *Chaos and Complexity*, edited by P. Berge, R. Conte, M. Dubois, and J. Van Thran Thanh, 331–32. Gif sur Yvette, France: Editions Frontieres.

Jervis, R. 1978a. "Cooperation under the Security Dilemma." *World Politics* 30 (2): 167–74.

Jervis, R. 1978b. *Perception and Misperception in International Politics*. Princeton, NJ: Princeton University Press.

Kaza, Nikhil, and Lewis Hopkins. 2009. "In What Circumstances Should Plans Be Public?" *Journal of Planning Education and Research* 28 (4): 491–502.

Kelly, K. 1994. *Out of Control*. New York: Addison-Wesley.

Koopmans, Tjalling C. 1967. "Objectives, Constraints, and Outcomes in Optimal Growth Models." *Econometrica* 35 (1): 1–15.

Krabs, W., S. Pickl, and Jürgen Scheffran. 2000. "An N-Person Game under Linear Side Conditions." In *Optimization, Dynamics and Economic Analysis*, edited by E. J. Dockner, R. F. Hartl, M. Luptacik, and G. Sorger, 76–85. Heidelberg, Germany: Springer Physica-Verlag.

Krawczyk, Jacek B., and Alastair Pharo. 2013. "Viability Theory: An Applied Mathematics Tool for Achieving Dynamic Systems' Sustainability." *Mathematica Applicanda* 41 (1): 97–126.

Kriesberg, Louis. 2009. "The Evolution of Conflict Resolution." In *The Sage Handbook of Conflict Resolution*, edited by Jacob Bercovitch, Victor Kremenyuk, and I. William Zartman, 15–32. Thousand Oaks, CA: SAGE.

Kuznets, Simon. 1955. "Economic Growth and Income Inequality." *The American Economic Review*, 1–28.

Link, P. Michael, Franziska Piontek, Jürgen Scheffran, and Janpeter Schilling. 2012. "On Foes and Flows: Vulnerabilities, Adaptive Capacities and Transboundary Relations in the Nile River Basin in Times of Climate Change." *L'Europe En Formation*, 365: 99–138.

Maoz, Zeev. 2010. *Networks of Nations: The Evolution, Structure, and Impact of International Networks, 1816–2001*, Vol. 32. Cambridge, England: Cambridge University Press.

Maxwell, J. W., and Rafael Reuveny. 2005. "Continuing Conflict." *Journal of Economic Behavior and Organization* 58: 30–52.

McHarg, Ian. 1969. *Design with Nature*. New York: John Wiley and Sons, Inc.

Mearns, Robin, and Andrew Norton. 2009. *Social Dimensions of Climate Change: Equity and Vulnerability in a Warming World*. New York: World Bank Publications.

Monod, Jacques. 1971. *Chance and Necessity*. New York: Alfred A. Knopff.

Nussbaum, Martha, and Amartya Sen, eds. 1993. *The Quality of Life*. Oxford, England: Oxford University Press.

Ostrom, Elinor. 1990. *Governing the Commons: The Evolution of Institutions for Collective Action*. Cambridge, England: Cambridge University Press.

Ostrom, Elinor. 2009. *Understanding Institutional Diversity*. Princeton, NJ: Princeton University Press.

Ostrom, Elinor, Joanna Burger, Christopher B. Field, Richard B. Norgaard, and David Policansky. 1999. "Revisiting the Commons: Local Lessons, Global Challenges." *Science* 284 (5412): 278–82.

Oubraham, Aïchouche, and Georges Zaccour. 2018. "A Survey of Applications of Viability Theory to the Sustainable Exploitation of Renewable Resources." *Ecological Economics* 145: 346–67.

Perrings, C. 2006. "Resilience and Sustainable Development." *Environment and Development Economics* 11: 417–27.

Petschel-Held, G., H. J. Schellnhuber, T. Bruckner, F. L. Toth, and K. Hasselmann. 1999. "The Tolerable Windows Approach: Theoretical and Methodological Foundations." *Climatic Change* 41 (3/4): 303–31.

Railsback, Steven F., and Volker Grimm. 2011. *Agent-Based and Individual-Based Modeling: A Practical Introduction.* Princeton, NJ: Princeton University Press.

Rasband, S. N. 1990. *Chaotic Dynamics of Nonlinear Systems.* New York: Wiley.

Reuveny, Rafael, and John W. Maxwell. 2001. "Conflict and Renewable Resources." *Journal of Conflict Resolution* 45 (6): 719–42.

Reuveny, Rafael, John W. Maxwell, and Jefferson Davis. 2011. "On Conflict over Natural Resources." *Ecological Economics* 70 (4): 698–712.

Richardson, Lewis Fry. 1960. *Arms and Insecurity: A Mathematical Study of the Causes and Origins of War.* Ann Arbor, MI: University of Michigan Press.

Sanchez, D. A. 1979. *Ordinary Differential Equations and Stability Theory: An Introduction.* N. Chemsford, MA: Dover Publications.

Saperstein, Alvin M. 1984. "Chaos – A Model for the Outbreak of War." *Nature* 309 (5966): 303–5.

Saperstein, Alvin M. 1986. "Predictability, Chaos, and the Transition to War." *Security Dialogue* 17 (1): 87–93.

Scheffran, Jürgen. 1989. *Strategic Defense, Disarmament, and Stability - Modeling Arms Race Phenomena with Security and Costs under Political and Technical Uncertainties (PhD Thesis; IAFA Publication Series No. 9).* Marburg, Germany: University of Marburg.

Scheffran, Jürgen. 1994. "Modelling International Security Problems in the Framework of the SCX Model." In *Mathematische Methoden in Der Sicherheitspolitik,* edited by J. Hermanns, B. v. Stengel, and A. Tolk, 103–20. Neubiberg, Germany: IASFOR.

Scheffran, Jürgen. 1996. "Modeling Environmental Conflicts and International Stability." In *Models for Security Policy in the Post-Cold War Era,* edited by R. K. Huber and R. Avenhaus, 201–20. Baden-Baden, Germany: Nomos.

Scheffran, Jürgen. 1999. "Environmental Conflicts and Sustainable Development: A Conflict Model and Its Application in Climate and Energy Policy." In *Environmental Change and Security,* edited by A. Carius and K. M. Lietzmann, 195–218. Berlin, Germany: Springer.

Scheffran, Jürgen. 2000. "The Dynamic Interaction between Economy and Ecology: Cooperation, Stability and Sustainability for a Dynamic-Game Model of Resource Conflicts." *Mathematics and Computers in Simulation* 53: 371–80.

Scheffran, Jürgen. 2001. "Stability and Control of Value-Cost Dynamic Games." *Central European Journal of Operations Research* 9 (7): 197–225.

Scheffran, Jürgen. 2002. "Economic Growth, Emission Reduction and the Choice of Energy Technology in a Dynamic-Game Framework." In *Operations Research Proceedings 2001: Selected Papers of the International Conference on Operations Research (OR 2001), Duisburg, September 3–5, 2001,* edited by Peter Chamoni, Rainer Leisten, Alexander Martin, Joachim Minnemann, and Hartmut Stadtler. Operations Research Proceedings. Berlin, Heidelberg: Springer-Verlag. Pages 329–336

Scheffran, Jürgen. 2003. "Calculated Security? Mathematical Modelling of Conflict and Cooperation." In *Mathematics and War,* edited by B. Booss-Bavnbek and J. Hoyrup, 390–412. Berlin, Germany: Birkhäuser.

Scheffran, Jürgen. 2006a. "The Formation of Adaptive Coalitions." In *Advances in Dynamic Games,* edited by A. Haurie, S. Muto, L. A. Petrosjan, and T. E. S. Raghavan, 163–78. Berlin, Germany: Birkhäuser.

Scheffran, Jürgen. 2006b. "Tools for Stakeholder Assessment and Interaction." In *Stakeholder Dialogues in Natural Resources Management,* edited by S. Stoll-Kleemann and M. Welp. Berlin, 153–85. Germany: Springer.

Scheffran, Jürgen. 2008a. "Adaptive Management of Energy Transitions in Long-Term Climate Change." *Computational Management Science* 5 (3): 259–86.

Scheffran, Jürgen. 2008b. "Preventing Dangerous Climate Change." In *Global Warming and Climate Change,* edited by Velma I. Grover, Vol. 2: 449–82. Boca Raton, FL: CRC Press.

Scheffran, Jürgen. 2013. "Conditions for Cooperation and Trading in Value-Cost Dynamic Games." In *Advances in Dynamical Games: Theory, Applications, and Numerical Methods (Annals of the ISDG, 13)*, edited by V. Krivan and G. Zaccour, 173–204. New York: Springer-Verlag.

Scheffran, Jürgen. 2016. "From a Climate of Complexity to Sustainable Peace: Viability Transformations and Adaptive Governance in the Anthropocene." In *Handbook on Sustainability Transition and Sustainable Peace*, 305–46. Cham, Switzerland: Springer.

Scheffran, Jürgen, and Todd BenDor. 2009. "Bioenergy and Land Use: A Spatial-Agent Dynamic Model of Energy Crop Production in Illinois." *International Journal of Environment and Pollution* 39 (1/2): 4–27.

Scheffran, Jürgen, and Bruce Hannon. 2007. "From Complex Conflicts to Stable Cooperation: Cases in Environment and Security." *Complexity* 13 (1): 78–91.

Scheffran, Jürgen, and M. Jathe. 1996. "Modeling the Impact of the Greenhouse Effect on International Stability." In *Supplementary Ways for Improving International Stability*, edited by P. Kopacek, 31–38. Oxford, England: Pergamon IFAC.

Scheffran, Jürgen, and Stefan Pickl. 2000. "Control and Game-Theoretic Assessment of Climate Change: Options for Joint Implementation." *Annals of Operations Research* 97 (1): 203–12.

Schlüter, Maja, and Claudia Pahl-Wostl. 2007. "Mechanisms of Resilience in Common-Pool Resource Management Systems: An Agent-Based Model of Water Use in a River Basin." *Ecology and Society* 12 (2): 4.

Schweitzer, Frank. 1997. *Self-Organization of Complex Structures: From Individual to Collective Dynamics.* Boca Raton, FL: CRC Press.

Shively, David. 2017. "Flood Risk Management in the USA: Implications of National Flood Insurance Program Changes for Social Justice." *Regional Environmental Change* 17 (6): 1663–72.

Sterman, John D. 2000. *Business Dynamics: Systems Thinking and Modeling for a Complex World.* New York: Irwin/McGraw-Hill.

Stern, David I. 2017. "The Environmental Kuznets Curve after 25 Years." *Journal of Bioeconomics* 19 (1): 7–28.

Susskind, Lawrence, and Elaine M. Landry. 1991. "Implementing a Mutual Gains Approach to Collective Bargaining." *Negotiation Journal* 7 (1): 5–10.

Thomas, J. A., M. G. Telfer, D. B. Roy, C. D. Preston, J. J. D. Greenwood, J. Asher, R. Fox, et al. 2004. "Comparative Losses of British Butterflies, Birds, and Plants and the Global Extinction Crisis." *Science* 303 (5665): 1879–81.

Olson, Mancur. 1971. *The Logic of Collective Action: Public Goods and the Theory of Groups (Revised Edition)*. Cambridge, MA: Harvard University Press.

Walker, Brian, and David Salt. 2012. *Resilience Practice: Building Capacity to Absorb Disturbance and Maintain Function*. Washington, DC: Island Press.

Wallace, M. D., and J. M. Wilson. 1978. "Non-Linear Arms Race Models." *Journal of Peace Research* 2 (15): 175–82.

Walras, Léon. 1900. *Elements of Theoretical Economics (translated by Donald Walker, and Jan van Daal)*. Cambridge, England: Cambridge University Press.

Walsh, John Raymond. 1935. "Capital Concept Applied to Man." *The Quarterly Journal of Economics* 49 (2): 255–85.

Weidlich, Wolfgang. 2006. *Sociodynamics: A Systematic Approach to Mathematical Modelling in the Social Sciences*. Mineola, NY: Dover Publications.

Part III

Applications of the VIABLE Model Framework

In Part III, we turn our focus to case studies of resource conflict applications using the VIABLE modeling framework that we introduced in Chapter 7. In Chapter 8, we offer an in-depth examination of a basic, multiagent fishery model. In Chapter 9, we distill the complexity of climate change policy into a dynamic-game model of emissions trading. In Chapter 10, we broaden our focus spatially by looking at a conflict over bioenergy crop introduction, which could exacerbate water and land use conflicts in a heterogeneous landscape.

Chapters 8–10 each have technical appendices with model information and additional mathematical details. Models are built in the NetLogo agent-based modeling platform, which is free, user-friendly, and has a large user base. We also maintain a website where readers can access and learn more about these models, as well as learn more about agent-based modeling (ABM) and conflict resolution: http://todd.bendor.org/viable.

Finally, in Chapter 11, we present a synthesis of our work building and applying the VIABLE conflict modeling framework. Much of this synthesis includes general themes of our findings and thoughts on the criteria for stability of social systems and the transition from conflict to cooperation. It is our hope that the reflections that we offer can provide guidance for decision making in future environmental conflicts, as well as ideas and opportunities for future scholarship in this area.

8

A Viability Approach to Understanding Fishery Conflict and Cooperation

It was a blustery, gray day in February 1986, and I was on a National Oceanic and Atmospheric Administration (NOAA) research ship in the middle of the commercial fleet fishing in the "Donut Hole," an international zone in the middle of the Bering Sea between the coastal waters of the U.S. and USSR… I counted 60 large factory trawlers around us belonging to four or five different nations. They lined up in a pattern of several rows to take turns dragging across a thin layer of Alaska Pollock at about 400 m depth, fishing with cavernous nets that opened 45 m high for durations of several hours. That year, the Donut Hole sustained a "reported" winter catch of about 1 million tons.

In hindsight, I was witnessing the [extermination] of pollock in the Aleutian Basin
> -Kevin M. Bailey (2011, p. 1), An Empty Donut Hole. The
> Great Collapse of a North American Fishery

The problem isn't so much with managing fish as it is with managing people having different needs, values, laws, institutions, and accessibility to the fish and the resources upon which they depend…. Solutions will require broader public education, involvement, and participation in the decision-making process.
> -William Ruckelshaus (1998, pg. xi), 1st Administrator of
> the US Environmental Protection Agency

Introduction

In this chapter, we turn our focus towards applying the Values and Investments for Agent-Based interaction and Learning in Environmental systems (VIABLE) model framework to conflicts over ecosystems and natural resources. We will examine a model of fishery conflicts that can be viewed as either an abstract representation of fishers involved in an ecologically and economically devastating conflict or as a starting point for the realistic representation of a specific, complex fishing conflict. While extensive research has employed modeling to better understand the status and management of fisheries worldwide, more limited work has focused on modeling the conflicts occurring within fisheries due to competition and overfishing. This is an area in need of improved understanding of conflict and its effects on the livelihoods of individual agents, and one where our VIABLE conflict modeling framework can provide insights (BenDor, Scheffran, and Hannon 2009).

Conflicts involving fisheries have been numerous (CEMARE 2002; Hendrix and Glaser 2011) and include disputes such as the landmark "Cod Wars" between Iceland and the UK in the North Atlantic during the late 1950s and early 1970s (Kurlansky 1998).

A more recent example is the "Turbot[1] War," which occurred in 1995 along the Grand Banks near Newfoundland and involved Canada, Spain, Ireland, the UK, and the early European Union (Missios and Plourde 1996). Fishery conflicts continue to be very common throughout the developing world, including areas such as Ghana, Bangladesh, and the Turks and Caicos Islands (Bennett et al. 2001).

This is not a declining problem. As recently as March 2013, an international team of fisheries scientists discovered that China was systematically and drastically underreporting its overseas fishing catch, particularly in fisheries near West Africa, some of the world's most fertile fishing grounds (Pauly et al. 2014). Underscoring the impact of this overfishing, the lead author of the study, Daniel Pauly, argued, "[i]t shows the extent of [China's] looting of Africa, where so many people depend on seafood for basic protein" (Pala 2013, p. 18). Although previous work had determined that China had underreported its catch (Watson and Pauly 2001), revelations about the sheer level of China's illegal fish catch in international waters may be a cause for future conflicts. Additional evidence indicates that China is not the only overfishing culprit (Pauly and Zeller 2016).

Although many studies focus on nations embroiled in international fishing conflicts (Missios and Plourde 1996), disputes also frequently occur through judicial actions, either between individual fishers (or companies) or between fishers and regulators. For example, Kearney (2002) summarized numerous conflicts in the 1980s and 1990s between recreational and commercial fishers over real and perceived problems with resource access in Australian waters of Victoria.

Likewise, Jentoft and Mikalsen (2004) and Healey and Hennessey (1998) explain that the complexity of fisheries contribute to conflict; for example, as regulations are made more equitable, they become more specialized for particular fishing groups, thereby increasing the difficulty in enforcement regulations and encouraging compliance. Uncertain and complex regulatory environments have increased the likelihood of litigation, leading to time lags in enforcement and judicial reversal of regulations. Gade et al. (2002) traced the history of legal challenges to US government fishery regulations, noting a rate of one to two rulings annually throughout the 1970s and 1980s. The annual rates of rulings rose to 10 per year in the late 1990s and exploded to over 20 per year in 2001.

While fishery problems bear considerable conflict potential (Charles 1992; Ruseski 1998), they also present promising avenues for market interventions, cooperation, and conflict avoidance. In this chapter, we delve into a case study involving a classic environmental problem: overfishing in the world's oceans. The case of fisheries represents an opportunity to apply agent-based modeling to better understand emerging disputes. We will begin by exploring how and why fishery declines, collapses, and conflicts occur, using fisheries as an analog for a broader set of commons-based natural resource conflicts. We will then discuss potential policy solutions (and their drawbacks) and some background on fishery modeling. Mirroring our introduction of the generalized VIABLE framework in the last chapter, we will begin introducing our fishery model as applied to a single agent, including discussing techniques for modeling fish stocks, harvest, and agent value. We will then extend our model to work with multiple agents, implementing it to explore two competitive scenarios, followed by a detailed exploration of cooperative institutional arrangements that could decrease conflict potential. Finally, we conclude the chapter by summarizing our findings, reflecting on the broader insights from our model, and offering ideas for model extensions and tinkering.

---

[1] Greenland halibut is also popularly known as "Greenland turbot."

Fisheries are frequently viewed as an analogy for many other complex resource problems, including other common pool (e.g., air pollution) and renewable resources (Ostrom 1990; Anderies, Janssen, and Ostrom 2004). Understanding the linkages between fishery disputes and other resource conflicts will also be an important theme in this chapter as we focus on the connection between individual decisions, the rules governing them, and the collective outcomes that occur when individuals interact.

The Economic and Ecological Nature of Fishery Conflicts

Fishery Decline and Collapse

Fisheries represent a significant basis of the world's food production and are a major income source in many coastal regions (FAO-FAD 2016). However, in a landmark study, Worm et al. (2006) found that fish population loss has accelerated over the last 50 years in marine ecosystems worldwide, particularly as fish diversity in fisheries declined. While their projection that *all* current fisheries will collapse

> Fishery *decline* often results in fishery *collapse*, which is defined as a decline in fish catch to 10% of the maximum previous level (Mullon, Fre, and Cury 2005).

by 2048 has been contested (Holker et al. 2007; Branch 2008), it is now almost universally acknowledged that global fisheries are continuing to decline and are experiencing a crisis (Gutierrez, Hilborn, and Defeo 2011). Dercole, Prieu, and Rinaldi (2010) summarize five general patterns observed in fisheries data (collected by Hutchings and Reynolds [2004] and Hutchings and Fraser [2008]), which suggest that (1) fish stock abundances tend to decline over time, (2) fishery stocks can collapse rapidly and continuously, (3) stocks recover very slowly after depletion, (4) fisheries become less resilient ("fragile") when fishing pressure increases, and (5) all of these patterns dramatically decrease the body size of fish.

Due to unsustainable fishing practices and rapid improvements in fishery technology (i.e., improving the efficiency of fishers in catching larger yields), many fishery resources are declining despite numerous attempts to improve scientific understanding and management practices (Myers and Worm 2003; Pinsky et al. 2013). Although overfishing has occurred for centuries (Jackson et al. 2001), only recently has the sheer scale of fishing directly led to overall ecosystem collapse.

Why is this happening? Many instances of fishery collapse directly due to overfishing have been recorded (Pauly et al. 2002). In just the last 20 years, overfishing has caused collapses around the world, including the Alaska pollock (*Theragra chaleogramma*), which is the world's largest fishery for human food (see Bailey 2011 for the historical account). Furthermore, over 33% of US commercial fish stocks (National Marine Fisheries Service 2002), including over 80 marine fish species and many top predator species like tuna (Myers and Worm 2003), are now considered to be "vulnerable, threatened, or endangered"[2] with anticipated extinction from North American waters (Musick et al. 2000; Dudley 2008).

Overfishing is commonly a cause of major fishing conflicts, such as those discussed by Missios and Plourde (1996), Bennett et al. (2001), CEMARE (2002), and Gutierrez, Hilborn, and Defeo (2011). Furthermore, the migratory nature of fish is commonly the cause of

---

[2] "Vulnerable," "threatened," and "endangered" are terms used to signal a continuum—moving from least to greatest—of threats to the long-term viability of a species.

transboundary disputes (Missios and Plourde 1996). Scholars that study fishery problems often blame competitive fishing practices as a primary reason for overexploitation and collapse. For example, when Link and Tol (2006) modeled overfishing, which often lowers growth rates and carrying capacities, they found that it can lead to smaller fish stock levels and consequently to smaller catches, which produces major long-term economic impacts. They also discovered that following regulatory interventions, resulting increases in fishing activity can often keep fish harvests at the same levels, even though stock sizes are still diminished.

So why does competitive fishing often create overfishing scenarios? Researchers have discussed many causes for overfishing throughout the world's fisheries including (1) the open access, or "common-pool," nature of many fisheries (Gordon 1954) that are not cooperatively managed; (2) fisheries that are solely owned and have high price-to-cost ratios (making them highly profitable); (3) the rapid replacement of small fishing vessels with larger, more efficient boats; and (4) large-scale government subsidies based on fish catch levels, which generate profits even when resources are overfished (Pauly et al. 2002). In open-access fisheries, the revenues from fish catch are private, while the costs induced by a reduced (or collapsed) resource stock are shared between all participants in the fishery.

Fishery Collapse and Conflict

Fishery collapses can lead to enormous economic and social problems. One of the most important and well-known examples has been the Northern Atlantic Cod (*Gadus morhua*) fishery off the Eastern coast of Canada (Myers 1993; Hutchings and Myers 1994; Power, Norman, and Dupré 2014), where declines in fish stocks were met with a moratorium on fishing. This moratorium was responsible for a loss of over 19,000 jobs of fishers and plant workers, as well as over 20,000 additional jobs with the ensuing general economic decline. This collapse hit rural Newfoundland the hardest, where the fishery industry had long been the largest employer and economic backbone of many communities. This collapse has been documented extensively, and more details on the social, environmental, and economic effects can be found in texts such as Kurlansky (1998) and Harris (1999).

Recent policy debates in the Massachusetts cod fishery underscore the dire situations that fishery collapses can cause in fishing communities. The Gulf of Maine cod population is currently at less than 19% of its sustainable target level, while the Georges Bank cod population is even lower, at approximately 7% (NOAA 2013). In January 2013, the New England Fishery Management Council proposed dropping allowable cod catch rates by 77% in the Gulf of Maine, a fishery running from Massachusetts's Cape Cod all the way north to Nova Scotia. The council also voted to decrease catches by 55% in Georges Banks, another expansive fishing zone off the coast of Cape Cod.

Although the decreases in allowable catch were in response to long-term declines in fish stocks, the reductions were extraordinarily controversial. Economists and residents expected the cuts to harm the economies of fishing communities throughout the region over the long term, potentially putting thousands of fishers, seafood processes, distributers, wholesalers, and retailers out of work. Although the cuts were absolutely necessary from an ecological perspective, the fear was that, over time, consumers would begin to look elsewhere for cod, importing the fish from other countries like Canada or Norway, prompting further destruction of local fishing capacity and culture. Additionally, interviews with local fishers put the conflict and its long-term cultural and economic implications in perspective (Ariosto 2013, pp. 2–3):

> "I call it the systematic castration of the groundfish fleet," said Al Cattone, a Gloucester fisherman. But the 47-year-old man ...also said he never gave his son—who's now in

college—the option to fish in the way that generations of his family had. "I was afraid he was going to fall in love with it the way I did," he said. "It's a shame, but it's something I knew I had to do."

Potential Solutions to Fishery Conflicts

A number of policy recommendations have been presented for reducing fishing capacity to more sustainable levels, including strong reductions of subsidies, catch restrictions, hatcheries (e.g., Paquet et al. 2011), mandatory gear modification, marine zoning, and fishing limitations in reserve areas (i.e., marine-protected areas; Pauly et al. 2002; Worm et al. 2009). Dudley (2008) discusses the dynamic problems associated with fisheries, whereby the consequences of management actions are affected by long lag times. He points out that fishery over-capacity occurs before any evidence of overfishing. Actions to lower excessive fishing capacity are problematic, especially given the economic and political realities of fishing communities. Moreover, after fish stocks return to sustainable levels, additional over-capacity develops again (Healey and Hennessey 1998). Fishery restrictions can also act counter-intuitively to stimulate policy resistance, whereby declining catch rates prompt increased regulatory violations as fishers maximize their margins in a declining industry. Violations include unreported catches, thereby decreasing the reliability of fishery data, which are the basis for management decisions. These feedback effects conspire to defeat the good intentions of fishery managers.

Looking at potential solutions from an analytical standpoint, we are left with a major question: what catch level can sustain a fish population? Or more accurately, what is the *total allowable catch* (TAC)? We can think of TAC as the optimal total fish harvest that will maximize the welfare of all the fishing agents, while ensuring that fish stocks will not fall below a critical level. Missios and Plourde (1996) note that even if this question can be answered through scientific studies, we need to understand how to allocate the TAC among fishing stakeholders (e.g., fishers, companies, countries).

Fisheries that require stock rebuilding often undergo mandated reductions in fishing rates (lowered TAC), meaning that fishers must make short-term sacrifices in the form of reduced catches. Scholars have discovered reluctance on the part of fishers to comply with a reduced TAC over the near term, especially when there is uncertainty about whether they will be able to eventually enjoy the benefits of increased catches in the future (Walters and Martell 2004). Therefore, additional questions emerge, including: how are any proposed solutions affected by the absence of monitoring and supervision of fishing intensity? And what will happen if the negotiating parties fail to reach any sort of actual, binding agreement?

These two questions are ongoing areas of research into fisheries, as are the optimal design of the TAC and its allocation among fishers (Otto and Struben 2004; McDonald et al. 2016). However, Kearney (2002) argues that successes in fishery conflict negotiations are largely facilitated by (1) identifying the sources of conflict, (2) gathering sufficient data to assess principal perceptions of fishers, (3) obtaining independent assessment of fish stock use, (4) creating forums for public debate of these assessments and their implications, and (5) having a broad set of interests represented during assessments and advisory processes. Fishery co-management has traditionally centered around the involvement of the fishing industry in the process of designing governmental fisheries management schemes (Figure 8.1).

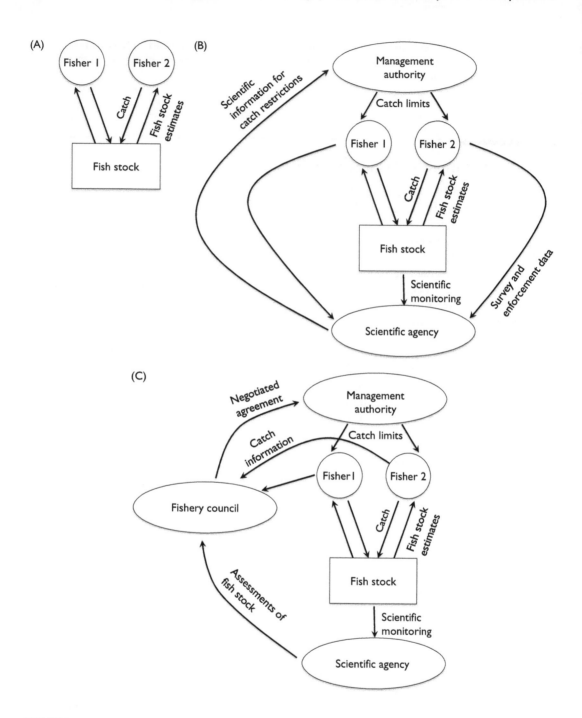

FIGURE 8.1

A comparison of fishery management schemes, adapted and modified from Eisenack, Scheffran, and Kropp (2006). (A) An unmanaged fishery, where fishers calculate investments independently and only related to the current stock status, as measured from previous year's catch. (B) A top-down strategy in which a monitoring organization collects data via scientific trawls and fisher surveys to estimate stock size. Management authorities then estimates to impose restrictions on fishers. (C) A fishery council is used in a co-management arrangement, where scientists and fisher representatives negotiate for catch or investment restrictions. The management authority approves subsequent plans that result from a negotiated agreement.

Individual Transferable Quotas (ITQs)

Although we will wait until the next chapter on carbon emissions to model a market structure for conflict resolution, it is important to note the explosion of recent research into "market-based" approaches that incentivize agents to follow TAC-related agreements and become more transparent about their catch intensities (Hilborn, Orensanz, and Parma 2005; Costello et al. 2010; Garrity 2011). Harlow (1974) provides a classic overview of how market mechanisms can increase economic efficiency and reduce multi-agent conflicts through *trading*, which amounts to a more flexible implementation of regulatory reforms. On the whole, market-based policies aim to create economic incentives for fishers to influence their behavior in a way that facilitates long-term stock viability (Garrity 2011).

One rights-based approach to fishery management is known as an *individual transferable quota* (ITQ) program (Costello, Gaines, and Lynham 2008; Costello et al. 2010), where catch shares (quotas) represent rights to catch a portion of the TAC, which can be traded among shareholders (fishers) through purchase and lease of quotas. The ITQ system is actually a form of property rights that is created after TAC limits are established. Since quota share-holders are guaranteed their particular share of the TAC, the "race to fish" behavior is (theoretically) eliminated, and fishers are afforded the freedom to pursue the timing and degree of effort (e.g., investment) that they wish to expend on catching fish.

Aided by a long history of theoretical and empirical modeling (Dewees 1989; Herrmann 1996; Weninger 1998; Fina 2005), the use of ITQs has been heralded as an economically efficient solution to fishery over-exploitation that effectively creates economic incentives for long-term fish stock stewardship. Costello et al. (2010) notably examined over 11,000 fisheries around the world, finding evidence that ITQ managed fisheries collapsed at half the rate of non-ITQ managed fisheries. Further successes in Canada, New Zealand, Australia, Iceland, Norway, and other locations have prompted changes in US fishery management (Garrity 2011), and as of 2017, approximately 10% of the total marine harvest was managed by some form of transferable quota system (Arnason and Gissurarson 2017).

Noteworthy Criticisms of ITQs

It is important to point out potential problems in using market mechanisms for conflict resolution. First, setting the TAC can itself be a highly politicized and difficult process (Garrity 2011). When faced with uncertain scientific advice indicating that a fish stock is at an unsustainable level, fishery managers may be reluctant to act. If managers lower the TAC immediately, then local fishers bear immediate financial impacts from reduced catch. The precautionary reduction of TAC can be damaging for long-run profitability and economic viability.

Instead, managers may decide not to reduce the TAC, hoping to receive improved scientific information at a later date that reveals a stock recovery. This yields a rational decision process, whereby fishery managers will consistently delay decisions based on uncertain information that have negative economic impacts on fishers. Garrity (2011) argues that the original scientific information is commonly more reasonable, meaning that the stock will continue to decline with the original TAC. These delays mean that managers are forced to adjust the TAC even further in the future to aid stock recovery, fueling further political resistance from fishers, more delays, and a dangerous vicious cycle.

ITQ systems have also been criticized based on their enduring susceptibility to being cheated by individuals (Hilborn, Parrish, and Litle 2005; Chu 2008). For example, so-called "quota busting" behavior occurs when fishers illegally catch above their purchased quota. As a fisher's catch becomes limited due to their quota, fishers are incentivized to maximize catch revenue by discarding lower valued fish that still count against their quota. Unfortunately, these discards can significantly reduce the fish stock. Scholars have theorized about the failure of some ITQs, including failures to maintain investments in monitoring and enforcement (i.e., making sure fishers do not cheat), as well as scientific assessment, which also involves the continuation of appropriate TAC levels, even in politicized environments (Chu 2008).

Perhaps the most alarming issue with ITQ adoption has been the concentration of quota shares in fewer hands, with larger-scale fishing operations typically controlling an increasing market share. This concentration can have major negative effects on fishing communities and their social equity (e.g., Dewees 1989). For example, Armstrong and Sumaila (2001) showed that the introduction of an ITQ system among Barents Sea cod trawlers and coastal vessels would not significantly improve economic output, because of the likelihood that one fleet (i.e., a single agent) could purchase all the quotas. A particularly large social impact can occur on community structure and employment if quota ownership and resource stewardship are decoupled. This issue may be exacerbated if quotas are bought up by non-local fishing company owners with no long-term interests in maintaining fish stock viability (Gibbs 2009).

Taking a Coupled Ecological-Economic Approach

To stay within limits that are ecologically sustainable, the general focus of many fishery policies has been on measuring and controlling fish populations, while increasingly taking into account uncertainties that are inherent in both ecological and economic systems (Whitmarsh et al. 2000; Davis and Gartside 2001). However, to resolve fishery conflicts, we need to understand the dynamics and interactions of the combined ecological–economic system, including those concerning the broader ecological (regeneration capacity of fish populations) and socioeconomic systems (profits, employment, and social cohesion) connected to fisheries.

Misperceptions and sensitivities surrounding fishery management are receiving growing attention in fishery policy analysis (Moxnes 2004, 2005). Increasingly, regulated fishing and compulsory cooperation among fishers have been employed in attempts to overcome the unsustainable results of competitive fishing practices (Roughgarden and Smith 1996; Pomeroy and Berkes 1997; Eisenack, Scheffran, and Kropp 2006). Co-management of marine resources and ITQs for fish catch have also been suggested and implemented to increase participation and strengthen compliance with regulatory constraints (Kearney 2002; Jentoft 2005).

However, the need for, and design of, sustainable fishery management policies raises many questions. How does the heterogeneity of individual fishers factor into fishery conflict? What are the conditions that could lead to long term viability of both fish stocks and fishers? Finally, how can cooperative fishery management policies be used to enhance both ecological and economic sustainability?

Fishery Modeling

Mathematical modeling has been used to understand fisheries for more than 170 years (e.g., Verhulst 1838), and numerous system dynamics (SD) approaches and other modeling studies have attempted to address a variety of fisheries questions and problems. For example, Otto and Struben (2004) described a group model building initiative that sought to understand how various actions could help revitalize the fishing industry in Gloucester, Massachusetts. More recently, game-theoretic and agent-based approaches have become useful for understanding the behavior of a large number of agents who can adapt to ecological and economic constraints through learning and negotiation processes (Little et al. 2009; Timmermann et al. 2014).

Castillo and Saysel's (2005) study of fishers and crab hunters in the Colombian Caribbean Sea is one example of agent-based modeling applied to fisheries. The authors develop an individual-level SD model, where five agents represent five real-life fishers. The authors then use experimental role-playing games to calibrate the model, exploring common pool resource settings in which groups of villagers (recruited from fisher and crab hunter communities) play against each other in different communication and regulation scenarios. Formulating causal relationships in the model based on collective action theory (defined in Chapter 7), Castillo and Saysel create feedback loops representing trust, reputation, and reciprocity interactions among actors. Through their experiments, the modelers demonstrated that Elinor Ostrom's (1998) approach to describing collective action is a plausible representation of fishermen and crab hunter behavior in the region.

Authors (e.g., Walters 1980; Dudley 2008) have long called for a more holistic examination of fishery systems (i.e., reframing fisheries as dynamic systems with interacting biological, political, social, and economic components). Today there remains a need for models that allow us to examine complex fishery issues, particularly management decisions and resulting fisher responses, in a transparent and understandable manner without becoming overly focused on details of population dynamics. This will be one of our goals in this chapter.

Defining Fishery Sustainability in Terms of *Viability*

Given that we are interested in the long-term functioning of fishery systems, it is first important to establish sustainability criteria for a complex nature–human interaction involving multiple actors and conflicting goals (Scheffran 2000). While definitions of *sustainability* have long been criticized as being overly broad and vague (Marcuse 1998; Vucetich and Nelson 2010), we define sustainability in our simulations in simple, yet very explicit terms. We define a future scenario to be *sustainable* if it neither degrades the natural resource stock of fish populations nor the economic capital stocks of agents over the long term. This approach is not necessarily new, and early sustainability studies also suggested a set of criteria that balanced several system variables in a dynamic environment (Daly 1991; Daily 1997).

To translate such conditions into dynamic system constraints, we call on *viability theory*, a branch of mathematics developed by Jean-Pierre Aubin (1991) and discussed in Chapter 7. Viability theory provides a framework for deriving whether the dynamics of an evolutionary system–which may be replete with policy controls and resilience mechanisms–will remain within constraints (given by critical thresholds or value-based judgments) over a given period of time. An evolution of a system through time is considered to be "viable" if it remains within the prescribed constraints (i.e., within the *viable region*, also called a *viability kernel* or a *viable corridor*) that we have defined for a prescribed period. In conceptual terms, these corridors can also be thought of as representing a subset of a system's possible "phase space," in which all possible states of a system are represented.

By combining viability theory with game-theoretic assumptions about the socio-economic behavior of fishery stakeholders, we can formally define the boundaries of viable corridors. These corridors are constrained by our knowledge of the conditions needed to achieve sustainability or prevent catastrophic developments. In a general sense, this is an appropriate concept for environmental problems where natural resource dynamics are constrained to a predefined set of possible system states. Viability theory is also a useful approach for exploring whether a complex system like a fishery, whose trajectory has moved outside of its viable corridor, can be "steered" back into it using policy controls that meet certain principles and criteria (e.g., adaptive fishery management).

Environmental applications of viability theory originally materialized in climate research, where researchers sought to define sustainable limits ("guardrails") for atmospheric greenhouse gas concentration (Bruckner et al. 1999, 2003; Petschel-Held et al. 1999). Early on, viability theory was applied to fisheries by Eisenack, Scheffran, and Kropp (2006) and Aubin and Saint-Pierre (2007) to determine "tolerable windows" (i.e., viability corridors) for fish catch (although these studies concerned a different context than the modeling framework we present here).

In focusing on fisheries, we seek to analyze whether control strategies, along with different fish harvest rates, can keep the system within its viable corridor or not. Determining the viability of a certain system trajectory or structure means that normative requirements (e.g., a company stays in business or resources are not depleted beyond a certain level) are translated into clear, mathematical boundary conditions that guide decisions, actions, and changes to system structure (Aubin, Bayen, and Saint-Pierre 2011). The choice of viability constraints cannot be justified by empirical considerations alone; these constraints involve *value-laden*, normative ideas (Eisenack, Scheffran, and Kropp 2006) that need to be justified to all actors involved in the system, which is the task of consensus processes between system stakeholders (discussed in Chapter 1).

The viability approach can assist in the definition and selection among long-term objectives at an ecosystem level, which can be derived from fishery objectives, conservation principles, or scientific results of modeling (Cury et al. 2005). Within this framework, stakeholders and managers can identify the controls that are necessary to maintain the system within sustainable limits (e.g., for fish catch), as well as to avoid nonviable states in which fish resources decline or fisheries become unprofitable. Considering the complexity of fishery systems and their management, we do not strive to determine *optimal* corridors for the coevolution of fish stocks and the fishing industry, but rather determine a set of dynamic corridors that meet the defined system constraints.

Several studies have used a viability approach to control the ecological and economic interactions of renewable marine resources within explicit viability constraints (e.g., Doyen et al. 2007; Martinet and Doyen 2007; Gourguet et al. 2013). While previous efforts (e.g., Cury et al. 2005; Eisenack, Scheffran, and Kropp 2006; Gourguet et al. 2013) have waded into the technical details of fishery modeling using viability approaches, our focus here is to build a model that demonstrates the basic behavior of fishery management schemes organized along different agent decision rules, including (1) competitive scenarios that test different representations of fisher behaviors and (2) a cooperative scenario where fishers work together in a coordinated system to increase their collective fishery returns. While we will represent fishers as having heterogeneous properties, for brevity we will assume that fishers use the same decision rules. This is an assumption that can be relaxed through the reader's own experimentation, or in real-world applications by determining actual fisher decision rules through participatory research and engagement processes (see Chapter 4).

As we set up the model of fishery conflicts, we will begin by describing a set of relatively basicmathematical representations of ecological–economic interactions in the fishing industry. Moreover, we will integrate fisher behavior into an SD simulation model and discuss three sets of decision rules that simulate the behavior of individual fisher agents under both competitive, cooperative, and trading fishery management policies.

Building an Agent-Based Model of a Fishery Conflict

BOX 8.1   VARIABLE DEFINITIONS

Given for each fisher $i = 1, ..., n$, unless otherwise specified

- V = agent profit (revenue minus cost).
- h^k = harvest (catch) of each fish species ($k = 1, ..., m$ fish species).
- x^k = size of the fish stock of each species.
- c^k = agent's unit cost of catching fish k.
- p^k = market price for fish k.
- a^k = initial price per unit of harvest for fish k.
- b^k = price elasticity for fish k.
- C^k = amount invested in catching each species.
- v_C = rate of change of the profit function V with respect to investment C, $v_c = \partial v / \partial c$ (marginal value).
- C^+ = upper investment limit of an agent, which is a fraction of the available capital K applied to fishing effort.
- q^k = an agent's catch efficiency or the amount of fish caught per unit of investment and per number of existing fish (also called "catchability").
- r^k = priority that a fisher has for catching a species of fish. This represents the fraction of total investment spent on harvesting fish species k.
- E^k = effective investment for fish harvest where investment is weighted by efficiency q^k.
- L^k = carrying capacity of environment for fish stock k (i.e., maximum fish stock supported).
- g^k = reproduction rate of fish k.
- K = capital stock of each agent.
- $\bar{C}(x)$ = rate of investment that yields *ecologically* viable conditions.
- $\hat{C}(x)$ = rate of investment that yields *economically* viable conditions.
- α = adaptation rate of a fisher in retooling their equipment and capital to catch a different fish species.
- β = adjustment rate at which agent adapts investment C to the changing state of value V (e.g., hiring or firing fishers, buying or selling boats or gear).

(Continued)

BOX 8.1 (*Continued*) VARIABLE DEFINITIONS

- γ^k = rate of adaptation of *effective investment* E^k toward the sustainable limit E^{k*} that restrains both the amount of investment and the catch efficiency.

- C^* = target investment rate toward catching each fish species that maximizes profit (used under optimizing rule).

- D = a decision rule.

- v_r^k = rate of change of profit V (i.e., marginal value) with respect to changes in allocation priorities r for given fish $k\left(v_r^k = \partial V / \partial r^k\right)$.

- E^k = "effective investment" of an agent, which is the product of the catch efficiency and the investment of each fisher in harvesting each fish species. This variable not only measures the investment that a fisher puts into into fish catch but also the sensitivity of the fish population to this investment (which is mediated by fisher efficiency).

- E^{k*} = steady-state (equilibrium) condition of effective investment where fish stocks are neither growing nor declining ($\Delta x^k = 0$).

- φ^k = fraction of the total change of effective investment assigned to every fisher in proportion to their share of actual effective investment of all fishers together.

Agent Value Functions

We begin by focusing on fishers as the key agents in altering fish stocks by catching fish as a source of income. Although several dynamic models (e.g., Dudley 2008) have explored how and when fishers enter and exit the market, for simplicity we will model a static number of fishers. Although many marine systems have a variety of other variables affecting fish stocks (e.g., climate change; Glantz 2005; Pinsky et al. 2013), here we will simplify this model by excluding those variables.

For a single fisher or a homogenous group of fishers (e.g., a single fishing company with multiple boats), we will start by representing the value gained from fishing through a net profit function, $V = ph - C$, where value (or profit) V accrues from selling fish catch (harvest) h at a market price p, which is diminished by the investment C associated with catching fish. Moreover, we can represent investments as $C = ch$, where c is the unit cost of harvesting h fish. Here, investment is a key variable that is synonymous with fishing "effort," which we represent as the financial capital expended during a given time period for harvesting, including costs for salaries, fuel, boats, and other equipment.

This is similar to the idea of "fishing intensity" suggested by Scott Gordon (1954) in his seminal work on economic theory of fisheries. We can think of "effort" as generally encompassing the *capability* that fishers actually use and invest in action. While, for the sake of simplicity, the model we present in this chapter limits this capability to the financial capital invested, a more extended concept could use other capability dimensions (e.g., labor, intellectual resources).

We estimate market price p by using a linear demand curve, $p = a - bh$, which declines with the total catch h, where a represents the initial price per unit of harvest and b the price elasticity. Again, there are other ways of representing the relationship between the amount of catch and the market price. While we have chosen the simplest possible linkage, this equation could become more sophisticated if used to specify the demand curve for a specific fishery; for example, see Copes (1970) for a classic analysis describing a complex, backward-bending supply curve linking price and fishing investment.

Now that we have a way of estimating profit, we can start to consider how profits affect fisher capital, which are the resources built up over time—essentially representing fisher's "net wealth"—that allows fishers to catch more or less fish. To do this, we can take each year's profit and compound it as financial capital K accumulated over time. So, in any given year, if profit is positive, it gets added to the stock of K, and if negative, it gets removed from K. This is a basic way of representing a year with net gains or losses. However, to make this more realistic, during each time step, we establish a maximum fraction of the total capital that can be used as an uppermost "investment limit" $C^+ = K$, which is the maximum effort a fisher is willing and able to invest in a year.

Modeling Fish Stocks

Having discussed some of the economic workings of the fisher agent, we can add some realism by extending this model to simulate the harvest of multiple types of fish, thereby modeling a more realistic, multistock fishery, such as the North Sea fisheries that produce cod, haddock, whiting, saithe [pollack], plaice, and sole (Vinther, Reeves, and Patterson 2004). While we will be making a lot of simplifying assumptions about this system, it is important to note the importance of information input from stakeholders regarding any specific fishery (Stöhr et al. 2014). With input from fish and wildlife experts, as well as local fishers, about the exact relationship between fish locations, populations, and the ability of fishers to catch those fish, we could be much more specific about how we represent fish harvest and how we model fish stock growth through re-population.

However, in this simplified model, we will assume that in a given area, a fisher can harvest from multiple fish stocks x^k; from now on, we will take variables to indicate a particular type of fish species by the superscript $k = 1, ..., m$. In cases where we have very small, incremental time steps, and tractable equations, the dynamic behavior of fish stocks could be represented by differential equations with continuous time (Sterman 2000). However, if we do not have easily tractable equations with closed-form solutions (see Chapter 5 for more details), we will instead represent fish stock dynamics using different equations, which change from one time step (year) to the next: $\Delta x^k(t) = x^k(t+1) - x^k(t)$.

To be general, simple, and yet "acceptably complex" (as Richard Dudley [2008] sensibly put it) in representing the structure of fish populations, we will rely on Schaefer's (1954) generalized dynamic model of biomass, which has found frequent use in fisheries models (e.g., Dudley 2008). This model represents fish populations as a logistic reproduction function, which is diminished by the yearly total harvest h^k:

$$\Delta x^k = g^k x^k (L^k - x^k) - h^k \tag{8.1}$$

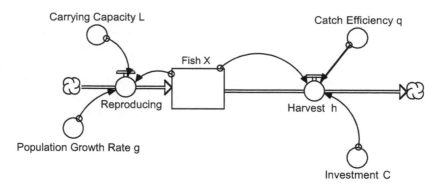

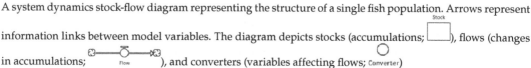

FIGURE 8.2

A system dynamics stock-flow diagram representing the structure of a single fish population. Arrows represent information links between model variables. The diagram depicts stocks (accumulations; ⬜), flows (changes in accumulations; �Flow⟩), and converters (variables affecting flows; Converter)

Here, g^k is the net rate of growth of the fish stock x^k; we will assume that this is the same across all fish age groups, although we could also consider the case that fish only reproduces above or below certain ages (as is the case of species such as sturgeon; [family Acipenseridae]). Additionally, L^k is the carrying capacity of the environment for fish x^k and h^k is the fisher's harvest. While we will assume that the carrying capacity L^k is constant, it is clear that certain fishing activities may decrease ecosystem-carrying capacity (trawling gear can damage sea floor habitat; Watling and Norse 1988). An equivalent SD stock-flow diagram representing Equation 8.1 is shown in Figure 8.2, where the fish stock (which is an accumulation) is decreased by harvest and increased by reproduction, which are both rates of change (flows).

Modeling Fish Harvest

Now that we have represented multiple fish stocks that dynamically change over time, we can start to represent how fishing cumulatively affects fish stocks. Let us start by thinking about the harvest of each fish species h^k as an outflow from each respective fish stock. Harvest can be calculated as $h^k = q^k x^k C^k$, which means that the harvest is taken to be proportional to the fish stock x^k and the amount invested C^k in catching each species. This method of linking harvest h^k and investment C^k implies that harvest increases with the productivity of fishers (high efficiencies q^k) and with more investment (e.g., labor, vessels, money), while the fish population represents an important mediating factor; if there are few fish, there will be lower harvest rates, whatever the investment.

For each agent's investment in fish species k, we can denote $C_i^k = r_i^k C_i$, which is the fraction r_i^k of the total investment C_i by each fisher to catch fish x^k. In this case, r^k actually represents the priority that a fisher has for catching a specific species of fish, which, as we will soon see, is a very important variable. Finally, the factor q^k is called the *catch efficiency* or the amount of fish caught per unit of investment and per number of existing fish. This efficiency, which is also known as *catchability* in much of the fishery economics literature (Arreguín-Sánchez 1996), is determined by the relative technological capability of the fisher and the catch methods that they apply. As the fisher becomes more sophisticated, perhaps using improved technology or techniques, their efficiency at catching fish using the same amount of general investment will rise.

Finally, the last key variable that we will introduce is the unit cost of each harvested fish species $c^k = C^k/h^k$, which we can calculate as the ratio of investment C^k to harvest h^k, which reduces to $c^k = 1/(q^k x^k)$ using the previous equation for h^k. This representation is somewhat intuitive in that as fish stock and the catch efficiency increase, the unit cost of catching fish decreases. Conversely, it becomes costlier to harvest additional fish as fish stocks decline.

Viability and Uncertainty

To specifically address conflicts surrounding fishery decline, it is important to now explore the long-term viability implications for this system. We begin by looking at the variables that control the dynamics of fish populations and fisher capital stocks.

For each fisher, the investment rate C is a key control variable that determines the dynamics of the fish stock as well as the fisher's harvest and income. The fish population x^k is unchanging when $\Delta x^k = 0$, which is equivalent to saying that harvest and reproduction are the same, a state known as "dynamic equilibrium" (Ford 2009). The mathematical representation of this can be seen if we set Equation 8.1 to zero (and divide out the x^k term on both sides), so that $q^k C = g^k (L^k - x^k)$ is the equilibrium condition. This is equivalent to taking both flows into the fish stock in Figure 8.2 and setting them equal.

Both sides of this equation represent preconditions for viability, balancing the "effective investment" for fish harvest $E^k = q^k C$, where investment is weighted by efficiency q^k (and thus is zero if either value is zero), with the ecologically viable supply rate without human intervention $\Delta x^k = x^k g^k (L^k - x^k)$.[3] Figure 8.3 depicts this relationship, with the right side indicating fish reproduction in the ecosystem and the left side indicating the effective effort of the fishers.

While in principle, the economic limits for $q^k C$ could be calculated from the ecological limits, each of the three parameters that simulate the fish ecosystem, including the reproduction rate g^k, the current fish stock x^k, and the carrying capacity L^k, are calculated under

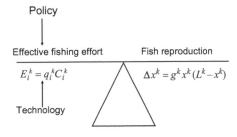

FIGURE 8.3
The balancing act between fish harvest and reproduction of fish stock.

---

[3] Regarding units: g is the growth rate [unit: 1/(fish * time)], q is catch efficiency [unit: percent fish catch/(cost unit * time)].

extreme uncertainty. Therefore, estimated static boundaries on $q^k C$ could vary significantly, based on worst-case or best-case assumption about the future of the fish population. More importantly, however, is the fact that fishery conflicts emerge in situations where many fishers act individually to maximize profit. This individual profit maximization, devoid of central organization, can lead to situations where the total effective investment in harvesting exceeds the sustainable limit. This, in turn, can lead to sustained overfishing, which adversely affects the interests of all fishers. Differently stated, this is a classic example of the tragedy of the commons (Hardin 1968), a common pool resource problem that is often present in open access fishing situations (Gordon 1954; Scott 1955; Berck and Perloff 1984; Homans and Wilen 1997) and many other environmental systems (Ostrom 1990; Ostrom et al. 1999).

To understand how fishery stocks can be maintained in a sustainable way, we need to develop a comprehensive understanding of the processes and mechanisms to hold total harvest within the ecological and economic boundaries that describe a "viability kernel." In terms of policy, these mechanisms can include top-down regulatory control, bottom-up cooperation on the part of the fishers, or some sort of trading program, as is increasingly proposed as a solution for overfishing (Costello et al. 2010). It is important here to consider the different sides of fishery conflicts, whereby "sustainability" of the fishery system also greatly concerns the long-term viability of fisher capital. That is, fishers must remain abundant and profitable for any fishery scenario to be termed "sustainable." This view incorporates a multi-pronged definition of "sustainability," incorporating economic, environmental, and cultural concerns (Campbell 1996; Berke et al. 2006).

Referring more directly to the VIABLE framework we presented in Chapter 7, we can focus this discussion to establish two viability criteria for sustainability: avoid degradation of *both* the fish stock x and fishers' capital stocks K, which implies that neither the fish population growth rate Δx nor profits V (which is essentially ΔK) should be negative over an extended period of time to maintain the capital stocks and avoid their destruction. In a viability context, this means that maintaining capital K and fish stocks x^k is the precondition for the existence of the system, and the ratios of profit to capital (V/K) and fish changes to stocks ($\Delta x/x$) are important metrics for considering the rate of degradation.

While these dual constraints may be relaxed during periods with high fish stocks and high capital, assuming that a short-term and limited drop in either variable could be acceptable, we take these conditions as the basic constraints for ecological and economic sustainability in our following analysis. Mathematically, we can consider these conditions more explicitly as:

1. Ecological viability of sustainable fish stocks: $x > 0$ (always) and $\Delta x > 0$ (except for short periods or minor changes relative to fish stock size). We can describe this as a boundary condition $\overline{C}(x)$, which relates investment costs C to the fish population x; and

2. Economic viability of nondeclining capital and sustained profit of a fisher: $K > 0$ (always), $V > 0$ (except for short periods or minor changes relative to total capital). We can describe this as another boundary condition $\hat{C}(x)$ for investment rates as a function of fish stock x. Details of both of the viability equations, $\overline{C}(x)$ and $\hat{C}(x)$, can be found in Appendix 8.1.

As long as the economic viability condition is satisfied, fishers can still increase their investment rate (i.e., $\Delta C > 0$). However, in the case where investment exceeds this threshold $C > \hat{C}(x)$, profit becomes negative ($V < 0$), which reduces the fisher's capital stock K. To avoid losing all capital, a fisher has to adjust by reducing the investment ΔC until again, $C \leq \hat{C}(x)$. The direction of fishery stock change (Δx) follows Equation 8.1 (increasing on the left side of the straight line and declining on the right side). The direction of investment change (ΔC) follows the decision rules given in Appendix 8.2 (declining for $C > C^*$, increasing for $C < C^*$).

The viability conditions (Equations 8.2 and 8.3 in Appendix 8.1) of the fish populations and profit can be visualized in Figure 8.4. Here, the curves divide the two-variable diagram (which is also called a "phase diagram" or "state space" connecting x and C) into a set of four qualitatively different situations in which ecological or economic sustainability conditions are either satisfied or not. The sustainability criteria are represented by boundary curves for both variables, the shape of which depend on numerous parameters, including g, q, a, b, and L. At the intersection of both viability limit curves, we find an equilibrium with $\Delta x = V = 0$. That is, fish population stocks are in a dynamic equilibrium state (same amount caught and harvested), while outside of the equilibrium, either of these variables is positive or negative.

With these rules in place, maintaining both the ecological and socioeconomic viability conditions leads to the most desirable outcome of "joint viability" (as denoted as Case 1 in Figure 8.4). If either viability condition is violated, the system moves toward decline of either the ecological or economic systems, or of both—which is the least desirable state (Case 4). By maintaining the viability of only one of the systems, we face the question of whether socioeconomic or ecological criteria are preferred (Cases 2 and 3).

In each of the four cases in Figure 8.4, the investment rate dynamics depend on the decision rule a fisher follows. Perhaps the simplest possible rule is to increase investment rate

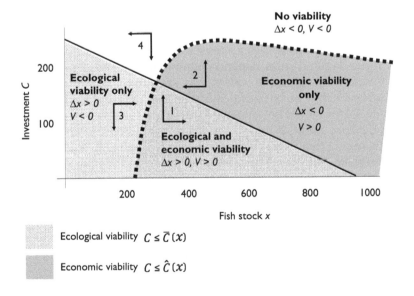

FIGURE 8.4
Phase diagram of investment rates C vs. fish stock x, showing the boundaries of economic viability and ecological viability. The zone of economic viability is depicted as the area under the dotted black line indicating where $V = 0$. The zone of ecological viability is shown in gray under the downward sloping solid line, indicating where $\Delta x = 0$. These two boundaries divide this phase space into four regions of economic or ecological viability: (1) dual economic/ecological viability, (2) economic viability only, (3) ecological viability only, and (4) dual nonviability. The arrows indicate the direction of adaptation dynamics in each of the regions and the intersection point of the two curves is the equilibrium point, whereby $\Delta x = V = 0$. This is a conflict that has been outlined within the welfare-based environmental management literature (e.g., see Farber 2000).

C when it is below the threshold investment and decrease the investment rate when it is above the target. Under this rule, in each of the four regions, we can establish a different combination of arrows, indicating the direction of the adaptation dynamics. In the represented case, a circular movement around the intersection point $\overline{C}(x) = \hat{C}(x)$ may result, which implies that the boundaries between viable and nonviable regions are repeatedly crossed and each of the viability conditions is temporarily violated.

The exact form of the trajectory depends on the adjustment rate β at which investment C adapts to the changing state of value V. For example, if the responses of a fisher are slow, the system state could remain in one of the nonviable regions for an extended period of time, leading to the strong decline of either the fish population or the capital stock. On the other hand, if fishers change their investment rate too quickly (the adjustment rate β is too high), the response could overshoot the target into nonviable regions. The analysis of our example model in Chapter 7 indicates that the speed of adaptation is very important.

Upon translating the boundaries of sustainability into these mathematical viability criteria, the question for the model becomes, how does this new understanding of long-term system viability guide the actions of fisher agents and translate into fisher decision rules for investing in fish catch? We will explore three of the many different types of decision rules that could simulate agent-fish stock relationships in specific fisheries around the world; each fishery would have a completely different viability graph and viability kernel boundaries that determine long-term system sustainability.

Decision Rules Describing the Behavior of Fishers

We will explore and compare two *competitive* decision rules that seek to increase fisher value. Both of these decision rules depend on some form of the partial derivative of the agent's value function v_C (marginal value) and are thus variations on the optimizing, rational-actor approach that we have critically discussed throughout this book. Under this design, fisher's decisions would be sensitive to all of the variables that affect profit, including cost, catch efficiency, and fish price.

We can start by saying that fishers can adjust their investments (within the upper investment limit C^+) between fish species in an attempt to generate more value that is added to their capital stock K. A general, adaptive view assumes that the increase or decrease in investment of a fisher is driven by $\Delta C(t) = \beta D(V(t))$, where $D(V(t))$ is a decision rule that drives changes in investment based on value considerations, and the adaptation rate β represents the speed at which a fisher is able to adapt to the investment strategy to the changing market or environment around them. This takes into account possible response delays as well as the fact that some investments cannot be easily changed from one year to the next, such as acquiring fishing boats or hiring and training more staff.

Model 1: Fishers as Global Optimizers

In our first case, we describe an "optimizer" as an agent who adjusts its investment in fish harvests based on the calculation of the global profit optimum. This is the classic representation of a rational actor that is able to account for all information in the system. If we imagine that seeking profit is analogous to climbing a mountain, then "optimizing" refers to a strategy where, during each time step, the agent knows where the mountain summit is, and is able to move directly toward it without delay. In calculus, this is analogous to finding the "global" maximum state associated with a given function.

This rule corresponds to the rational optimizing behavior described in classic economics approaches, in which agents singularly aim to maximize profits over a certain time period by either maximizing utility or minimizing cost. This *optimizing decision rule* requires knowledge about the global optimum (e.g., how much money they could be making if they performed optimally) and assurance that the path leading toward it is feasible. To determine the peak investment strategy, which maximizes the efficiency of investment C^k in each fish species, fishers could optimize their strategy such that the rate they gain value for additional investment (marginal value) drops to zero (e.g., determining where the partial derivative $\partial V/\partial C = 0$). In this case, investment rates are adapted in proportion to the distance from this target according to the decision rule $D = C^* - C$, which leads to a linear investment increase when the fisher is below the optimum, and to an investment reduction when the fisher is above the optimum for a given fish species. See Appendix 8.2 for more information on this decision rule.

Model 2: Fishers as Local "Satisficers"

In contrast to the optimizing rule, we can consider behavior that gradually seeks to increase profit in a local manner, rather than a global manner. Here, we can explore what we call a *gradient decision rule*, whereby fishers increase or decrease investment in catching a given species of fish in proportion to the incremental change in profit they experience. This approach means that changes in agent decisions are proportionate to locally visible gains in profit; to return to the mountain climbing analogy, here an agent climbs the mountain without knowing the location of the summit, and thus has limited knowledge of possible better decision options. While the gradient strategy is aimed at optimizing profit, in the long run, it may (but not with certainty) lead to an optimal strategy.

Rather, harvest strategies are characterized by *adaptation*, which adjusts fishers' fleet utilization for catching each type of fish based on a comparison of the profit and catch priority with previously calculated values. Because of this, the gradient decision rule represents the behavior of agents who are constrained by "bounded rationality," wherein a fisher's decision making is limited by their information, cognitive limitations, and the time that they have to make investment allocation decisions (Simon 1991). This mirrors behavior known as "satisficing" (also discussed in Chapter 3), a term combining "satisfy" and "suffice" that was famously coined by Herbert Simon (1947, 1956) to describe how agents search through available alternatives until they meet one that is acceptable (e.g., follow your neighbor's actions if they seem better than yours). As the criteria for adjusting fish species depend on the ability to earn incrementally more profit by moving up the gradient toward a higher target value (a local maximum of the profit function), we would expect that the profits from fishing obtained by following the gradient strategy will almost always be smaller than profits that would have been obtained under the profit-optimizing harvest strategy. See Appendix 8.2 for more information on this decision rule.

Adaptation Delays and Priorities

Under both of these decision rules, adaptation processes depend on the response parameter β, which defines how fast fishers adapt to changes in price and profit. For instance, given $\beta = 1$, an optimizer would reach their target investment, given as C^*, in one time step if nothing else changes. However, given a $\beta > 1$, agents would overshoot their target, being overly aggressive in adjusting their investment rates. The idea that a fisher can "overshoot" their target investment rate suggests that changes in investment can be either negative or positive; if the investment rate is higher than their target investment rate, then fishers reduce investments according to their decision rule.

Adaptation processes under both decision rules can also be applied to the allocation of catch investment to multiple fish species (represented by superscript k), given by the allocation priorities r^k, which represent the fraction of total effort spent on harvesting each fish species. We can alternatively think of r as a fisher's "priority" for each fish species; these priorities must sum to one $\sum_k r^k = 1$. Under the optimizing decision rule, we can calculate optimal allocation priorities r^{k*} for each fish species k using the optimality condition that seeks maximum profits, whereby we take the derivative of profit V as a function of priority r^k, and set it equal to zero to find the global maximum, as we did before (e.g., $v_r^k = \partial V / \partial r^k = 0$).[4]

To actually modify the priorities that fishers have for each fish species, we can use a simple goal-seeking function, where the change in priority for a fish species is the difference between the optimal priority and the current priority, multiplied by a rate of adaptation α, which represents the "adaptation" rate of a fisher in retooling their equipment and capital to catch a different fish species; much like the adaptation rate β seeks to adapt investments, α seeks to adapt the use of capital in actually catching fish. Mathematically, this means changes in fish species priority are $\Delta r^k = \alpha(r^{k*} - r^k)$ under the optimizing decision rule (see Appendix 8.2 for derivation of the optimal priority r^{k*} and investment C^* under the optimizer approach). While, for simplicity, we assume α to be the same for each fish k, high variation in α could represent situations where it is very easy or very difficult to switch catch investments between specific fish species.

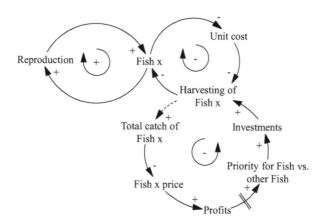

FIGURE 8.5
The feedback relationships within the single-agent model of fisher–fish stock interactions. The dotted line between harvest and total catch is a link between individual fishers through the shared fish stocks x^k, which represent the "environmental system" x we defined in Chapter 7. Causal relationships are denoted on each arrowhead: [–] inverse causal relationships (e.g., increase in supply causes decrease in price), and [+] proportional causal relationships (e.g., higher price means more profit). The hashes represent a delay in the link between profits V and changes in fish priority r^k. The clockwise and counterclockwise symbols indicate the presence of feedback loops and whether the loop exhibits re-enforcing (+) or balancing (–) behavior.

---

[4] Because this derivative is linear with respect to r^k, this is all that needs to be done. However, if it were not linear, then additional, higher order derivatives would be necessary to determine the global maximum.

To represent changes to fish species priority under the gradient approach, we hypothesize—again, in the absence of stakeholder input—that priority changes under this rule will be proportional to (1) the adaptation rate α, (2) the current fish priority r^k, and (3) the difference between the marginal value v_r^k for fish k (e.g., the change in profit from a change in priority for that fish, $v_r^k = \partial V / \partial r^k$) and the weighted average of the marginal values across all fish species ($\sum_l r^l v_r^l$; weighted by priority), which ensures that the priorities sum to one, $\sum r^k = 1$.

As we demonstrated while introducing the VIABLE framework in Chapter 7, we can mathematically represent proportional, incremental changes in profit (i.e., the gradient approach) as the partial derivative of the profit function V with respect to investment C, $v_C = \partial V / \partial C$. Differently stated, by taking this derivative, we can learn how profit changes with respect to small changes in investment rates for each species of fish. Therefore, the associated gradient decision rule, $D(V) = v_C$, responds to the immediate impact of increasing or decreasing investment on profit, and depending on the adaptation rate β for new investment strategies, acts to continuously improve the mix of fish species for each fisher.

While these may seem like relatively undeveloped ways of representing the decisions to change allocation of investment between different fish species, these variables and their interactions can become very complex due to the nonlinear feedback loops shown in Figure 8.5.

Multiagent and Multifish Stock Interactions

While the model so far has been developed for a single fisher, we can now extend it to multiple agents $i = 1, ..., n$. Fish species $x^{k = 1, ..., m}$ are diminished by the total harvest from all fishers $\sum_i h_i^k$, where $h_i^k = q_i^k x^k r_i^k C_i$ with catch efficiencies q_i^k and allocation priorities r_i^k of the efforts of each fisher and to a particular fish species k. Equations for both decision rules have the form of evolutionary dynamic games among competing priorities r_i^k for different fish species (Hofbauer and Sigmund 1998).

What is a *simulation*? Endy and Brent (2001, p. 391) define simulation as a representation embodying the information contained within, and created by, a *model* (e.g., model output). Simulations provide access to the model by allowing computation of system behavior. That is, while the model represents the system, a simulation is an examination—compressed over space and time—of the output of the model based on distinct inputs and parameters.

Economic Competition: Satisficers and Optimizers

Now that we have developed a multi-agent model of a fishery under several different decision rules, we can explore *simulations* that (1) describe competition between fishers who individually behave either according to the optimizing or gradient decision rules or (2) analyze coalition formation whereby fishers cooperate to jointly determine optimal and sustainable investment rates. Most importantly, the VIABLE framework allows us to observe the interactions between fishers whose abilities and resources are not homogenous. To explore differences between fishers, we vary a set of hypothetical parameter values for fish price, price elasticity, and fish "catchability" (catch efficiency).

As an example, we will simulate six fishers, who could represent individual fishers or large fishing companies, each centrally coordinating dozens of boats (BenDor, Scheffran, and Hannon 2009). These fishers will compete for two fish species, which we will call "x" and "y." Although we set the initial fish prices (a) to be different ($a^y = 2$ and $a^x = 1$), and the price elasticity of y ($b^y = 0.01$) to be twice as high as for x ($b^x = 0.005$), we set both fish species to have the same carrying capacity, $L^x = L^y = 1000$ fish, the same growth rate, $g^x = g^y = 0.2$ (20%/year), and the same initial population $x(0) = y(0) = 500$ fish "units" (e.g., biomass measurements). Additionally, the initial capital endowment for all fishers is set at $K_i(0) = 100$ cost units, and the initial investment rate is also uniformly set to $C_i(0) = 10$. The maximum investment rate C_i^+ is initially set to 50% ($C_i^+ = 50$) of the available capital K_i, although this may later change for each fisher based on their accumulation of positive or negative profits V_i. Finally, we initialize this simulation so that the allocation priorities (r^k; the amount of investment allocated between species) are equal for both fish, $r_i^x = r_i^y = 0.5$.

Most of the dynamics in this system will emerge from the important factor that governs the heterogeneity of fishers, which we characterize through different efficiencies (catchability)

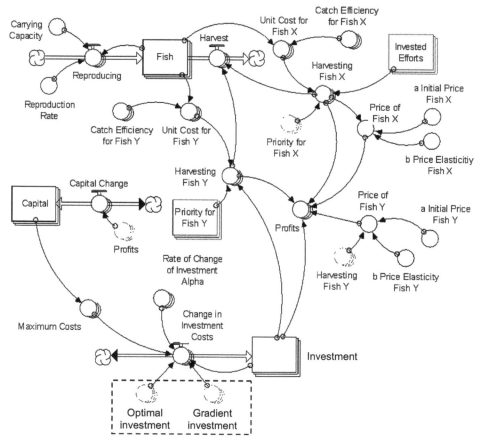

FIGURE 8.6
STELLA diagram of the model structure under competition with optimizing and gradient decision rules. *Optimizing* and *gradient* rules are shown as different possible inputs into the calculation of the "Change in Investment Costs" at the bottom of the diagram. The *optimal investment rule* alters investment rule costs based on a calculated global optimal cost, while the *gradient investment rule* alters investment costs using the local derivative of the fisher's profit (value) function.

for each fisher. For the purpose of illustration, we give efficiencies for fishers pursuing fish y as a linearly decreasing function between Fisher 1 and Fisher 6, where $q^y = 0.012$ for Fisher 1 down to $q^y = 0.003$ for Fisher 6 (0.0018 increments). Doing this simulates Fisher 1 as having a substantial competitive advantage in catching fish y over Fisher 6 (with everyone else in between). Conversely, the catch efficiency for fishers pursuing fish x *increases* linearly from $q^x = 0.002$ for Fisher 1, to $q^x = 0.008$ for Fisher 6 (0.0012 increments), giving Fisher 6 a substantial technical advantage over Fisher 1. Given these values, Fisher 6 is the most efficient and Fisher 1 is the least efficient in harvesting x, with the reverse being true for fish y. These hypothetical values were chosen to demonstrate differences in cost and technical efficiency for catching both fish species, as well as allow us to explore the impact of widening gaps in technical efficiency between fishers. A more sophisticated structure could incorporate changes in catch efficiency given a framework for generating individual fisher experience (i.e., agent learning).

Figure 8.6 depicts the model's competitive market structures using an SD stock-flow diagram, whereby fishers use optimizing and gradient decision rules (one of the two inputs into investment in the gray dotted box) to alter their priorities and investment decisions. These different decision rules govern the way in which fishers react to changes in their profits with regard to their level of investment in catching Fish x or Fish y. This books' website—http://todd.bendor.org/viable—an in-depth description of this model implemented in the NetLogo simulation software (see this chapter's additional resources section for more NetLogo information and resources)

Optimizing and Gradient Decision Rules: The Fishery Decline

In the first scenario, all fishers use the optimizing decision rule, while in the second scenario, all fishers use the gradient decision rule. It is important to note, however, that when using NetLogo, it is not difficult to define a mix of decision rules for specific fishers, which is possible by explicitly redefining the *delta-fish-priority* variable for specific fishers (using the **ask turtle** command; Wilensky and Rand 2015).

The initial total harvest for both fish species is around 100 fish units per year under both optimizing and gradient approaches. Since these harvest rates exceed reproduction rates, both fish species decline and fish y, which is caught more efficiently, comes close to extinction (Figure 8.7). Because of the declining fish stocks, fish scarcity and unit costs rise, thereby increasing market prices (in fact, doubling market prices for fish y) along with unit costs. With the upsurge in unit costs, harvest decreases until it equals the reproduction rate and both fish stocks stabilize.

Under the optimizing rule, fishers experience a rapid decline in profits since optimizers, by their very nature, are driven to heavily exploit the fish stock of the species that they are best at catching to the maximum possible degree. Motivated by their initially optimal investments, which rise over twice as high as under the gradient rule (see Figure 8.8), optimizers are myopic and invest large amounts of capital in the early years, resulting in high harvests (20%–30% higher than the gradient rule simulation) and rapid depletion of fish stocks, thus ignoring long-term sustainability issues. As a result, optimal investment rates, and eventually harvests, fall rapidly in later years, and even become negative for some fishers before year 10. Although all fishers end up with a higher capital stock than under the gradient decision rule, fish stocks are more rapidly depleted.

This behavior demonstrates that except for those fishers that are most efficient in catching one of the two fish species and can sustain their investments, most fishers using globally optimizing investment strategies actually stop fishing due to nonprofitable fishing investments

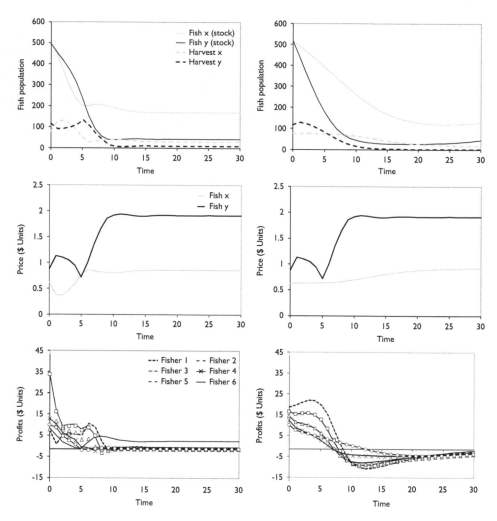

FIGURE 8.7

Fish populations x^k, harvest rates h^k, price p^k, and agent profits V under competition with agents using (left column) optimizing and (right column) gradient decision rules.

after about 10 years. Given their strong technical efficiencies and resulting competitive advantages, Fisher 1 and Fisher 6 continue with relatively high rates of fishing investment. One important insight: by continuing to exploit their competitive advantages, fishing activities by Fishers 1 and 6 act to keep unit costs high for other fishers, thereby shutting them out of the market and effectively creating monopolies for each of the fish species.

Under this optimizing rule, capital for all fishers stabilizes at a level that is higher than in the beginning (Figure 8.8). However, this only occurs once Fishers 2-5 have effectively ended their fishing efforts and lowered their harvests to zero, which signals that they have been driven out of the market. Why does this happen? Each of these fishers sought to maximize their individual profits and accumulate capital over time. However, as we see, their actions lead to a collective action problem: as all fishers aim for their own optimum harvests, together they overharvest the fish population until it declines to the point where unit costs are high and profit is no longer possible. By virtue of exploring this model, we can understand the causes of massive exploitation of fish stocks, the violation of the ecological viability criterion, and the resulting long-term economic viability problems for two-thirds of the fishers, who are displaced by their now more capitalized and specialized competitors.

Contrasting fisher behavior under the optimizing rule, the gradient decision rule initially exhibits a smooth, steady response to these complex interfisher dynamics. Here, all fishers gradually increase their investment, while some experience increasing profits. This occurs most notably for Fishers 1 and 6, who are most easily able to catch fish due to their high catch efficiencies for one of the given fish species. However, after approximately 7 years in both scenarios, fishers slowly begin to reduce their investment in response to declining or even negative profits, which contribute to major capital declines. After 30 years, most fishers have low investment rates and profits close to zero, implying that they are no longer able to compete in the market. Again, the exceptions to this are Fishers 1 and 6, who are the most specialized in catching one of the two fish species.

Stated in economic terms, these fishers have substantial competitive advantage over the other fishers against whom they are competing. In fact, our simulations show that only these two fishers are able to achieve and maintain capital growth and positive profits. While Fisher 1 initially profits from overfishing y, Fisher 6 is the only agent capable of sustaining capital growth (Figure 8.8) from the larger fish population x (Figure 8.7). This shows that even though we have only defined a single variable delineating differences among the fishing agents, this variable creates significant heterogeneity that has major effects on the system's behavior.

Under the gradient rule, starting from the same priority for both fish species, Fishers 1–4 initially increase priority toward fish y due to high catch efficiency, but only Fishers 1 and 2, being specialized in y, continue this in the long run. Likewise, Fishers 5 and 6 move toward their own specialization and prioritize their investments toward fish x. Fisher 3, having no specialization in harvesting either species, returns to giving equal priority to each of the fish species (Figure 8.8). A similar behavior is reproduced in the interaction among optimizers, even though the changes are more rapid. Although fishers exhibit individual rationality using both decision rules, they deplete both fish populations and thus degrade the economic basis upon which they rely.

"Policy resistance" is the tendency for policy interventions to be defeated by the system's response to the intervention itself (Sterman 2000, 2006). Examples include increasing congestion on highways after lane expansion, overuse of antibiotics that lead to resistant pathogens, and increased workforces that actually lower overall productivity.

Economic Cooperation for Sustainable Fishery Management

Creating Cooperative Institutions and Policies

Classic restrictions on fishing, such as marine zoning or required gear modifications (Pauly et al. 2002; Worm et al. 2009), are often perceived as constraining economic opportunity for the fishing industry and damaging local economies (Eisenack, Scheffran, and Kropp 2006). Management regimes that consider only biological metrics of fish stocks as policy targets have been called "ichthyocentric," indicating problems when scientific input places emphasis entirely on fish stock biomass, lending little effort to examine the

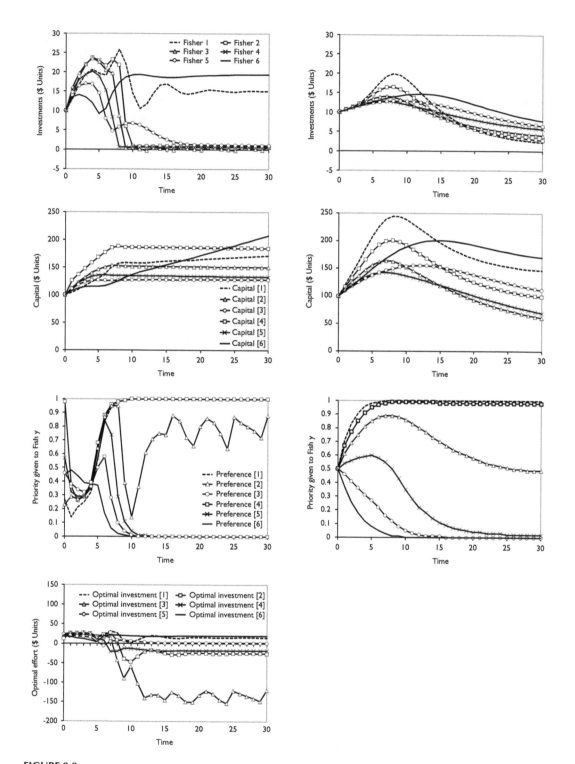

FIGURE 8.8

Investments *C*, Capital *K*, and priorities *r** under competition with agents using *optimizing* (left column) and *gradient* decision rules (right column).

behavior of fishers and their economic necessities (Lane and Stephenson 2000; Davis and Gartside 2001). As a result, fishing firms often act as opponents to management authorities, resulting in *policy resistance* in the form of illegal landings or unreported catches.

One solution to avoid these problems is through co-management schemes (e.g., Jentoft, McCay, and Wilson 1998; Jentoft 2005), whereby the fishing industry is included in policy and decision-making processes. Self-governance within an agreed-upon legal framework is a basic principle of this strategy. By involving fishers, it is assumed that economic imperatives of industry complement the conservation goals of regulatory agencies. This type of strategy aims to have all fishers participate actively with respect to the overall target in order to keep the utilization of marine resources sustainable. Actually, achieving participation is the topic of negotiation, consensus building, and joint fact-finding efforts described in resources, such as work by Susskind, McKearnen, and Thomas-Lamar (1999), and more fully reviewed in Chapter 1. Co-management may be exercised via quasi-governmental organizations such as fishery councils, where the representatives of fishing firms, processing firms, scientific institutions, and regulators negotiate the TAC and other management actions for each different species.

The dynamics considerably change in our model when fishers draw on co-management structures that promote cooperation or coalition formation as a way to maintain a sustainable limit for catch efforts. As we saw in the last section, under unfettered, competitive market structures, fishers define their investments based on individual evaluation and profit maximization goals, leading to overfishing (e.g., a "race to fish"), and the subsequent ecological and economic decline seen in fisheries around the world. This situation contracts with a cooperative case, where fishers have, in theory, engaged in a dispute resolution process to collectively define their capabilities and goals for the system as a whole (e.g., the type of negotiated conflict resolution discussed in Susskind, Levy, and Thomas-Larmer 2000). In the cooperative case, we can model the implementation of a negotiated sustainable collective target, which can be used to determine individual harvest and investments rates based on negotiated distribution rules.

The design, implementation, and enforcement of these allocation mechanisms, which diverge from the individually preferred decision rules, require effective institutional procedures that provide either incentives to join (or penalties in case of violation of) an agreed-upon or imposed framework. Elinor Ostrom's (1990, 2009) Nobel Prize-winning work on social institutions has been key to understanding how institutions can successfully create collaborative frameworks for implementing consensus resolutions. A critical factor is the selection of distribution rules that are agreeable, verifiable, and enforceable for all fishers involved in the coalition. Rules that attract opposition by major fishers or that cannot be monitored and verified are less likely to be successful (Gutierrez, Hilborn, and Defeo 2011).

Reconsidering Investment and Competitive Advantages

We begin to modify the model by defining the *effective investment* E_i^k of fisher i devoted to fish species k as $E_i^k = q_i^k r_i^k C_i$, which is the product of the catch efficiency q_i^k and the investment $C_i^k = r_i^k C_i$ of each fisher for each fish species; $E^k = \sum_i E_i^k$ is taken as the joint effective investment of all fishers i with regard to fish k. Effective investment not only measures the investment that a fisher invests into fish catch but also the sensitivity of the fish population to this investment, which is mediated by fisher efficiency. Combined, these two factors determine the total impact on fish k in terms of the harvested fraction of fish stock

h_i^k / x^k induced by fisher i. Thus, efforts to control effective investment can be closely related to quotas of fish catch.

In building the coalition that will function together and jointly assess the collective effort to sustainably harvest fish, the catch efficiencies q_i^k of all of the fishers involved must be known to the group. While catch efficiencies were each fishers' private information in the competitive cases, in this cooperative scenario, globally known catch efficiencies are key to determining the possible joint effective investment that can be allocated as well as the optimal allocation of sustainable investment rates among fishers and species.

Next, we can define the steady-state (equilibrium) condition $\Delta x^k = 0$ (where fish stocks are neither growing nor declining) of effective investment E^{k*} for fish stock x^k as $E^{k*} = h^k / x^k = g^k (L^k - x^k)$, as was calculated under the competitive scenarios. If all agents jointly aim to restrain their effective investment effort to within the sustainable limit E^{k*}, given by the fishery ecosystem conditions, one way of adapting effective investment effort over time is

$$\Delta E^k = \gamma^k (E^{k*} - E^k)$$

where γ^k is the rate of adaptation of E^k toward the sustainable limit E^{k*} that restrains both the amount of effort and the catch efficiency. This rate of adaptation, which is analogous to α^k in the competitive scenarios (γ^k takes a value $0 \leq \gamma^k \leq 1$), is the speed that the *collective group* adapts its joint effective investment rate toward the sustainable goal E^{k*}.

To consider the broader system structure for a moment, we can think of E^k as the joint effective effort invested by all fishermen; rather than acting according to independent decision rules for changing their individual investment effort E_i^k, which created the problems of overfishing described previously. A potential solution is to treat the group of fishers as a collective agent that merges their individual effective effort to E^k and then adjust it to their sustainable limit for efforts E^{k*} according to the adaptation rule ΔE^k. This implies that through their coordinated collective action, they will not exceed the ecological limit (which serves as a quota of sustainable effective effort for each fish type), and they will also not be chased out of the industry as a result of their lowered investment rates.

At this point, an important question comes to the surface: which share, denoted by φ_i^k, of the changing joint effective effort ΔE^k should be assigned to each fisher i for each fish species k? In other words, how much should each fisher individually contribute to the changes of joint investment effort by all fishers at each point in time?

Negotiations and Cooperative Fishing Arrangements

The answer to this question depends on distribution rules that are established during negotiation processes (Susskind, Levy, and Thomas-Larmer 2000). In many cases of managed fisheries, quotas have been distributed on the basis of historical catch, which can be interpreted as a measure of fisher capital acquired in the process of harvesting fish (e.g., the biggest boats get the highest quotas) and reflects the level of investment corresponding to a system of ITQs. Another plausible distribution rule would be to assign every fisher a fraction of the total change of effective investment that is proportionate to their actual effective effort that they are capable of expending within the group:

$$\varphi_i^k (t) = \frac{E_i^k (t)}{E^k (t)}$$

This distribution rule, which is one of the many possible negotiated distribution designs, translates into a decision rule for the effective effort of each fisher:

$$\Delta E_i^k = \varphi_i^k \Delta E^k$$

This rule implies that fishers with higher effective efforts are allocated a higher share of the increases or decreases in the TAC of the entire group (BenDor, Scheffran, and Hannon 2009). Differently stated: under this approach, larger fishing operations are effectively granted a higher share of the sustainable fishing limit.

While this is not the only plausible rule for distributing harvest share among fishers, it is a simple example of a "fair" distribution rule, which allocates quotas based on expertise and actual investment. Another plausible rule, implemented by Garrity (2011), assumes an equal distribution of the total catch among fishers. As mentioned earlier, limits on effective effort can be directly translated into quota on fish catch and thus connected to an ITQ framework. We explore analogous allocation issues with regard to carbon trading more in the next chapter.

The initial allocations in an ITQ system are often highly contentious, as are the initial allocation of shares under any cap-and-trade framework (Ellerman and Buchner 2007; Ellerman, Convery, and Perthuis 2010). As described in many case studies in Shotton (2001), establishing an ITQ program or other comanagement system absolutely necessitates the presence of well-developed participatory processes that involve fishers in the initial allocation process. In the cooperative scenario that we describe in this chapter, the allocation priorities r_i^k that are integral to the competitive scenarios are transformed into the product of the distribution mechanism created by a negotiation process. As a result, each fisher cannot freely and dynamically choose these priorities, as they did in the competitive case. While ITQ systems distribute shares of the output (harvest), the system structure modeled here instead distributes shares of the input (effective effort).

In the cooperative scenario, fisher value functions are unchanged from the competitive scenarios, and the net values (profits), $V = ph - C$, for each fisher i accrue from selling the fish catch (harvest) h_i at a market price p, which is diminished by the investment C associated with catching fish. Now, considering that each fisher i no longer internally determines their own allocation priorities r_i^k for dividing their total investment cost among each fish species k, the former representation of harvest rate, where every fisher decides individually on their investment effort, changes from $h_i^k = C_i^k / c_i^k$ for individual action to $h^k = \sum_i h_i^k = \sum_i C_i^k / c_i^k$ for a collective group harvest, where c_i^k is the unit cost and C_i^k is the investment allocated to fisher i to catch fish of species k.

Assuming a fishery coalition agrees on a total harvest quota h^k, the challenge is to find the *allowable* investment C_i^k or investment quota for each fisher i regarding a given species k. Changes in these investment allocation priorities ΔC_i^k for each fisher, and each fish species, are determined jointly as $\Delta C_i^k = (\Delta E^k \varphi_i^k)/q_i^k$, which is a function of the total joint effective investment, the allocation share φ_i^k, and the catch efficiency of each fisher for each fish species. We will assume that the share of the joint effective effort apportioned to each fisher, and each species, is calculated as $\varphi_i^k(t) = E_i^k(t)/E^k(t)$. This means that the share allocated to each fisher is proportional to the effective investment of each fisher, divided by the effective investment for the whole group. That is, the more a fisher invested in the beginning, and the more efficient their investment was, the more a fisher can increase effort in the next time step.

Finally, as stated earlier, the joint effective effort of this coalition of fishers is constantly seeking the sustainable goal $E^{k^*} = g^k(L^k - x^k)$, through the goal-seeking function that we defined earlier, where changes in the group's joint effective effort occur gradually, $\Delta E^k = \gamma^k(E^{k^*} - E^k)$.

Simulating a Cooperative Fishing Scenario

The SD articulation of this model is shown in Figure 8.9. When implemented in NetLogo, the dynamic simulation in the cooperative case completely differs from the competitive cases: both fish populations stabilize at considerably higher levels. This allows for greater sustained harvests and profits, which are distributed in proportion to investment, thus sustaining capital growth for all fishers (Figure 8.10).

The price for fish y again increases by almost a factor of two, while the priority for fish x—now the same for all agents—increases compared to y. When compared with the competitive scenarios, we see that the price for x is comparable to the price seen in both competitive scenarios, while the price for y is consistently lower. This is due to the stabilization of the fish y population at much higher levels due to sustainable cooperation, which avoids the price effects of fish y scarcity.

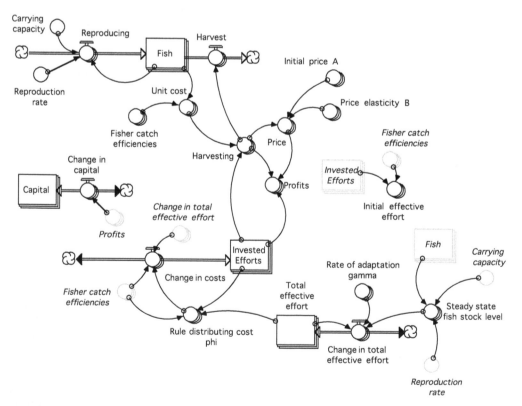

FIGURE 8.9
A system dynamics stock-flow diagram representing cooperation among fishers.

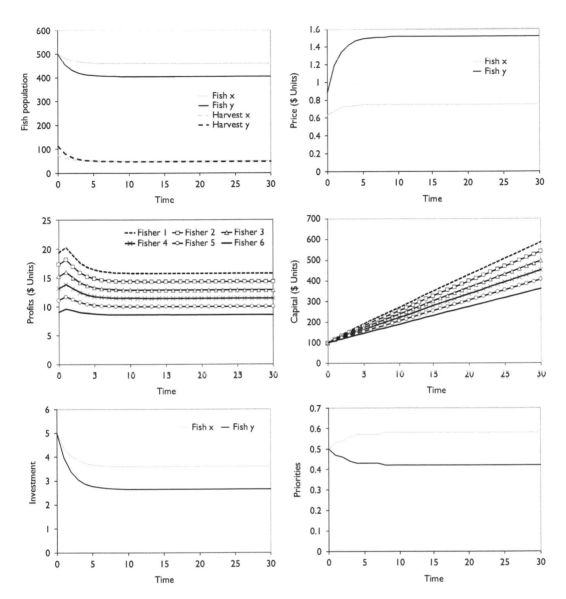

FIGURE 8.10

Fish populations x^k, harvest rates h^k, fish price p^k, agent profits V, capital K, investments C, and priorities r^k under a scenario where agents are in a cooperative agreement.

Summary

In this chapter, we aimed to model the underlying reasons for the tragedy of the commons that characterizes overfishing, which has been observed empirically as the cause of decline and conflict in many of the world's fisheries (Roughgarden and Smith 1996). We have discussed the underlying reasons for overfishing and have applied the VIABLE agent-based conflict modeling approach to the assessment of fishery co-management regimes, while at the same time demonstrating a method for modeling heterogeneous agents that change their behavior and investment patterns over time.

The agent-based capabilities of the VIABLE framework create a straightforward modeling approach that incorporates some of the complexity found in real fisheries. Unlike aggregate modeling techniques, such as SD (Otto and Struben 2004; Moxnes 2005; Dudley 2008), we can explicitly represent agent heterogeneity to explore competitive advantages in fisheries, demonstrating ways in which ecological and economic well-being of fishery systems can suffer at the expense of competitive efficiency. We have explored how heterogeneity in catch efficiency is important among the fishers because it can critically determine who survives the competition, as well as who can gain access to catch shares under co-management strategies.

We also explored two competitive decision rules dictating how fishers formulate fishery investment strategies. We found a tendency for fishers with technological specialization in catching a certain fish species to over-exploit that species while driving other fishers out of business. This creates monopolies for each of the two fish species, whose populations only stabilize at substantially smaller populations. This pattern of overexploitation supports fewer fishers and leads to economic decline and instability, which cannot be compensated for by either more investment or more efficient fishing techniques (i.e., better technology).

In a purely competitive fishery environment, ecological and socioeconomic sustainability can suffer, leading to major environmental damage and conflict. However, our analysis clearly showed that cooperation and sustainability can complement and strengthen each other. In a cooperative structure, we found that limiting investments in specific ways can prevent the negative side effects of uncontrolled competitive interactions between fishers that tend to degrade both the ecological and the economic system.

By testing a cooperative mechanism that serves the viability of both the ecological and the socioeconomic systems, we found vastly different results in terms of sustainability and market stability. Here, our cooperative market structure requires a mechanism for setting catch targets and distributing the right to invest in actually catching fish. Creating and implementing such a mechanism requires institutional procedures, including negotiation frameworks, management authorities, and input from scientific institutions (Eisenack, Scheffran, and Kropp 2006).

Fisheries around the world are still facing serious ecological and economic challenges, with ongoing debate over how to implement adequate, sustainable management strategies. With no "silver bullet" solutions available, there still exists the need for integrated assessments in fisheries around the world (Eisenack, Scheffran, and Kropp 2006). Likewise, developing adaptive management frameworks that allow for social learning will be essential to overcome the tragedy of the commons in fisheries around the world.

Involvement of key stakeholders is as important as incorporating the best available scientific information about the complex interactions and the data that feed into models. Sustainable market structures must be implemented along with policy initiatives that balance the interests of individual fishers and support negotiation of quotas within sustainability limits. However, understanding the economic realities of fisheries is essential to design effective incentive systems; any policy decisions will require fisher "buy-in" for creating a successful policy framework. Here, buy-in implies that fishers have been included in the policy-making process to an extent that they feel a sense of ownership of the policy and will accept—and possibly even advocate for—final policy decisions. We discuss this concept at length in Chapter 4.

Appendix 8.1: Ecological and Economic Viability Conditions

The ecological viability condition $\Delta x > 0$ can be expressed as a linearly declining function $\bar{C}(x)$ that relates investment costs C to the fish population x:

$$\text{Ecological viability condition} \quad C \le \frac{g(L-x)}{q} = \bar{C}(x) \tag{8.2}$$

The economic viability condition, with profit $V = ph - C > 0$, where price $p = a - bh$ and harvest $h = qxC$, leads to a quadratic equation that describes V in terms of harvest or cost: $V = h(a - bh) - h/(qx) = qxC(a - bqxC) - C$. This can be resolved into another boundary condition $\hat{C}(x)$ for investment:

$$\text{Economic viability condition} \quad C \le \frac{aqx - 1}{bq^2 x^2} = \hat{C}(x) \tag{8.3}$$

Appendix 8.2: More Details on the Multi-Agent Fishery Model

The fishers' value functions are the net profits from harvest h_i^k of fish $k = 1, ..., m$ for a market price p^k, diminished by their investments C_i. Due to the linear price–supply function, we can re-state fisher value as a function of harvest and cost of all fishers, where

$$V_i = \sum_k p^k h_i^k - C_i = \left(u_i - \sum_j w_{ij} C_j \right) C_i$$

$$u_i = \left[\sum_k \frac{a^k}{c_i^k} r_i^k \right] - 1 \quad w_{ij} = \left[\sum_k b^k \frac{r_i^k r_j^k}{c_i^k c_j^k} \right] \tag{8.4}$$

Here, ij represents the mutual couplings between investment of fisher j and value of fisher i, as articulated by the priorities and unit costs of both fishers, while u_i indicates the direct positive marginal value (benefit) of each fisher's investments. Due to the now-quadratic form of the value function, the partial derivative $v_{C_i} = \partial V_i / \partial C_i$ is a linear function of the investments of all fishers, and the decision rules $\Delta C_i = \beta_i D_i(V_i)$ are linear difference equations describing this adaptation of investments.

Optimizers

In the two decision rules we design, we have an optimization case (a global optimizer) and a satisficing case (local optimizer; "gradient" method). In the optimization case, we will take the derivative of the value function (Equation 8.4) and find the global maximum:

$$v_{r_i}^k = \frac{\partial V_i}{\partial r_i} = \left(\frac{\partial u_i}{r_i^k} - \sum_j \frac{\partial w_{ij}}{r_i^k} C_j \right) C_i = 0 \tag{8.5}$$

Since the general case of $k = 1, \ldots, m$ fish species is quite complicated, we limit the analysis here to $k = 2$ fish species. In this case, we define $r_i^1 \equiv r_i$ and $r_i^2 \equiv 1 - r_i$ and, from Equation 8.4 obtain

$$u_i = \frac{a^1}{c_i^1} r_i + \frac{a^2}{c_i^2}(1 - r_i) - 1$$

$$w_{ij} = b^1 \frac{r_i r_j}{c_i^1 c_j^1} + b^2 \frac{(1 - r_i)(1 - r_j)}{c_i^2 c_j^2}$$

Therefore,

$$\frac{\partial u_i}{\partial r_i} = \frac{a^1}{c_i^1} - \frac{a^2}{c_i^2}$$

$$\frac{\partial w_{ij}}{\partial r_i} = b^1 \frac{r_j}{c_i^1 c_j^1} - b^2 \frac{(1 - r_j)}{c_i^2 c_j^2}$$

$$\frac{\partial w_{ii}}{\partial r_i} = 2b^1 \frac{r_i}{c_i^1 c_i^1} - 2b^2 \frac{(1 - r_i)}{c_i^2 c_i^2}$$

which leads to (according to equation 8.4):

$$\frac{\partial V_i}{\partial r_i} = \left(\frac{\partial u_i}{r_i} - \sum_j \frac{\partial w_{ij}}{r_i} C_j \right) C_i = \left\{ \frac{a^1}{c_i^1} - \frac{a^2}{c_i^2} - \sum_{j \neq i} \left[\left(b^1 \frac{r_j}{c_i^1 c_j^1} - b^2 \frac{(1 - r_j)}{c_i^2 c_j^2} \right) C_j \right] \right.$$

$$\left. - \left(2b^1 \frac{r_i}{c_i^1 c_i^1} - 2b^2 \frac{(1 - r_i)}{c_i^2 c_i^2} \right) C_i \right\} C_i = 0.$$

This is a linear equation in r_i and can be resolved for:

$$r_i = \frac{z_i + \sum_j m_{ij}^2 C_j - \sum_{j \neq i} \left(m_{ij}^1 + m_{ij}^2 \right) r_j C_j}{\left(m_{ii}^1 + m_{ii}^2 \right) C_i} = r_i^*$$

where

$$z_i = \frac{a^1}{c_i^1} - \frac{a^2}{c_i^2}$$

$$m_{ij}^1 = \frac{2b^1}{c_i^1 c_j^1}, \quad m_{ij}^2 = \frac{2b^2}{c_i^2 c_j^2}$$

Optimizers seek this global value of r_i^* as determined by the rate α, such that changes to priority for a given species takes on the form of

$$\Delta r_i^k = \alpha\left(r_i^{k*} - r_i^k\right)$$

Here, the agent is simply closing the gap between their current priority and their target priority.

The interaction matrix F^C and its stability—described in Chapter 7 and Appendix 7.4—is determined by partial derivatives as a function of C and r:

$$f_{ii}^C = \frac{\partial V_i}{\partial C_i} = u_i - \sum_j w_{ij}C_j - w_{ii}C_i \text{ and } f_{ij}^C = \frac{\partial V_i}{\partial C_j} = -w_{ij}C_i < 0$$

Setting $\partial V_i / \partial C_i = 0$ leads to the optimal values V_i^* and the related optimal costs C_i^* as sought by each agent:

$$C_i^* = \frac{u_i - \sum_{j \neq i} w_{ij}C_j}{2w_{ii}}$$

where changes to investment are given as $\Delta C_i = \beta\left(C_i^* - C_i\right)$

Satisficers (Gradient Decision Rule)

For the gradient decision rule, or satisficing approach, we can represent the change in priority Δr^k as linear functions of the priorities of all fishers ($i = 1, ..., n$) and the rates of change (partial derivatives) of each fisher's value V_i, with respect to its priorities for different fish types, $v_{r_i}^k = \partial V_i / \partial r_i^k$. This leads to a decision rule that is a nonlinear difference equation of these priorities for all fishers:

$$\Delta r_i^k = \alpha_i^k r_i^k \left(v_{r_i}^k - \sum_l r_i^l v_{r_i}^l \right)$$

where, as previously stated earlier in Equation 8.5:

$$v_{r_i}^k = \frac{\partial V_i}{\partial r_i^k} = \left(\frac{\partial u_i}{r_i^k} - \sum_j \frac{\partial w_{ij}}{r_i^k} C_j \right) C_i \tag{8.6}$$

Given that we are not globally optimizing, in this case, we simply find $v_{r_i}^k = \partial V_i / \partial r_i^k$ for Equation 8.6 using Equation 8.4:

$$v_{r_i}^k = \frac{\partial V_i}{\partial r_i^k} = \left(\frac{a^k}{c_i^k} - \sum_{j \neq i} \left[b^k \frac{r_j^k C_j}{c_j^k c_i^k} \right] - \frac{b^k 2 r_i^k}{(c_i^k)^2} C_i \right) C_i = \left(\frac{a^k}{c_i^k} - \sum_j \left[\frac{b^k r_j^k}{c_j^k c_i^k} C_j \right] - \frac{b^k r_i^k}{(c_i^k)^2} C_i \right) C_i$$

In this case, those agents with the highest marginal value $v_{r_i}^k$ have the highest value growth rate, while the values of those agents with below average marginal values experience decline (Scheffran 2000; Eisenack, Scheffran, and Kropp 2006). Changes to the investment rate are given as $\Delta C_i = \beta(\partial V / \partial C) = \beta v_C$.

Questions for Consideration

1. In the last 30 years, overfishing has caused fishery collapses around the world, including the Alaska pollock (*Theragra chaleogramma*) fishery, which is the world's largest for human consumption. *Check out Kevin Bailey's (2011) fascinating, firsthand account of this fishery collapse. Why did this fishery collapse? What could have been done? What efforts were made to prevent it?*

2. We have mentioned several times the immense importance of negotiation processes between stakeholders in this system, such as fishers. The ability of fishery systems to remain on viable system trajectories depends partly on the ability of agents to individually or collectively adapt to conditions that they find acceptable and even desirable (Susskind, Levy, and Thomas-Larmer 2000). *What other relationships—among fisher agents, nonfisher agents, or exogenous actions—could affect fishery viability? What else could you theorize would be economic to connect agents to each other and to fish stocks? What are other types of viability conditions for individual agents that we have not discussed? What other types of agents exist in fishery systems?*

3. Although long-run sustainability is a primary concern of management agencies and the public, Garrity (2011) has argued that ITQ programs can give fishers continued incentives to compete and remain profitable, elevating short-term goals to the forefront of their decision making. *Find and read Garrity's paper (open-access journal). What does he mean by short-term agent goals vs. long-term fishery goals? How do you think this issue could be resolved in ITQ designs? What would need to happen?*

4. We referred to "sustainability" within our broader discussion of "viability" throughout this chapter and the last. *What is the difference between "sustainability" and "viability?" What are the assumptions made by each approach or paradigm?*

5. Competing fishing firms base their decisions on their available capital, the required effort to catch fish, and on the observed state of their target species; this is important since catch is commonly used as an estimator for the abundance of fish biomass. *What are other, perhaps better metrics for determining proper catch?*

6. Costello et al. (2010) has estimated that ITQs have been implemented in nearly 150 fisheries around the world, including fisheries for Bering Sea crabs, British Columbia halibut, mid-Atlantic surf clams, many New Zealand fisheries, and others. Find and read his paper. How did some of these quotas work? Did they have similar rules or arrangements? *Find and read Costello et al.'s (2010) paper. How did some of these quotas work? Did they have similar rules or arrangements? Find other literature tracking ITQ programs through resources such as the U.S. National Marine Fisheries Service (NMFS 2018) Catch Share Program (within its Office of Science and Technology). What are the trends in ITQ policies, designs, and implementations? How are they changing over time?*

7. **There are many ways to improve this model**. Future work on this model could go in a number of directions, potentially incorporating capabilities to consider additional fish species, multiple types of fishing gear, technological development, various forms of uncertainty, and political factors (Dudley 2008). One idea is that we could make this model more sophisticated to address the effects of climate change and invasive species introduction. We could do this by explicitly representing non-fishing-related fish deaths rates as $d^k \cdot x^k$, where d^k is either a crude death rate or

a complex function that determines death rate. We also assume that the fish stock is not affected by migration patterns (e.g., there is a balance between movements into and out of the area of concern).

Evolving Fishing Technology

Model modifications could also focus on interesting trade-offs; *for example, how can agent-based conflict modeling address situations where limits on catch investment are offset by improved fishing technology?* In theory, technological improvements could increase effective investment efforts and create counter-productive behavior. There are other behavior fishery models, work by including Dercole, Prieu, and Rinaldi (2010) and Dudley (2008), which include evolving fishing technology, a factor that can vastly increase model complexity. Dercole, Prieu, and Rinaldi (2010) justify the adaptive technological features of their model by referring to the impressive improvements in fishing technology that occurred over the last several decades (Salthaug 2001). In their approach, technological changes in fishing are the endogenous result of innovation and competition processes.

This contrasts our approach, where technology does not evolve at all[5] and with bioeconomic modeling techniques, where fishing fleet technology increases at an exogenously determined rate.

Transferable Allocations

Reducing the problems of overfishing requires the ability to control fishing effort (Worm et al. 2009). However, monitoring and controlling dispersed fishing activity is difficult, and finding efficient mechanisms to mitigate the adverse impacts of overfishing on fisheries and economies remains a major challenge. Model expansions could also explore solutions that involve ITQ frameworks, which can be helpful for reducing fishing capacity, effort, and exploitation rates.

However, fishery scientists are increasingly recognizing that ITQs may not be the panacea for overfishing, and instead have begun recommending a combination of incentive- and non incentive-based approaches to mitigate the impacts of fishing on ecosystems (Arnason and Gissurarson 2017; Costello et al. 2010). These approaches include the use of marine-protected areas (i.e., no fishing zones), incentives to reduce by-catch (e.g., dolphins and other accidentally caught species), and restrictions on fishing gear (Worm et al. 2009).

Try out some of these changes; start by going to this: book's website—http://todd.bendor. org/viable—and tinker around with the model. What other ideas do you have to modify the model? We invite you to contact the authors (http://todd.bendor.org/contact) to forward along your modifications so that we can add them to this website so others can learn from your work!

---

[5] In our case, the fishing effort is adjusted to produce economic optimization.

Additional Resources

Interested in fishery declines or collapse? In addition to Worm et al.'s (2006) landmark study about fish population loss, in another study of predatory fish communities, Myers and Worm (2003) estimated that large predatory fish biomass today is less than 10% that of preindustrial levels, with industrialized fisheries reducing biomass an average of 80% within 15 years of initial exploitation. In two well-known examples, devastating effects were seen in Monterey, California's cannery row after the Pacific sardine's (*Sardinops sagax*) collapse in the 1940s (Mangelsdorf 1986), as well as local economic stress created by the decline of the Peruvian anchoveta (*Engraulis ringens*) in the 1970s (Glantz 1979). Bailey (2011) also points out that irregular periods of abundance of the Bohuslän herring have led to cycles of feast and famine in the nearby Swedish coast over the last 500 years (Smith 1994).

Interested in learning more about fishery modeling? Mathematical modeling has been used to understand fisheries for more than 170 years (Verhulst 1838). Numerous SD studies (Ruth 1995; Yndestad and Stene 2002; Otto and Struben 2004; Moxnes 2005; Arquitt and Johnstone 2008; Dudley 2008; Garrity 2011; Stouten et al. 2007) have attempted to address a variety of fisheries questions and problems. We also provide links to more than a dozen additional studies in this chapter's references.

Perhaps the most well-known dynamic fishery model is *Fishbanks*, which was originally created as a board game (Meadows, Fiddaman, and Shannon 1986). Today, Fishbanks can be played online, and is described by its host as "a multiplayer web-based simulation in which participants play the role of fishers and seek to maximize their net worth as they compete against other players and deal with variations in fish stocks and their catch" (see Meadows, Sterman, and King 2018).

We should note that ideally, a fishery model should capture the evolution of fish stocks by completely modeling the interactions between harvesting and biological renewal ("recruitment" in biological terms) of the population (we do not do this in this chapter). Yet, in practice, stakeholders may not be able to agree on either the structure or the impact of these dynamics (Otto and Struben 2004). Although it is difficult to identify the exact critical thresholds and sensitivities of particular fisheries, like other conflict modeling efforts, our goal is to understand the dynamic, nonlinear feedback structure of this fishery system. If we sought enough stakeholder input, and were so inclined, this structure could include the complex socioeconomic interactions between the community and the fishers, as well as more complex structural considerations around the fish population.

Among our additional simplifying assumptions that could be relaxed or explored include our omission of non-fishing-related death rates, as well as our exclusion of any interactions between fish species. We did this because we were more interested in focusing on the interaction among fisher agents, but also because of the complexity that emerges when accounting for how separate fish species might compete for resources (known as "interspecific competition" in ecology; Larkin 1956). Some species even have predator–prey interactions with each other. However, Brewer (2011) argues that it is imperative to consider multiple species in policy making and quota design, otherwise attention is diverted away from broader ecosystem considerations, including habitat, migratory patterns, and trophic relationships. More sophisticated models, such as that by Armstrong and Sumaila (2000), even explore impacts of factors like fish "cannibalism" (e.g., fish eating their own species) on fish population viability. These are all issues that are open for exploration!

Interested in learning more about the NetLogo agent-based modeling platform? We implement this book's models in the NetLogo agent-based simulation package. NetLogo is a widely used platform with strong support documentation, a large user base, and it is freely available for most desktop computer operating systems, including Microsoft Windows, Apple, and Linux. NetLogo models are actually programmed in the NetLogo modeling language, which builds on the older "Logo" language and is a variant of "StarLogo." The NetLogo language is an efficient tool for our purpose, because its "vocabulary" of commands is specifically designed for agent-based model creation. For example, many commonly used modeling elements are prebuilt, eliminating the need to code them from scratch. In comparison to general-purpose languages such as C++ or Java, it is much easier to write agent-based simulation modeling code in NetLogo. A relatively short bit of code in a NetLogo model can efficiently create powerful simulation models. All details on the NetLogo language and platform can be found at the platform's website (http://ccl. northwestern.edu/netlogo/).

NetLogo is also an advantageous modeling platform because it affords a high level of accessibility and a shallow learning curve for novice modelers. This platform also supports a wide array of powerful tools for researchers who wish to extend NetLogo output by linking it to a geographic information system, statistical analysis software, or additional specialized modeling environments. This includes "BehaviorSpace," a feature (accessible via the Tools menu) that allows NetLogo models to run multiple simulation experiments (e.g., changing parameter values to compare alternate scenarios) concurrently on multiple processors, which are now ubiquitous.

Although NetLogo models can be rapidly developed, the NetLogo system does not support the enormous data sets or agent populations often necessary to fully implement some models. In response, an on-going project, known as ReLogo, has extended NetLogo by providing a method for importing NetLogo models into Repast (Ozik et al. 2013), an agent-based modeling platform that, while much more sophisticated for advanced research applications, is also much more complex and requires JAVA programming expertise (Lytinen and Railsback 2012).

The NetLogo interface is straightforward, with three work areas organized into tabs at the top of the screen, including an *Interface* tab, an *Information* tab, and a *Procedures* tab. When beginning model development (as well as parameterizing, running, and experimenting with existing models), it is often useful to start with the Interface tab, which allows users to setup user-defined parameters and visualizations of the model. The *Information* tab provides an area for storing "metadata" i.e., (data about data) about the model, including model documentation (very useful when using NetLogo's sample models), running instructions, limitations, and caveats. Finally, the *Procedures* tab contains the actual modeling workspace, where the model's source code is actually developed and stored. The *Procedures* tab is simply a specialized text editor designed for writing and revising code in the NetLogo programming language.

While we use NetLogo throughout Part III of this book, we refer interested readers with no previous NetLogo experience to Stigberg's (2012) in-depth introduction to the NetLogo modeling environment, as well as Railsback and Grimm's (2011) and Wilensky and Rand's (2015) comprehensive expositions and guides to the NetLogo language and use.

We recommend that users wishing to gain familiarity with NetLogo and agent-based simulations begin by experimenting with the large and highly instructive library of sample models that is included with the software, which can be downloaded (https://ccl. northwestern.edu/netlogo/) or used in a web browser (www.netlogoweb.org/). Users can tinker with sample models at their leisure or follow Netlogo's established web tutorials to immediately construct their first model.

Resources for Learning Netlogo Modeling Software

Railsback, Steven F., and Volker Grimm. 2011. *Agent-Based and Individual-Based Modeling: A Practical Introduction*. Princeton, NJ: Princeton University Press.

   Wilensky, Uri, and William Rand. 2015. *An Introduction to Agent-Based Modeling: Modeling Natural, Social, and Engineered Complex Systems with NetLogo*. Cambridge, MA: MIT Press.

References

Anderies, John M., Marco A. Janssen, and Elinor Ostrom. 2004. "A Framework to Analyze the Robustness of Social-Ecological Systems from an Institutional Perspective." *Ecology and Society* 9 (1): 18.

Ariosto, David. 2013. "Historic Cod Fishing Cuts Threaten Centuries-Old Industry in New England," April. Atlanta, GA: Cable News Network. www.cnn.com/2013/01/31/us/northeast-cod-fishing-cuts

Armstrong, Claire W., and Ussif Rashid Sumaila. 2000. "Cannibalism and the Optimal Sharing of the North-East Atlantic Cod Stock: A Bioeconomic Model." *Journal of Bioeconomics* 2: 99–115.

Armstrong, Claire W., and Ussif Rashid Sumaila. 2001. "Optimal Allocation of TAC and the Implications of Implementing an ITQ Management System for the North-East Arctic Cod." *Land Economics* 77 (3): 350–59.

Arnason, Ragnar, and Hannes H. Gissurarson, eds. 2017. *Individual Transferable Quotas in Theory and Practice*. Reykjavik, Iceland: University of Iceland Press.

Arquitt, Steve, and Ron Johnstone. 2008. "Use of System Dynamics Modelling in Design of an Environmental Restoration Banking Institution." *Ecological Economics* 65 (1): 63–75.

Arreguín-Sánchez, Francisco. 1996. "Catchability: A Key Parameter for Fish Stock Assessment." *Reviews in Fish Biology and Fisheries* 6 (2): 221–42.

Aubin, Jean-Pierre. 1991. *Viability Theory*. Berlin, Germany: Birkhäuser.

Aubin, Jean-Pierre, Alexandre M. Bayen, and Patrick Saint-Pierre. 2011. *Viability Theory: New Horizons* (2nd Edition). Berlin, Germany: Springer-Verlag.

Aubin, Jean-Pierre, and Patrick Saint-Pierre. 2007. "An Introduction to Viability Theory and Management of Renewable Resources." In *Advanced Methods for Decision Making and Risk Management in Sustainability Science*, edited by Jürgen P. Kropp and Jürgen Scheffran, 56–95. New York: Nova Science Publishers.

Bailey, Kevin M. 2011. "An Empty Donut Hole: The Great Collapse of a North American Fishery." *Ecology and Society* 16 (2): 28.

BenDor, Todd, Jürgen Scheffran, and Bruce Hannon. 2009. "Ecological and Economic Sustainability in Fishery Management: A Multi-Agent Model for Understanding Competition and Cooperation." *Ecological Economics* 68 (4): 1061–73.

Bennett, Elizabeth, Arthur Neiland, Emilia Anang, Paul Bannerman, A. Atiq Rahman, Saleemul Huq, Shajahan Bhuiya, Mark Day, Michelle Fulford-Gardiner, and Wesley Clerveaux. 2001. "Towards a Better Understanding of Conflict Management in Tropical Fisheries: Evidence from Ghana, Bangladesh and the Caribbean." *Marine Policy* 25 (5): 365–76.

Berck, Peter, and Jeffrey M. Perloff. 1984. "An Open-Access Fishery with Rational Expectations." *Econometrica* 25 (2): 489–506.

Berke, Philip R., David R. Godschalk, Edward J. Kaiser, and Daniel A. Rodriguez. 2006. *Urban Land Use Planning* (5th Edition). Urbana, IL: University of Illinois Press.

Branch, Trevor A. 2008. "Not All Fisheries Will Be Collapsed in 2048." *Marine Policy* 32: 38–39.

Brewer, Jennifer F. 2011. "Paper Fish and Policy Conflict: Catch Shares and Ecosystem-Based Management in Maine's Groundfishery." *Ecology and Society* 16 (1): 15.

Bruckner, Thomas, Georg Hooss, Hans-Martin Fussel, and Klaus Hasselmann. 2003. "Climate System Modeling in the Framework of the Tolerable Windows Approach: The ICLIPS Climate Model." *Climactic Change* 56: 119–37.

Bruckner, Thomas, G. Petschel-Held, F. L. Tóth, H. M. Füssel, C. Helm, M. Leimbach, and H. J. Schellnhuber. 1999. "Climate Change Decision-Support and the Tolerable Windows Approach." *Environmental Modeling and Assessment* 4 (4): 217–34.

Campbell, Scott. 1996. "Green Cities, Growing Cities, Just Cities: Urban Planning and the Contradictions of Sustainable Development." *Journal of the American Planning Association* 62 (3): 296–312.

Castillo, Daniel, and Ali Kerem Saysel. 2005. "Simulation of Common Pool Resource Field Experiments: A Behavioral Model of Collective Action." *Ecological Economics* 55 (3): 420–36.

CEMARE. 2002. "The Management of Conflict in Tropical Fisheries, Final Technical Report." Portsmouth, England: Centre for the Economics and Management of Aquatic Resources. www.gov.uk/dfid-research-outputs/the-management-of-conflict-in-tropical-fisheries-final-technical-report

Charles, A. T. 1992. "Fishery Conflicts: A Unified Framework." *Marine Policy* 16 (5): 379–93.

Chu, C. 2008. "Thirty Years Later: The Global Growth of ITQs and Their Influence on Stock Status in Marine Fisheries." *Fish and Fisheries* 10: 1–14.

Copes, Parzival. 1970. "The Backward-Bending Supply Curve of the Fishing Industry." *Scottish Journal of Political Economy* 17 (1): 69–77.

Costello, Christopher, Steven D. Gaines, and John Lynham. 2008. "Can Catch Shares Prevent Fisheries Collapse?" *Science* 321 (5896): 1678–81.

Costello, Christopher, John Lynham, Sarah E. Lester, and Steven D. Gaines. 2010. "Economic Incentives and Global Fisheries Sustainability." *Annual Review of Resource Economics* 2 (1): 299–318.

Cury, Philippe M., Christian Mullon, Serge M. Garcia, and Lynne J. Shannon. 2005. "Viability Theory for an Ecosystem Approach to Fisheries." *ICES Journal of Marine Science* 62 (3): 577–84.

Daily, Gretchen C. 1997. *Nature's Services: Societal Dependence on Natural Resources.* Washington, DC: Island Press.

Daly, Herman E. 1991. *Steady-State Economics: Second Edition With New Essays.* Washington, DC: Island Press.

Davis, D., and D. Gartside. 2001. "Challenges for Economic Policy in Sustainable Management of Marine Natural Resources." *Ecological Economics* 36: 223–36.

Dercole, Fabio, Charlotte Prieu, and Sergio Rinaldi. 2010. "Technological Change and Fisheries Sustainability: The Point of View of Adaptive Dynamics." *Ecological Modelling* 221 (3): 379–87.

Dewees, C. M. 1989. "Assessment of the Implementation of Individual Transferable Quotas in New Zealand's Inshore Fishery." *North American Journal of Fisheries Management* 9: 131–39.

Doyen, L., M. De Lara, Jocelyne Ferraris, and Dominique Pelletier. 2007. "Sustainability of Exploited Marine Ecosystems through Protected Areas: A Viability Model and a Coral Reef Case Study." *Ecological Modelling* 208 (2–4): 353–66.

Dudley, Richard G. 2008. "A Basis for Understanding Fishery Management Dynamics." *System Dynamics Review* 24 (1): 1–29.

Eisenack, Klaus, Jürgen Scheffran, and J. Kropp. 2006. "Viability Analysis of Management Frameworks for Fisheries." *Environmental Modeling and Assessment* 11 (1): 69–79.

Ellerman, A. Denny, and Barbara K. Buchner. 2007. "The European Union Emissions Trading Scheme: Origins, Allocation, and Early Results." *Review of Environmental Economics and Policy* 1 (1): 66–87.

Ellerman, A. Denny, Frank J. Convery, and Christian de Perthuis. 2010. *Pricing Carbon: The European Emissions Trading Scheme.* Cambridge, England: Cambridge University Press.

Endy, D., and R. Brent. 2001. "Modelling Cellular Behaviour." *Nature* 409 (6818): 391–95.

FAO-FAD. 2016. *The State of World Fisheries and Aquaculture (SOFIA).* Geneva, Switzerland: United Nations Food and Agriculture Organization.

Farber, Stephen. 2000. "Welfare-Based Ecosystem Management: An Investigation of Trade-Offs." *Environmental Science and Policy* 3: S491–98.

Fina, Mark. 2005. "Rationalization of the Bering Sea and Aleutian Islands Crab Fisheries." *Marine Policy* 29 (4): 311–22.

Ford, Andrew. 2009. *Modeling the Environment* (2nd Edition). Washington, DC: Island Press.

Gade, M. A., T. D. Garcia, J. B. Howes, T. M. Schad, and S. Shipman. 2002. *Courts, Congress, and Constituencies: Managing Fisheries by Default*. Washington, DC: National Academy of Public Administration.

Garrity, Edward J. 2011. "System Dynamics Modeling of Individual Transferable Quota Fisheries and Suggestions for Rebuilding Stocks." *Sustainability* 3: 184–215.

Gibbs, M. T. 2009. "Individual Transferable Quotas and Ecosystem-Based Fisheries Management: It's All in the T." *Fish and Fisheries* 10: 470–74.

Glantz, Michael H. 1979. "Science, Politics and Economics of the Peruvian Anchoveta Fishery." *Marine Policy* 3: 201–10.

Glantz, Michael H. 2005. *Climate Variability, Climate Change, and Fisheries*. Cambridge, England: Cambridge University Press.

Gordon, H. Scott. 1954. "The Economic Theory of a Common-Property Resource: The Fishery." *The Journal of Political Economy* 62 (2): 124.

Gourguet, Sophie, Claire Macher, Luc Doyen, Olivier Thébaud, Michel Bertignac, and Olivier Guyader. 2013. "Managing Mixed Fisheries for Bio-Economic Viability." *Fisheries Research* 140: 46–62.

Gutierrez, Nicolas L., Ray Hilborn, and Omar Defeo. 2011. "Leadership, Social Capital and Incentives Promote Successful Fisheries." *Nature* 470 (7334): 386–89.

Hardin, G. 1968. "The Tragedy of the Commons." *Science* 162: 1243–47.

Harlow, Robert L. 1974. "Conflict Reduction in Environmental Policy." *Journal of Conflict Resolution* 18 (3): 536.

Harris, Michael. 1999. *Lament for an Ocean: The Collapse of the Atlantic Cod Fishery, A True Crime Story*. Toronto, ON: McClelland & Stewart.

Healey, M. C., and T. Hennessey. 1998. "The Paradox of Fairness: The Impact of Escalating Complexity on Fishery Management." *Marine Policy* 22 (2): 109–18.

Hendrix, Cullen S., and Sarah M. Glaser. 2011. "Civil Conflict and World Fisheries, 1952–2004." *Journal of Peace Research* 48 (4): 481–95.

Herrmann, Mark. 1996. "Estimating the Induced Price Increase for Canadian Pacific Halibut with the Introduction of the Individual Vessel Quota Program." *Canadian Journal of Agricultural Economics* 44 (2): 151–64.

Hilborn, Ray, J. M. Orensanz, and Ana M. Parma. 2005. "Institutions, Incentives and the Future of Fisheries." *Philosophical Transactions of the Royal Society B: Biological Sciences* 360 (1453): 47–57.

Hilborn, Ray, J. K. Parrish, and K. Litle. 2005. "Fishing Rights or Fishing Wrongs?" *Reviews in Fish Biology and Fisheries* 15: 191–99.

Hofbauer, J., and K. Sigmund. 1998. *Evolutionary Games and Population Dynamics*. Cambridge, England: Cambridge University Press.

Holker, Franz, Doug Beare, Hendrik Dorner, Antonio di Natale, Hans-Joachim Ratz, Axel Temming, and John Casey. 2007. "Comment on 'Impacts of Biodiversity Loss on Ocean Ecosystem Services'." *Science* 316 (5829): 1285.

Homans, Frances R., and James E. Wilen. 1997. "A Model of Regulated Open Access Resource Use." *Journal of Environmental Economics and Management* 32 (1): 1.

Hutchings, Jeffrey A., and Dylan J. Fraser. 2008. "The Nature of Fisheries- and Farming-Induced Evolution." *Molecular Ecology* 17 (1): 294–313.

Hutchings, Jeffrey A., and R. A. Myers. 1994. "What Can Be Learned from the Collapse of a Renewable Resource? Atlantic Cod, Gadus Morhua, of Newfoundland and Labrador." *Canadian Journal of Fisheries and Aquatic Sciences* 51: 2126–46.

Hutchings, Jeffrey A., and J. D. Reynolds. 2004. "Marine Fish Population Collapses: Consequences for Recovery and Extinction Risk." *Bioscience* 54: 297–309.

Jackson, Jeremy B. C., Michael X. Kirby, Wolfgang H. Berger, Karen A. Bjorndal, Louis W. Botsford, Bruce J. Bourque, Roger H. Bradbury, et al. 2001. "Historical Overfishing and the Recent Collapse of Coastal Ecosystems." *Science* 293 (5530): 629–37.

Jentoft, S. 2005. "Fisheries Co-Management as Empowerment." *Marine Policy* 29: 1–7.

Jentoft, S., B. J. McCay, and D. Wilson. 1998. "Social Theory and Fisheries Co-Management." *Marine Policy* 22: 423–36.

Jentoft, S., and K. H. Mikalsen. 2004. "A Vicious Circle? The Dynamics of Rule-Making in Norwegian Fisheries." *Marine Policy* 28 (2): 127–35.

Kearney, R. E. 2002. "Co-Management: The Resolution of Conflict between Commercial and Recreational Fishers in Victoria." *Australia Ocean & Coastal Management* 45 (4–5): 201–14.

Kurlansky, Mark. 1998. *Cod: A Biography of the Fish That Changed the World.* New York: Penguin Press.

Lane, D. E., and R. L. Stephenson. 2000. "Institutional Arrangements for Fisheries: Alternate Structures and Impediments to Change." *Marine Policy* 24: 385–92.

Larkin, P. A. 1956. "Interspecific Competition and Population Control in Freshwater Fish." *Journal of the Fisheries Board of Canada* 13 (3): 327–42.

Link, P. Michael, and Richard S. J. Tol. 2006. "Economic Impacts of Changes in the Population Dynamics of Fish on the Fisheries of the Barents Sea." *ICES Journal of Marine Science* 63 (4): 611–25.

Little, L. Richard, André E. Punt, Bruce D. Mapstone, Gavin A. Begg, Barry Goldman, and Ashley J. Williams. 2009. "An Agent-Based Model for Simulating Trading of Multi-Species Fisheries Quota." *Ecological Modelling* 220 (23): 3404–12.

Lytinen, Steven L., and Steven F. Railsback. 2012. "The Evolution of Agent-Based Simulation Platforms: A Review of NetLogo 5.0 and ReLogo." In *Proceedings of the Fourth European Meeting on Cybernetics and Systems Research*, edited by Wolfgang Hofkirchner. Vienna, Austria : Bertalanffy Center for the Study of Systems Science.

Mangelsdorf, T. 1986. *History of Steinbeck's Cannery Row.* Santa Cruz, CA: Western Tanager Press.

Marcuse, Peter. 1998. "Sustainability Is Not Enough." *Environment and Urbanization* 10 (2): 103–11.

Martinet, Vincent, and Luc Doyen. 2007. "Sustainability of an Economy with an Exhaustible Resource: A Viable Control Approach." *Resource and Energy Economics* 29 (1): 17–39.

McDonald, Gavin, Tracey Mangin, Lennon R. Thomas, and Christopher Costello. 2016. "Designing and Financing Optimal Enforcement for Small-Scale Fisheries and Dive Tourism Industries." *Marine Policy* 67: 105–17.

Meadows, Dennis, Thomas Fiddaman, and Diana Shannon. 1986. *Fish Banks, Ltd.* Durham, NH: Loboratory for Interactive Learning.

Meadows, Dennis L., John D. Sterman, and Andrew King. 2018. "Fishbanks: A Renewable Resource Management Simulation." 2018. Cambridge, MA: MIT Sloan School of Management. https://mitsloan.mit.edu/LearningEdge/simulations/fishbanks/Pages/fish-banks.aspx

Missios, Paul C., and Charles Plourde. 1996. "The Canada-European Union Turbot War: A Brief Game Theoretic Analysis." *Canadian Public Policy* 22 (2): 144–50.

Moxnes, E. 2004. "Misperceptions of Basic Dynamics: The Case of Renewable Resource Management." *System Dynamics Review* 20 (2): 139–62.

Moxnes, E. 2005. "Policy Sensitivity Analysis: Simple versus Complex Fishery Models." *System Dynamics Review* 21 (2): 123–45.

Mullon, C., P. Fre, and P. Cury. 2005. "The Dynamics of Collapse in World Fisheries." *Fish and Fisheries* 6: 111–20.

Musick, J. A., M. M. Harbin, S. A. Berkeley, G. H. Burgess, A. M. Eklund, L. Findley, R. G. Gilmore, et al. 2000. "Marine, Estuarine, and Diadromous Fish Stocks at Risk of Extinction in North America (Exclusive of Pacific Salmonids)." *Fisheries* 25 (11): 6–30.

Myers, Norman. 1993. *Ultimate Security: The Environmental Basis of Political Stability* (1st Edition). New York: W.W. Norton.

Myers, Ransom A., and Boris Worm. 2003. "Rapid Worldwide Depletion of Predatory Fish Communities." *Nature* 423 (6937): 280–83.

National Marine Fisheries Service. 2002. *Annual Report to Congress on the Status of U.S. Fisheries 2001.* Silver Spring, MD: National Marine Fisheries Service, NOAA, U.S. Department of Commerce.

National Marine Fisheries Service. 2018. "Catch Share Programs." Catch Share Programs. 2018. Silver Spring, MD: National Marine Fisheries Service, NOAA, U.S. Department of Commerce. www.st.nmfs.noaa.gov/economics/fisheries/commercial/catch-share-program/overview

NOAA. 2013. "Northeast Regional Stock Assessment Workshop (SAW) 55," April. Woods Hole, MA: National Oceanic and Atmospheric Administration. www.nefsc.noaa.gov/saw/

Ostrom, Elinor. 1990. *Governing the Commons: The Evolution of Institutions for Collective Action.* Cambridge, England: Cambridge University Press.

Ostrom, Elinor. 1998. "A Behavioral Approach to the Rational Choice Theory of Collective Action." *American Political Science Review* 92 (1): 1–22.

Ostrom, Elinor. 2009. *Understanding Institutional Diversity.* Princeton, NJ: Princeton University Press.

Ostrom, Elinor, Joanna Burger, Christopher B. Field, Richard B. Norgaard, and David Policansky. 1999. "Revisiting the Commons: Local Lessons, Global Challenges." *Science* 284 (5412): 278–82.

Otto, P., and J. Struben. 2004. "Gloucester Fishery: Insights from a Group Modeling Intervention." *System Dynamics Review* 20 (4): 287–312.

Ozik, Jonathan, Nicholson T. Collier, John T. Murphy, and Michael J. North. 2013. "The ReLogo Agent-Based Modeling Language." *In* 2013 *Winter Simulations Conference (WSC)*, 1560–68. Washington, DC: IEEE.

Pala, Christopher. 2013. "Detective Work Uncovers Under-Reported Overfishing." *Nature* 496 (7443): 18.

Paquet, P. J., T. Flagg, A. Appleby, J. Barr, L. Blankenship, D. Campton, M. Delarm, et al. 2011. "Hatcheries, Conservation, and Sustainable Fisheries—Achieving Multiple Goals: Results of the Hatchery Scientific Review Group's Columbia River Basin Review." *Fisheries* 36 (11): 547–61.

Pauly, Daniel, Dyhia Belhabib, Roland Blomeyer, William W. W. L. Cheung, Andrés M. Cisneros-Montemayor, Duncan Copeland, Sarah Harper, Vicky W. Y. Lam, Yining Mai, and Frédéric Manach. 2014. "China's Distant-Water Fisheries in the 21st Century." *Fish and Fisheries* 15 (3): 474–88.

Pauly, Daniel, Villy Christensen, Sylvie Guenette, Tony J. Pitcher, U. Rashid Sumaila, Carl J. Walters, R. Watson, and Dirk Zeller. 2002. "Towards Sustainability in World Fisheries." *Nature* 418 (6898): 689–95.

Pauly, Daniel, and Dirk Zeller. 2016. "Catch Reconstructions Reveal That Global Marine Fisheries Catches Are Higher than Reported and Declining." *Nature Communications* 7.

Petschel-Held, G., H. J. Schellnhuber, T. Bruckner, F. L. Toth, and K. Hasselmann. 1999. "The Tolerable Windows Approach: Theoretical and Methodological Foundations." *Climatic Change* 41 (3/4): 303–31.

Pinsky, Malin L., Boris Worm, Michael J. Fogarty, Jorge L. Sarmiento, and Simon A. Levin. 2013. "Marine Taxa Track Local Climate Velocities." *Science* 341 (6151): 1239–42.

Pomeroy, R. S., and F. Berkes. 1997. "Two to Tango: The Role of Government in Fisheries Co-Management." *Marine Policy* 21 (5): 465–80.

Power, Nicole, Moss Norman, and Kathryne Dupré. 2014. "'The Fishery Went Away': The Impacts of Long-Term Fishery Closures on Young People's Experience and Perception of Fisheries Employment in Newfoundland Coastal Communities." *Ecology and Society* 19 (3).

Railsback, Steven F., and Volker Grimm. 2011. *Agent-Based and Individual-Based Modeling: A Practical Introduction.* Princeton, NJ: Princeton University Press.

Roughgarden, Jonathan, and Fraser Smith. 1996. "Why Fisheries Collapse and What to Do about It." *Proceedings of the National Academy of Sciences* 93 (10): 5078–83.

Ruckelshaus, William D. 1998. "Foreword: Managing Commons and Community: Pacific Northwest People, Salmon, Rivers, and Sea." In *Managing the Commons* (2nd Edition), edited by J. A. Baden and D. S. Noonan, ix–xiv. Bloomington, IN: Indiana University Press.

Ruseski, G. 1998. "International Fish Wars: The Strategic Roles for Fleet Licensing and Effort Subsidies." *Journal of Environmental Economics and Management* 36 (1): 70–88.

Ruth, Matthias. 1995. "A System Dynamics Approach to Modeling Fisheries Management Issues: Implications for Spatial Dynamics and Resolution." *System Dynamics Review* 11 (3): 233–43.

Salthaug, Are. 2001. "Adjustment of Commercial Trawling Effort for Atlantic Cod, Gadus Morhua, Due to Increasing Catching Efficiency." *Fishery Bulletin* 99: 338–42.

Schaefer, M. B. 1954. "Some Aspects of the Dynamics of Populations Important to the Management of Commercial Marine Fisheries." *Inter-American Tropical Tuna Commission Bulletin* 1: 27–56.

Scheffran, Jürgen. 2000. "The Dynamic Interaction Between Economy and Ecology: Cooperation, Stability and Sustainability for a Dynamic-Game Model of Resource Conflicts." *Mathematics and Computers in Simulation* 53: 371–80.

Scott, Anthony. 1955. "The Fishery: The Objectives of Sole Ownership." *The Journal of Political Economy* 63 (2): 116.

Shotton, Ross, ed. 2001. *Case Studies on the Allocation of Transferable Quota Rights in Fisheries (FAO Fisheries Technical Paper 411)*. Rome, Italy: United Nations: Food and Agriculture Organization (FAO).

Simon, Herbert A. 1947. *Administrative Behavior: A Study of Decision-Making Processes in Administrative Organization*. New York: Macmillan Co.

Simon, Herbert A. 1956. "Rational Choice and the Structure of the Environment." *Psychological Review* 63 (2): 129–38.

Simon, Herbert A. 1991. "Bounded Rationality and Organizational Learning." *Organization Science* 2 (1): 125–34.

Smith, T. D. 1994. *Scaling Fisheries*. Cambridge, England: Cambridge University Press.

Sterman, John D. 2000. *Business Dynamics: Systems Thinking and Modeling for a Complex World*. New York: Irwin/McGraw-Hill.

Sterman, John D. 2006. "Learning from Evidence in a Complex World." *American Journal of Public Health* 96: 505–14.

Stigberg, David. 2012. "An Introduction to the NetLogo Modeling Environment." In *Ecologist-Developed Spatially-Explicit Dynamic Landscape Models*, edited by James D. Westervelt, Gordon L. Cohen, and Bruce Hannon, 27–41. Modeling Dynamic Systems. New York: Springer.

Stöhr, Christian, Cecilia Lundholm, Beatrice Crona, and Ilan Chabay. 2014. "Stakeholder Participation and Sustainable Fisheries: An Integrative Framework for Assessing Adaptive Comanagement Processes." *Ecology and Society* 19 (3): 14.

Stouten, Hendrik, Kris Van Craeynest, Aimé Heene, Xavier Gellynck, Jochen Depestele, Els Vanderperren, Bart Verschueren, and Hans Polet. 2007. "A Preliminary Microworld to Gain Insights in Belgian Fishery Fleet Dynamics". In *Proceedings of the 25th International Conference of the System Dynamics Society*, July 29–August 2, 2007. Boston, MA: International System Dynamics Society. www.systemdynamics.org/assets/conferences/2007/proceed/index.htm

Susskind, Lawrence, Paul Fidanque Levy, and Jennifer Thomas-Larmer. 2000. *Negotiating Environmental Agreements: How to Avoid Escalating Confrontation, Needless Costs, and Unnecessary Litigation*. Washington, DC: Island Press.

Susskind, Lawrence, Sarah McKearnen, and Jennifer Thomas-Lamar. 1999. *The Consensus Building Handbook: A Comprehensive Guide to Reaching Agreement*. New York: SAGE Publications.

Timmermann, Karen, Grete Dinesen, Stiig Markager, Lars Ravn-Jonsen, Marc Bassompierre, Eva Roth, and Josianne Støttrup. 2014. "Development and Use of a Bioeconomic Model for Management of Mussel Fisheries under Different Nutrient Regimes in the Temperate Estuary of the Limfjord, Denmark." *Ecology and Society* 19 (1).

Verhulst, P. F. 1838. "Notice Sur La Loie Que La Population Suit Dans Son Accroissement." *Correspondence Mathematique et Physique* 10: 113–21.

Vinther, Morten, Stuart A. Reeves, and Kenneth R. Patterson. 2004. "From Single-Species Advice to Mixed-Species Management: Taking the next Step." *ICES Journal of Marine Science: Journal Du Conseil* 61 (8): 1398–409.

Vucetich, John A., and Michael P. Nelson. 2010. "Sustainability: Virtuous or Vulgar?" *BioScience* 60: 539–44.

Walters, Carl J. 1980. "Systems Principles in Fisheries Management." In *Fisheries Management*, edited by R. T. Lackey and R. A. Nielsen, 167–83. New York: Wiley.

Walters, Carl J., and Steven J. D. Martell. 2004. *Fisheries Ecology and Management*. Priceton, NJ: Princeton University Press.

Watling, L., and E. A. Norse. 1988. "Spcial Section: Efforts of Mobile Fishing Gear on Marine Benthos." *Conservation Biology* 12 (6): 1178–79.

Watson, Reg, and Daniel Pauly. 2001. "Systematic Distortions in World Fisheries Catch Trends." *Nature* 414 (6863): 534–36.

Weninger, Quinn. 1998. "Assessing Efficiency Gains from Individual Transferable Quotas: An Application to the Mid-Atlantic Surf Clam and Ocean Quahog Fishery." *American Journal of Agricultural Economics* 80 (4): 750–64.

Whitmarsh, D., C. James, H. Pickering, and A. Neiland. 2000. "The Profitability of Marine Commercial Fisheries: A Review of Economic Information Needs with Particular Reference to the UK." *Marine Policy* 24 (3): 257–63.

Wilensky, Uri, and William Rand. 2015. *An Introduction to Agent-Based Modeling: Modeling Natural, Social, and Engineered Complex Systems with NetLogo*. Cambridge, MA: MIT Press.

Worm, Boris, Edward B. Barbier, Nicola Beaumont, J. Emmett Duffy, Carl Folke, Benjamin S. Halpern, Jeremy B. C. Jackson, et al. 2006. "Impacts of Biodiversity Loss on Ocean Ecosystem Services." *Science* 314 (5800): 787–90.

Worm, Boris, Ray Hilborn, Julia K. Baum, Trevor A. Branch, Jeremy S. Collie, Christopher Costello, Michael J. Fogarty, et al. 2009. "Rebuilding Global Fisheries." *Science* 325 (5940): 578–85.

Yndestad, Harald, and Anne Stene. 2002. "System Dynamics of the Barents Sea Capelin." *ICES Journal of Marine Science: Journal Du Conseil* 59 (6): 1155–66.

9

An Adaptive Dynamic Model of Emissions Trading

There's one issue that will define the contours of this century more dramatically than any other, and that is the urgent and growing threat of a changing climate.

—Barack Obama, 44th President of the United States of America, 2014 U.N. Climate Change Summit

We stand now where two roads diverge. But unlike the roads in Robert Frost's familiar poem, they are not equally fair. The road we have long been traveling is deceptively easy, a smooth superhighway on which we progress with great speed, but at its end lies disaster. The other fork of the road—the one less traveled by—offers our last, our only chance to reach a destination that assures the preservation of the earth.

—Rachel Carson, American biologist, writer, and leader of the 1960s US environmental movement Silent Spring (1962, p. 277)

Introduction

In this chapter, we will shift our focus to conflict and cooperation surrounding climate change. While this is perhaps the world's most complex and controversial environmental conflict (Scheffran, Brzoska, Kominek, et al. 2012; Hsiang, Burke, and Miguel 2013; Ide and Scheffran 2014), we will show how it is possible to distill much of the complexity embedded in this conflict into a straightforward dynamic adaptation model of regions aiming to achieve negotiated emissions targets. In doing so, we will develop a modeling framework to analyze the decision making and interaction processes of agents[1] engaged in emission reductions that involve carbon trading.

The linkages between emissions, increased atmospheric variability, weather changes (National Assessment Synthesis Team 2001), habitat changes (Parmesan and Yohe 2003), and a litany of other consequences of climate change (IPCC 2014) are difficult to determine with complete certainty. As a result, substantial controversy remains around climate change and its problematic effects on the environment and society at large (Scheffran, Brzoska, Brauch, et al. 2012; Scheffran, Brzoska, Kominek, et al. 2012).

Following and adapting the fishery model presented in Chapter 8, we may initially imagine modeling the conflict in a similar manner; we can begin by thinking about the structure of the system and how the conflict over a resource actually damages the resource base.

---

[1] We will continue to use the words "stakeholder," "actor," "party," and "agent" interchangeably. In this case, nations, or regional groups of nations, function as the agents in our model.

In the fishery case, there is a renewable resource that is being depleted through fish harvest and consumption. In the case of climate change, it is the atmosphere that is polluted with greenhouse gases (GHGs) as a result of a variety of economic activities (e.g., manufacturing, transportation, etc.). Therefore, if we were to pursue the same modeling approach as in the last chapter, we could imagine the entirety of the earth's biosphere as a series of "resource stocks" that are depleted as emissions increase and natural resources are damaged or changed in a particular region, inducing numerous risks to the natural and social environment.

"Greenhouse gases" (GHGs)—which include carbon dioxide, methane, nitrous oxide, and water vapor—are gases that absorb infrared radiation and contribute to the "greenhouse effect" (Karl and Trenberth 2003; IPCC 2014). The discovery of the greenhouse effect dates back to the 19th century, with major contributions to the theory made by Joseph Fourier (1824), John Tyndall (1869), and Svante Arrhenius (1896). The analogy drawn with greenhouses arose from the erroneous perception that these gases acted like greenhouse glass, which traps warm air primarily by reducing airflow to the outside (limiting "convective cooling").

However, given that we already pursued this line of thinking in our fisheries model, in this chapter, we instead focus directly on the mechanism surrounding emissions trading, a proposed *solution* to mitigate climate change and its impending damage (i.e., a conflict *resolution* scenario). This is an important distinction to make; we could alternatively model the collective action problem that dominates the lack of major action toward addressing climate change by many major emitters. In doing so, we could model how different types of interactions and gaming situations would change agents' behavior and interactions with other agents. However, in this chapter, we are not modeling agents interacting as they attempt to come to an agreement. Instead, we model the active progression of an already agreed-upon climate solution, which as we will see, has its own complex viability and stability issues.

We will begin this chapter discussing climate change and its impact on environmental conflict. Next, we will discuss strategies for mitigation of climate change, focusing on how economists have endeavored to price carbon. We will go on to introduce aspects of our use of the VIABLE modeling approach, including strategies for defining emissions baselines and targets, and the initial allocation of permits. We will follow this by creating an individual agent model, performing a viability analysis of that agent within their constraints, and extending the model to multiple agents. Finally, we will describe and model scenarios designed to reveal different trading outcomes that could lead to value gain or loss for different agents (potentially feeding or soothing conflict). We will close the chapter by summarizing this VIABLE application and offering several takeaways for readers interested in building on this model and extending it for their own purposes.

Climate Change and Conflict

Throughout the world, there is a growing concern about our changing climate and the impact it will have on the world's population and the ecosystems on which we depend. While atmospheric CO_2 concentration exceeded 400 parts per million (ppm) in 2015, up from about 280 ppm in preindustrial times, temperatures have risen 0.87°C relative to the

1951–1980 average, and may continue to rise by 4°C relative to the 1986–2005 average by the end of this century (IPCC 2014).

Recent research has demonstrated that patterns of global warming will cause substantial changes in ecosystems, sea level, ice cover, habitat distributions, and many other impacts (Parmesan and Yohe 2003). This can already be seen above the Arctic Circle where ecosystems and ice cover are quickly changing. It has been theorized that water resources will feel the most immediate impacts of climate change, including dramatic changes in precipitation and runoff, increased demand for water use, increased frequency in extreme floods and droughts, and declines in water quality (Pittock 2011).

Among the geographically specific impacts are water scarcity, droughts, and heat waves in Southern Europe, Central Asia, Northern Africa, the Sahel and the Middle East; floods and storms in Southern/Eastern Asia and the American Gulf region; glacier melting in the polar regions, the Himalayas, and Andes mountains; sea-level rise on many of the world's coasts; and biodiversity loss in various ecosystems, in particular in tropical forests, mountain regions, and coral reefs (see WBGU 2008; Scheffran and Battaglini 2011; Scheffran, Brzoska, Brauch, et al. 2012; IPCC 2014; Ide et al. 2016). In recent years, increasing attention has been paid to the potential security risks and conflicts associated with global warming (for an overview see Scheffran and Battaglini 2011; Scheffran, Brzoska, Brauch et al. 2012). For instance, a US-based panel of experts has portrayed global warming as "one of the greatest national security challenges" that could breed new threats and conflicts (Campbell et al. 2007). Similarly, the German Advisory Council on Global Environmental Change (WBGU 2008) expressed concerns that without an effective climate policy, the consequences "…could well trigger national and international distributional conflicts and intensify problems already hard to manage such as state failure, the erosion of social order, and rising violence."

Despite such sweeping statements, the empirical evidence remains ambiguous; research on the link between climate change and conflict has been inconclusive about the empirical significance of causal mechanisms (Barnett 2003; Barnett and Adger 2007; Nordås and Gleditsch 2007; Scheffran, Brzoska, Brauch et al. 2012; Scheffran, Brzoska, Kominek et al. 2012). While this may be partly due to the fact that climate change is still in its early stages, it may also be possible that in coming decades, the conflict impacts of climate change may emerge on a larger scale, while becoming more visible in the social sphere.

Beyond generalized statements, climate change is likely to affect security and conflict along multiple pathways, such as the degradation of freshwater resources, the decline in food production, increasing storm and flood disasters, and environmentally induced migration (WBGU 2008). For instance, due to water scarcity and soil degradation, agricultural yields could drop, diminishing food supply and driving migration. Extreme weather events put economic institutions and infrastructure at risk, including industrial sites and production facilities, as well as networks for transportation and supply of goods. Other conflict pathways include sea-level rise, energy insecurity, deforestation, loss of biodiversity, and fishery collapses. Acting as a threat multiplier, climate change may also interfere in complex ways with political, economic, and social conflict factors, including population growth, distribution of resources, or already questionable political legitimacy of governments.

For all these concerns, there are major uncertainties leaving room for worst-case scenarios, including economic damage and risk to coastal cities and critical infrastructure; loss of territory, border disputes and environmentally-induced migration; and tensions over energy and resource supplies. Some regional hot spots are seen as more fragile and vulnerable due to their geographic and socio-economic conditions and the lack of adaptation capabilities. Most vulnerable to climate change are those living in poverty, especially in vulnerable areas with poor governance and conflict-prone societies, where living conditions are already precarious (Reuveny 2007, 2008). Exposed to these risks, "failing states" have inadequate management and problem-solving capacities and cannot guarantee the core functions of government. In the worst affected regions, climate change conflicts could spread to neighboring states through refugee flows, ethnic links, environmental resource flows or arms exports. Such spillover effects can expand the geographical extent of a crisis and intensify the erosion of social stability elsewhere. Although industrialized countries will also be exposed to climate stress, they are usually less susceptible and have better capacities to cope with the challenge.

A cycle of environmental degradation, economic decline, social unrest, and political instability could destabilize affected societies. When tipping points are reached, abrupt changes and cascading sequences in the climate system could have incalculable consequences (Lenton et al. 2008). Worst-case scenarios may exceed adaptive capacities even in the wealthiest countries, as demonstrated by the 2003 heat wave in Europe, or by Hurricanes Katrina, Sandy, Harvey, and Irma in the United States in 2005, 2012, and 2017. Overall, it has become increasingly clear that the interconnections between environmental conditions, resource scarcity, human well-being, and possible societal instability are complex and difficult to assess quantitatively.

Strategies for Mitigation of Climate Change

A significant body of literature has developed detailed recommendations for swift and deep reductions in GHG emissions as an important step in avoiding the devastating impacts of climate change (Metz et al. 2007). According to the European Commission (EDGAR 2017), between 1990 and 2015, the world's total CO_2 emissions increased from 22.7 to 36.3 billion metric tons (Gigatons; GT). However, this increase has not been uniform across the globe. During this period, total American and Chinese emissions increased from 5.0 to 5.2 GT, and 2.3 to 10.6 GT, respectively, while European and Russian emissions decreased from 4.4 to 3.5 GT, and 2.4 to 1.8 GT, respectively. During the same time period, the distribution of per-capita emissions has ranged widely between nations (Figure 9.1), increasing in China (2 to 7.7 tons), while declining in the United States (from 19.8 to 16.1 metric tons), Europe (9.2 to 6.9 tons), and Russia (16.2 to 12.3 tons).

Throughout all of these regions, energy "productivity"—economic output per energy unit consumed—has grown (in China, energy productivity has nearly doubled). Similarly, the carbon intensity of production (carbon emissions per wealth unit) declined throughout these regions, including in North America (0.623 metric tons of carbon dioxide per thousand 2005 US Dollars gross domestic product [GDP] in 1990 to 0.429 in 2009), Europe (0.433 to 0.280), Russia (2.948 [1992] to 1.800), and China (4.317 to 2.219; EIA 2015). This means that production is getting more efficient from an emissions perspective. However, these improvements are not occurring fast enough to offset the accelerating total rate of global emissions (EDGAR 2017).

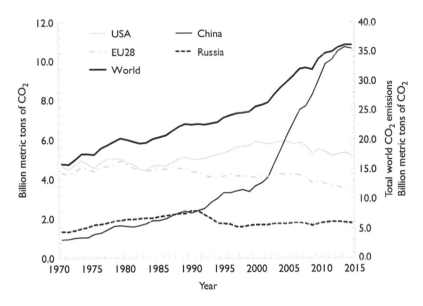

FIGURE 9.1
CO_2 emissions 1970–2015. Data source: EDGAR (2017).

The Promise of Climate Solutions

International climate negotiations, including those in Rio (1992), Kyoto (1997), Copenhagen (2009), Paris (2015), and beyond, attempt to converge on amicable limits to GHG emissions rates that will prevent dangerous climate change. This is one of the goals explicitly set forth in Article 2 of the UN Framework Convention on Climate Change (UNFCCC; Philibert and Pershing 2001; Ott et al. 2003). In particular, these negotiations seek to establish emission reduction pathways that determine reductions in relation to an emissions baseline (Ipsen, Rösch, and Scheffran 2001; Philibert and Pershing 2001). These future emissions trajectories are akin to the "viability corridors" that we discussed in Chapters 7 and 8. During this negotiation process, one of the most important issue involves finding mechanisms that can establish emissions limits and distribute rights among the many agents involved in the negotiation, at regional, national, and local levels, and possibly even at the level of individual firms and consumers (Morthorst 2003).

At the 2014 Conference of Parties (COP-20) in Lima, Peru, over 190 countries initiated new steps toward a global 2015 climate change agreement in Paris (UNFCCC 2015). The conference explored how loss and damage due to climate change affects most vulnerable developing countries and populations. Signatories at COP-20 also gave new urgency toward fast tracking climate adaptation and building resilience initiatives (Scheffran 2016), as well as strengthened efforts to develop and finance national climate adaptation plans. Several initiatives were launched or continued, and several nations announced scaling-up of climate interventions, including the European Union, China, and the United States.

Although we can think of this conflict as only involving a few "agents" at the global level (e.g., dividing the world into groups of industrialized and developing countries, a dozen major regions, or approximately 200 nation states), at the local level the number of individual agents can swell to millions, if not billions. It is not possible to organize a complex, negotiation process to find accordance with so many agents. Instead, we can imagine that a key issue in this process involves determining a fair emission allocation mechanism among parties, based on equity principles such as equivalent per-capita emissions (see Leimbach 2003). *Fairness* is very important to the global success of climate negotiations; although developing countries are less responsible for the current state of the climate, they are likely to be more strongly affected by damage from climate changes and less capable of taking effective counter-measures to mitigate climate-related damage (Scheffran 2008).

Putting a Price on Carbon

What can societies around the world do to slow or stop climate change? A huge range of solutions have been discussed, from local interventions by cities (Beatley 2009) to global mitigation options to avoid climate change (Newman, Beatley, and Boyer 2009), to adaptation strategies of local or national governments to minimize climate impacts (Carmin, Anguelovski, and Roberts 2012). However, the primary discussions regarding solutions to the climate problem often revolve around some sort of government economic intervention that strives to place a price on carbon, thereby creating economic incentives to lower emissions, improve technologies, and reduce fossil fuel usage (e.g., Ellerman, Convery, and Perthuis 2010).

Economic theory has focused on two primary ways of establishing a price for carbon (Lippke and Perez-Garcia 2008). Governments, or other institutions with enforcement authority, can either set a price on the cost that is born by society for each unit of carbon emitted (a "tax"), or they can establish a maximum quantity of carbon that may be emitted (a "cap"). In a perfect market, economic theory predicts that whether we choose a socially desirable emissions quantity or emissions price, these policies create equivalent results (Cleetus 2011). There is a continuing economic debate over whether the increased certainty over price or emissions quantities will be more important for policy in reality (Lippke and Perez-Garcia 2008; Avi-Yonah and Uhlmann 2009).

By determining a tax-induced price for carbon, regulators are essentially allowing the market to determine the corresponding level of emissions (Metcalf and Weisbach 2009). Emissions taxes have been proposed a number of times as a stable method of internalizing the costs of carbon output and climate change into economic production (Elkins and Baker 2001; Scheffran 2002). Alternatively, if we cap the quantity of emissions, then market transactions between polluters will determine the resulting price—a policy framework known as an *environmental market*. At the moment, these types of policies appear to be winning the policy debate and are now being implemented around the world (Kossoy and Guigon 2012; The World Bank 2016; Hamrick and Gallant 2017; Pizer and Zhang 2018; Figure 9.2).

The concept of market-based approaches to environmental policy was originally introduced in the 1960s by Thomas Crocker (1966) for air pollution and by John Dales (1968) for water pollution. Market approaches like emissions trading were meant to draw contrast with inflexible "command-and control" regulations, such as the permitting processes for large-scale, stationary source polluters under the 1990 US Clean Air Act requirements (42 U.S. Code §7661) and point source polluters under the US Clean Water Act (33 U.S. Code §1342). In comparison, market-based approaches are commonly viewed as providing a more efficient and cost-effective way of reducing pollution and resolving pollution-related

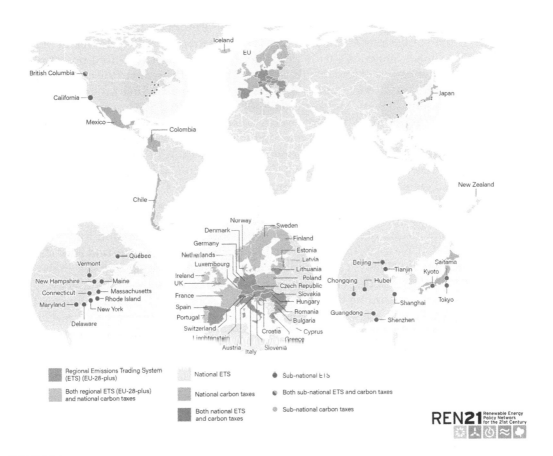

FIGURE 9.2

Carbon pricing policies (2016), including regional, national, and subnational emission trading systems and carbon taxes. Source: REN21 Renewables 2017 Global Status Report, Renewable Energy Policy Network for the 21st Century (REN21 2017, Figure 3).

conflicts (Harlow 1974; Seifert, Uhrig-Homburg, and Wagner 2008; Haya et al. 2016).[2] We discuss the role of markets as conflict resolution mechanism more extensively in Chapter 1 and market-based solutions in the context of individual transferable quota systems for fisheries in Chapter 8.

The most popular emissions trading system—called "cap and trade"—is an environmental policy regime in which governments set a maximum emissions level ("cap") and create permits (i.e., the rights to pollute) that polluters are allowed to buy or sell to other polluters (Boyd et al. 2003; Lippke and Perez-Garcia 2008). Polluters then determine their emissions based on the market price of pollution and the relative ease and expense at which they can reduce their pollution (i.e., their "abatement" cost, as it is commonly referred to in environmental economics; Fisher 1981). Under an emissions trading scheme, polluters can buy or sell permits to increase or reduce their allowed emissions at a market price that is determined by the interplay of supply and demand for emissions permits. This supply and demand depends on the benefits and costs of emission reductions for other polluters in the market.

---

[2] This is not always the case, as there have been major controversies over the use of environmental markets (CEJAC 2017).

Cap-and-trade markets have been implemented in a number of places, including the European Union's Emissions Trading Scheme (EU-ETS) for carbon (Ellerman and Buchner 2007), and the United States' SO_2 (sulfer dioxide; a byproduct of coal-fired power plants) market to reduce the effects of acid rain (Rico 1995). The establishment of the EU-ETS proved that major progress on carbon emissions could be made in a trading program involving 30 nations (Ellerman, Convery, and Perthuis 2010). In the United States, cap-and-trade programs have been proposed as the controversial centerpiece of federal climate legislation, including the American Clean Energy and Security Act (the "Waxman-Markey" bill), the American Power Act (the "Kerry-Lieberman" bill), and the Carbon Limits and Energy for America's Renewal Act (the "Cantwell-Collins" bill). In California, the California Air Resources Board created a cap-and-trade program under the California Global Warming Solutions Act of 2006 (AB 32), which requires the state to reduce GHG emissions to 1990 levels by 2020 (Chen, Liu, and Hobbs 2011; CARB 2014, 2018).

Given the recent history of national government convergence on cap-and-trade solutions for carbon pollution, we focus this chapter on how such a solution can be understood, modeled, and analyzed. By explicitly representing an emission trading system, we analyze the emission paths for nations that change their behavior in order to reach their goals (given by value functions). Continuing the analogy with the fisher agents in the last chapter, these value functions take into account net benefits of economic growth (increased revenue) as well as the marginal damages caused by climate change (decreased fish stocks), the costs for emission reduction (investment costs), and, as a new concept, the sale and purchase of emission permits.

This chapter builds on previous work developed to analyze dynamic games in climate policy (Scheffran and Pickl 2000; Ipsen, Rösch, and Scheffran 2001; Scheffran 2004). Although we will discuss the construction of this model in general terms, we again must note that it remains imperative to include expert interviews, stakeholder dialogs, and experimental gaming to capture the full extent of the dynamics of climate trading. Only with proper participatory input can we make an agent-based model of climate trading a useful tool for understanding emissions trading schemes and their role in improving climate negotiations.

Modeling Emissions Trading and Policy

Numerous models have been applied in the field of emission trading (see surveys in Springer 2003; Antes, Hansjürgens, and Letmathe 2006; Sathaye and Shukla 2013), some of which combine general equilibrium models (i.e., models that use economic data to estimate how economies could react to changes in policy) with the selection of policy instruments (Jensen and Rasmussen 2000; Springer and Varilek 2004). Understanding the underlying linkages between the macro-economy and emission trading programs remains a challenge for emission trading theories as well as practical implementation policies. However, like the last chapter, we are methodologically focused on how we can translate this complex system into an understandable model of the behavior of conflicting agents.

Significant work devoted to modeling climate solutions has also focused on "climate games," which are essentially game-theoretic representations of climate-related conflict and cooperation and are structured using game theory, the pioneering theoretical framework for studying conflict (more thoroughly discussed in Chapters 2 and 3).

Game theory is useful in providing the terminology and theoretical framework for analyzing interdependent decision making, negotiations and coalition formation in climate policy (Svirezhev, von Bloh, and Schellnhuber 1999; Finus 2001; Grundig, Ward, and Zorick 2001; Kemfert 2001; Scheffran 2006). As a result, these climate games have been particularly useful for understanding decision-making processes and interactions among multiple parties at the global, regional, and local levels of climate policy.

However, research has shown that for a large number of agents, constraints, or complex interactions, other methods are appropriate, such as optimal control models, multi-criteria decision analysis, and agent-based modeling (Brassel et al. 2000; Scheffran and Pickl 2000; Pickl 2001; Krabs and Pickl 2003; Scheffran 2004; Weber, Barth, and Hasselmann 2005). All of these additional methodologies help to provide real-world links for climate modeling and remain as key tools for integrated climate trading assessments (Moss 2002).

A key issue in establishing an emissions trading system is to define indicators, such as the levels of CO_2 emissions, atmospheric concentration of CO_2, and global average temperature that would likely lead to dangerous levels of climate change. More importantly, these indicators would need to be defined in such a way that they can be translated into emissions trajectories that can actually be used in a negotiated climate solution. One established mechanism is the "tolerable windows" approach, which defines "guardrails" (limits) for temperature changes that, when observed, lead us to emissions paths that reduce the negative effects of climate change (Bruckner et al. 1999; Petschel-Held et al. 1999). This is compatible with our approach in the VIABLE framework to identify viability limits in the natural and socio-economic systems at stake.

A rigorous mathematical way to do this uses viability theory (discussed in detail in Chapters 7 and 8) to create a series of limits to ensure that the dynamic climate system stays within viable constraints (Aubin, Bernado, and Saint-Pierre 2005; Aubin, Chen, and Durand 2012). In such an analysis, agents involved in negotiations aimed at creating an emissions trading system choose emissions targets as well as their individual actions set at achieving those targets. Here, specific target-setting actions are the product of joint evaluation processes that consider the selection or exclusion of certain sets of system states or trajectories. These evaluations may find that certain trajectories are intolerable to certain parties, such as asking a nation to cut 90% of its emissions in 5 years. These intolerable targets are unlikely to be considered successful outcomes of a group-centered negotiation process.

Once the group has negotiated a way forward (including a series of baseline emissions values and emissions goals), the interesting facets of the emissions trading program will vary based on the strategies that agents (countries or regions) use to achieve given targets. For modeling purposes, we must assume that this behavior will, in some way, be rule based, meaning that in many cases, agents may be functioning in a way that intrinsically seeks to optimize[3] the value function that defines their goals. In other words, the agents seek strategies to maximize their own welfare.

It is at this stage that we will focus our model of emissions trading on addressing three questions: How do agents strategize their emissions trading behavior? Do they buy or sell credits? Is this trading system likely to break down and incentivize agents to leave, or will agents likely remain? We will employ the VIABLE modeling framework that we developed in Chapter 7 to understand how agents iterate their reactions to the actions of other agents in an evolutionary, adaptive way in order to maximize their own welfare and achieve their goals.

---

[3] In Chapters 6 and 7, we discuss the rational agent model in economics, as well as extensive critiques of the idea that agents act rationally or seek to "optimize" outcomes or "maximize" their value.

Defining Emission Baselines, Targets, and Reduction Goals

We begin creating a model of an emissions trading system by considering the process by which two regions negotiate their way through an emissions trading program and How do they each determine if they should reduce their emissions and sell permits, or purchase emissions permits and continue to emit at current levels? This part of our model is derived from a model presented in Scheffran (2004) and Scheffran and Leimbach (2006) and is articulated in Figure 9.3.

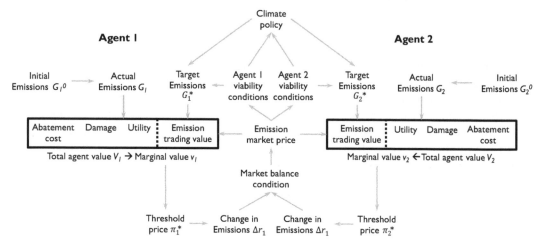

FIGURE 9.3

Flow chart representing the decision-making process of two agents as they consider changes to their emissions based on emission trading price.

BOX 9.1   VARIABLE DEFINITIONS

V_i	= Value function for each agent (country/region $i = 1,...,n$).
$G_i(t)$	= Emissions of individual regions $i=1,...,n$ in period t (typically 1 year).
$G(t)$	= $\sum_i G_i(t)$ = Total flow of emissions generated in time period t.
$G^*(t)$	= Collective emissions limit for all agents (possibly the whole world) in time period t. This is the *trajectory* of emissions targets through time.
G_i^0	= Baseline (initial) emissions for each agent (e.g., "1990 emissions levels").
$N_i(t)$	= Human population for agent i at time t.
$R_i(t)$	= Emission reduction of agent i relative to its baseline $R_i(t) = r_i G_i^0 = -\Delta G_i$.
$r_i(t)$	= Percentage of emissions reductions from the baseline emissions G_i^0.

(Continued)

BOX 9.1 (*Continued*)   VARIABLE DEFINITIONS

α_i	=	Response parameter that determines the speed of each agent in changing their emission reduction and adapting to new system states.
$v_{r_i} = \partial V_i / \partial r_i$	=	Rate of change of profit function V_i with respect to changes in emissions reduction percentage r_i (partial derivative or marginal value of emissions reduction)
π	=	Market or trading price for a single emissions unit.
π_i^*	=	Threshold price where an agent i is equally likely to buy or sell permits. This threshold price corresponds to the viability condition whereby, at this price, small increases or decreases of small emissions have no effect on an agent's marginal value.
$\tilde{\pi}_i$	=	Threshold price at which agent value is zero: $V_i = 0$. This threshold price defines the "economic" viability condition.
$\hat{\pi}$	=	Threshold price to meet the collective emission limit G^* of all agents together.
Π_i	=	Total increase or decrease in value for agent i by emission trading, where agent i deviates from the target emissions level G_t^*. This is the value V_i gained from selling emissions permits, or lost due to required investment in buying additional permits $-\Pi_i$.
$R^*(t)$	=	Total emission reduction requirement $G^0 - G^*(t)$.
Q_i	=	Production of economic goods (output).
$C_i(Q)$	=	Investment C into production of economic goods Q.
$U_i(Q)$	=	Utility generated from production Q.
F_i	=	Total abatement costs for each agent.
$D_i(G)$	=	Environmental damages D as a result of emissions G. Defined as a function of total emissions $G = \sum_i G_i$.
u_i	=	U/G = Utility U per emission unit G, which combines production Q and emissions G in a single utility function.
d_i	=	D/G = Damage D per emission unit G, typically represented by a nonlinear function.
c_i	=	Abatement cost (investments) per unit of emission reduction (mitigation) r.

The task of setting emissions targets is to ensure that the total flow of emissions $G(t)$ generated at time t by $i = 1,...,n$ agents, each with individual emissions $G_i(t)$, should not exceed an agreed-upon total limit $G^*(t)$ in any given year (see Box 9.1 for all variable definitions). In lieu of an individual agent limits $G^*(t)$, we can think of as a well-defined "collective emissions limit" that would be created through a negotiated allocation process or other policy. We can define this mathematically as $G(t) = \sum_i G_i(t) \le G^*(t)$. A key concept here is that total emissions $G(t)$ and target emissions $G^*(t)$ are both flow variables, representing the amount of gases emitted *per year*.

Any global target trajectory $G^*(t)$—a series of emissions targets through time—must be produced from consensus-based negotiation processes that produce value judgments on the meaning of dangerous climate change, as well as agreed-upon estimates of emissions modifications that are actually achievable. Emissions targets would certainly need to be set to avoid carbon concentration exceeding critical thresholds that lead to dangerous climate change (which is prohibited according to Article 2 of the United Nations Framework Convention on Climate Change; United Nations 1992).

One way to construct this target trajectory $G^*(t)$ is to derive "guardrails" for admissible temperature changes that yield a tolerable level of climate change, and then reverse-engineer a total emissions goal G^* that would be compatible with this limit (Scheffran 2008). This method is known as the "tolerable windows approach," and is described in WBGU (1995), Petschel-Held et al. (1999), and Bruckner et al. (2003). Unfortunately, this approach can be complicated by a range of different views of emissions that are seen as either tolerable or dangerous (Ott et al. 2003).

Perhaps the most prominent example has been the 2°C guardrail that was agreed upon at the Copenhagen climate summit (COP-15) in 2009, which is based on a maximum "world budget" of CO_2 emissions that would prevent a 2°C rise in global average temperature by end of the century. According to WBGU (2009), for a maximum atmospheric release of approximately 750 billion metric tons (Gigatons; GT) CO_2, there is a probability of 67% that this guardrail is not exceeded. If we raise the probability of success to 75%, cumulative emissions within this period would need to remain below 600 GT CO_2. In any case, after 2050, only a small amount of CO_2 may be emitted worldwide.

While Copenhagen had set a nonbinding target of 2°C, the 2015 Treaty of Paris (COP-21) made it mandatory under international law (but without any time limit) to constrain global warming to well below 2°C, and if possible, to under 1.5°C, relative to preindustrial level (UNFCCC 2015). Moreover, under the Paris Treaty, after reaching its maximum, the world's net emissions would need to be reduced to zero in the second half of the 21st century. While no binding targets for CO_2 reduction were set for individual countries, *Intended Nationally Determined Contributions* (INDC) were set by 186 States, who will verify and report progress every 5 years.

Aggregate GHG emission levels in 2025–2030 resulting from the INDCs will not likely fall within the <2°C scenario, but rather will lead to a projected level of 55 GT of emissions. Therefore, much greater emission reduction efforts will be required than those associated with the INDC to hold the increase in the global average temperature to below 2°C. To achieve less than a 1.5°C increase, these nations will need to reduce their 2030 emissions to a total of only 40 GT (Scheffran 2016).

Rather than fixing constant emissions, countries may define a time-dependent target trajectory that connects the current emission rate with the set lower emission target at a later date (e.g., an 80% emission reduction by 2050). There are multiple possible target trajectories connecting the initial emissions G^0 (at an initial time $t = 0$) and the target emissions $G^*(T)$ at a given final time T. One reasonable choice, which we will use in this model (but others can modify), is to take a linear emissions pathway as represented by a linear decline of annual emissions toward a lower bound G^* by a particular date:

$$G^*(t) = G^0 - \left[G^0 - G^*(T)\right]\frac{t}{T}.$$

There are many different complex ways that we think about emissions reductions: relative to the previous year, a reference year, or relative to some projected trajectory that agents have agreed upon. Here, we focus on how we could adapt baseline emissions levels to a worldwide emissions target, using the baseline year (e.g., the starting year) as the reference. Therefore, we assume that each agent will take efforts to reduce emissions, where the emissions reductions $R_i(t) = r_i(t)G_i^0$ for each agent i are simply a percentage of emissions reductions $r_i(t)$ from the "baseline" emissions G_i^0, the emissions in the initial year of the program, $\Delta G_i(t) = -r_i(t)G_i^0 = -R_i(t)$. We define the global condition when emissions $G(t)$ of all actors together stay below the target as

$$G(t) = G^0 - R(t) = \sum_i \left[G_i^0 - R_i(t) \right] = \sum_i G_i^0 \left[1 - r_i(t) \right] \leq G^*(t) \tag{9.1}$$

Therefore, efforts to seek the total, worldwide emissions target where $G(t) = G^*(t)$ translates into a worldwide emission reduction requirement or limit $R^*(t)$, which can be disaggregated for agents as the gap (difference) between the baseline G^0 and the goal $G^*(t)$,

$$R^*(t) = \sum_i R_i^*(t) = G^0 - G^*(t).$$

Initial Allocation of Permits

Allocation Philosophy

Assuming that negotiation processes converge on some type of aggregate cap (target) on emissions $G^*(t)$, it is a major challenge to find institutional mechanisms to ensure that a large number of agents jointly pursue this target while avoiding *collective action problems* through cooperative and regulative measures (Ellerman and Buchner 2007). Therefore, the next step in designing an emissions trading program is to find mechanisms that allocate initial emission permits from global levels down to individual agents. These initial allocations are often the subject of significant debate in the design and administration of trading programs (Ellerman, Convery, and Perthuis 2010).

Mancur Olson (1971 [1965]) first developed the concept that individuals in any group pursue *"collective action,"* as they undertake a joint effort to achieve a common objective. *Collective action problems* are situations in which many individuals would benefit from an action, but the action's associated cost makes it implausible that any individual will undertake it alone. The ideal solution is for the group to perform this action together, thereby distributing the cost. See Hopkins (2001) for a discussion of the link between collective goods and collective action, and Fisher (1981) for a discussion of the resource economics context of collective action problems.

In our framework, we can think about this allocation as a question of how to initially distribute the total reduction requirements $R^*(t)$ to the individual agents i, depending on the global target $G^*(t)$ and baseline emissions G_i^0. Measuring the differences between baseline G_i^0, actual emissions $G(t)$, and target emission trajectories $G^*(t)$, and estimating the costs of emission reduction $R_i(t)$ are not easy tasks, and as of this book's publication, no *global* agreement has been achieved to perform them.

All of these issues are further complicated by the fact that emissions reductions are not likely be achieved simply through technological advances alone (Edenhofer, Bauer, and Kriegler 2005). The most realistic scenarios typically involve closing down historically large emitters, while pursuing carbon capturing and sequestration facilities (Stefan 2000).

Thus, investments in new, and often more costly, low-emission technologies and facilities are accompanied by closure of older facilities in order to provide a net emission reduction. This process implies a loss of production and invested capital (sunk costs; Heal and Kristom 2002), which can become significant not only at the macro-economic level, but also at the scales of firms, specific products, and technologies.

Initial allocation approaches considered include permit auctions, which require agents to purchase all necessary emissions permits, as well as free initial allocations to agents based on an allocation rule, such as size- or efficiency-dependent mechanisms, or through an equal emissions per capita standard (Jensen and Rasmussen 2000). These free allocations are made to reduce economic impacts and financial hardship due to the start of the emissions trading program. This issue was quite contentious during early discussions of the EU ETS (Ellerman, Convery, and Perthuis 2010), as major equity considerations, economic impacts, and political decisions helped to determine the complex initial allocation of permits. Similar issues arose during the initial allocation of the United States' SO_2 market (Johnston, Sefcik, and Soderstrom 2008).

Rather than simulating a permit auction, we freely allocate emission permit pathways (i.e., requirements for reductions over time)—given as $G_i^*(t)$—to agents to explore different ways of structuring the start of an emissions trading program. Summing individual national or regional "free" emissions permits leads to a joint emission limit, $G^*(t) = \sum_i G_i^*(t)$, which is typically the starting point for an emissions reduction program. These allocations act as a political starting point for moving actual emissions $G_i(t)$ toward the global target trajectories $G^*(t)$. We must emphasize that these negotiated initial allocations are different than the baseline emissions levels G_i^0, which act as our reference for measuring emissions reductions.

Allocation Mechanisms to Test

Although there are numerous types of initial permit allocations, for our purposes, we will explore the impacts of alternatively setting emissions targets $G_i^*(t)$ using two allocation principles of global targets $G^*(t)$ to each region.[4]

Case 1. Allocation in proportion to the amount initially emitted, $G_i^*(t) = \dfrac{G_i^0}{\sum_i G_i^0} G^*(t)$.

Under this scenario, each region i receives a share that is proportional to their baseline (initial year) emissions rates G_i^0, meaning that the required reduction *percentage* is equal for all regions. This allocation principle follows the argument that larger emitters demand more emissions and thus have the right to a bigger share. With the need for emissions reduction, however, larger emitters would also have to take on a greater share of *absolute* reduction and thus would be more severely affected in terms of total costs (similar statements apply for other size-dependent mechanisms). This is a simplified way of representing the major tension between developed and developing countries after the Kyoto Protocol negotiations, and is the primary reason for the refusal of the United States to formally ratify the agreement (see Sauquet [2014] for more detail on this conflict).

Case 2. Allocation in proportion to an agent's population N_i, such that $G_i^*(t) = \dfrac{N_i}{\sum_i N_i} G^*(t)$, where $\sum_i N_i$ is the world population.

---

[4] An important assumption here is that we do not use projections to grow the population by the target year and instead keep the population share static. This is a relatively easy assumption to update in order to derive yield a per-capita emissions target trajectory.

In both cases, we will assume that previous efforts to decrease emissions intensity are not considered in the negotiated emissions reductions for each country. Contrast these arrangements with the agreement made under the Paris Accord (COP-21; UNFCCC 2015), where no binding CO_2 reduction targets were set for individual nations, but instead 186 nations agreed to reductions in the form of voluntary INDC.

Modeling Conflict Potential in Emissions Trading

We employ the VIABLE modeling framework to understand the decision-making processes and agent interactions that accompany emission reductions through an emissions trading system. Understanding the dynamics of agents in an emissions trading scheme has major implications for improving the current debate over climate actions, including more certainty about trade-offs among investments made in purchases of emissions permits, economic damages, and investments in new technology. To do this, we start by considering that a particular party (or region, as we will consider) can invest $C(Q)$ into production of economic goods Q, thereby creating economic benefits measured by a utility function $U(Q)$. During production, emissions G are released, leading to emissions-related environmental damages $D(G)$ as a negative side effect (e.g., sea level rise, flooding damage, increased natural hazards; Anthoff, Hepburn, and Tol 2009; National Academies of Sciences 2017). Finally, in this system, agents can buy (or sell) emission permits, thereby generating additional costs (or income) Π when their emissions are above (or below) target emissions G_i^*.

Elements of Agent Goals or Values

As we did in Chapters 7 and 8, we create a "value function" V_i for each agent, taking into account the utility $U(Q)$ and investments $C(Q)$ associated with their additional production Q, as well as the damage D each agent experiences associated with emissions, the payment for purchasing or selling additional emissions Π, and the emission reduction cost F, where

$$V_i(t) = U_i(Q_i) - C_i(Q_i) - F_i - D_i - \Pi_i \tag{9.2}$$

In the following equations, we define functions for these value terms:

- The benefits generated from economic output Q can be measured by a utility function $U(Q)$, which indicates the positive values generated by production associated with emissions. We will focus only on the utility generated from production pathways associated with GHG emissions G, while other pathways based on zero emissions (e.g., with renewables) are not considered here. This reflects the assumption that with emissions G declining toward zero, the associated utility $U(G)$ generated from emissions also declines and is not replaced by carbon-free production.[5]

- Due to the relationship between production and emissions, the investments $C(Q)$ ("production costs") are associated with emissions G that are needed to

---

[5] This is a simplifying assumption responsible for much of the behavior in this model. We discuss ways of relaxing this assumption in this chapter's *Questions for Consideration* section.

generate economic output Q. We consider the difference between utility and costs $U(G) - C(G) = (u_g - c_g)G = uG$ as the net utility gain (relative benefit) associated with emissions (where u_g and c_g are the marginal benefits and costs of emissions, respectively). For the rest of this chapter, $u = u_g - c_g$ is taken as the net unit value of emissions, thereby defining net utility as $uG(t)$.

- It is important to note that we distinguish between the investments $C(G)$ for generating emissions and the abatement (mitigation) costs $F(R)$ for emission reductions (Khanna 2001). We are going to assume that costs are proportionate to the magnitude G^0 to which reduction is applied, but that the effect of percentage reduction r is nonlinear. This nonlinearity is represented by an exponent γ, which indicates the relationship between costs and emission reduction, where $\gamma = 1$ is the linear case, $\gamma > 1$ represents increasing marginal abatement costs (i.e., due to technological limits) and $\gamma < 1$ represents decreasing marginal abatement costs (i.e., due to learning). We will use $\gamma = 2$ as a standard case. Therefore, the abatement costs or the costs to reduce (mitigate) emissions are specified separately as reduction costs $F = cG^0r^\gamma = cG^0r^2$. This relationship is modified when the total emissions reduction falls below zero ($R < 0$) (i.e., emissions increase), such that the abatement costs F are zero ($F = 0$).

- The damages in terms of utility losses $D = dG^\delta$ induced by emissions and resulting climate change can be similarly calculated, where exponent δ indicates a potential nonlinear relationship, with increasing marginal damage if $\delta > 1$ and decreasing marginal damage if $\delta < 1$. We will use $\delta = 2$ as a standard case, much in the same way as we treat cost as a quadratic function of emissions reduction. Therefore, we can take emission-related damage as $D = dG^2$.

- To add our final term, we consider the role of emissions trading. Emissions are only traded when regions either emit more or less than their admissible emission goals G_i^*. Since only these "deviations" $G_i(t) - G_i^*(t)$ between the actual and target emissions in each period t are traded, we can calculate the value of the emissions trading market for each agent as $\Pi_i = \pi\left(G_i - G_i^*\right)$, which is a function of the price per emission unit π and the amount of emissions traded $(G_i - G_i^*)$. Here, Π_i also represents a way of calculating the total decrease or increase in value for actors who deviate from the target emissions level G_i^*.

Taking all terms together, for $i = 1,..., n$ agents, we have a value function of emissions G_i and emission reduction fractions r_i:

$$V_i(t) = U_i - C_i - D_i - F_i - \Pi_i = u_iG_i(t) - d_i[G(t)]^2 - c_iG_i^0r_i^2 - \pi\left(G_i - G_i^*\right) \tag{9.3}$$

For simplicity, we are going to assume that these exponents are equal for all agents in the model. However, this could be changed to represent increased heterogeneity in agents. Note that damage D is defined explicitly as a function of total emissions $G = \sum_i G_i$, since damage is a function of the *global* emissions, not just those given off by an individual emitting agent.

Based on this model, we can start trade-offs by considering the interplay of emissions targets G^*, market price π, and the value function V. For instance, take the case where an emissions target is very ambitious (i.e., low) and G_i^* moves toward zero. In this case, we see agent value (as given in Equation 9.3) diminishing in two ways; first, if the market price for emissions rises high enough, purchasing additional allowances will overwhelm any effort to create utility from economic production (arising from emissions) and agent value drops.

If an agent instead decided to reduce emissions in response to these prices, their utility as a whole may drop, diminishing value as well. If damage from climate change D_i is also high, then forced reductions may partially compensate for value losses. In the opposite case, if we set the emissions target G^* too high, then there may be no incentive to cut emissions, and the emissions price π remains low. That is, the emissions price has no effect on agent value or behavior.

Would agents leave the system if the emissions price spiralled out of control? Countries control the emission target G^* because they agree to it during climate negotiations (countries are not forced to enter into an agreement). However, if they feel threatened by excessive reductions, they would likely not agree to participate in the trading system any longer. On the other hand, excessive climate damage provides economic incentives for emission reductions. Ultimately, as we will mention in our suggestions for model expansions in this chapter's *Questions for Considerations* section, a country may instead switch to economic production that has low or zero net-emissions (e.g., low-carbon energy production from renewables).

Pricing Carbon through Emissions Trading

In our model of fishers depleting a fishery stock in Chapter 8, we designed a decision rule whereby agents acted as "satisficers," following their preferences and searching for gradual improvements to their values, until they eventually reached a local optimum. In a similar way, we can create a decision rule for emitting agents who adjust their relative rate of emissions reductions Δr_i toward increasing values. As we have earlier used the VIABLE framework, we first take the derivative of the agents' value function V_i with respect to the action pathway (emissions reduction fraction r_i), yielding $v_{r_i} = \partial V_i / \partial r_i$. This indicates the condition under which further emission reductions tend to increase or possibly maximize their target (values) based on their emission reductions. Thus, it is reasonable to change r_i proportionate to this partial derivative, which indicates the direction of growing value:

$$\Delta r_i = \alpha_i \frac{\partial V_i}{\partial r_i} = \alpha_i v_{r_i} \tag{9.4}$$

Under this decision rule, the reactivity parameter α_i accounts for the rate at which agents can actually change their strategies, which determines the speed of each agent in changing their emissions reduction and adapting to new system states (responses may be delayed by a variety of factors). We can think of this as the rate that an agent changes their emissions in a direction that will increase their value function. We should note that seeking the set target G_i^* does not necessarily mean the agent is seeking their *optimum* emissions level. Any target can be set, including the 2°C target, which is not based on optimization calculations (Jaeger and Jaeger 2011).

In Appendix 9.1, as we have in Chapters 7 and 8, we find v_{r_i} by taking the derivative of the value function with respect to the emissions reduction fraction r_i. This leads to the definition of the threshold price π_i^* at which an agent will be indifferent to buy or sell permits. We derive the (approximate) threshold price π_i^* in Appendix 9.1, which is a function of emission reduction r_j of all agents (G is a function of the reductions of other agents r_j):

$$\pi_i^* = u_i - 2d_i G + 2c_i r_i \tag{9.5}$$

If the actual market price π of emissions is higher than this value, then it makes sense for the agent to *reduce* its emissions and *sell* permits. If the market price is *below* this threshold price, then this agent will *increase* their emissions and *buy* permits to maximize their value function. Using this concept, we can collapse the equation for the rate of change of value with respect to emissions reduction (derived in Appendix 9.1):

$$v_{r_i} = \frac{\partial V_i}{\partial r_i} = \left(\pi - \pi_i^*\right)G_i^0 \tag{9.6}$$

Consider that the market price π results as a function of each agent's threshold price for buying or selling permits. In the process of trading credits, the total changes in emissions reductions ΔR_i (again, this is relative to an agent's initial emissions G_i^0) would equal zero to achieve the market clearing condition of emissions trading. We assume that (1) the total number of purchased permits and sold permits will sum to zero and (2) the negotiated emissions target trajectory $G^*(t)$ remains the same trajectory as originally negotiated (i.e., the trajectory is not renegotiated along the way, whereby G^* would be reduced midstream as the program became increasingly stringent).

Therefore, if we take our previous representation in Equation 9.4 of a single agent's decision rule, $\Delta r_i = \alpha_i \dfrac{\partial V_i}{\partial r_i}$, and integrate it with Equation 9.6, we now have

$$\Delta r_i = \alpha_i v_{r_i} = \alpha_i\left(\pi - \pi_i^*\right)G_i^0 \tag{9.7}$$

Thus, the change in emission reduction between two time steps depends on the difference between total price and threshold price, which determines when countries buy or sell permits. As buyers and sellers transact permits, the market price π adjusts to this demand-supply interaction, facilitating agents meeting the global emissions target. When the collective change in total emissions reduction equals the politically planned (i.e., negotiated) emission reduction $\Delta G^*(t)$, we can use Equation 9.7 to aggregate all of the agents:

$$\sum_i \Delta R_i(t) = \sum_i \Delta r_i(t)G_i^0 = \sum_i \left\{\alpha_i\left(\pi(t) - \pi_i^*(t)\right)\left[G_i^0\right]^2\right\} = \Delta G^*(t)$$

If we solve for π, we get the market price of emissions (for simplicity, not including time t):

$$\pi = \frac{\Delta G^* + \sum_i\left(\pi_i^*\alpha_i\left[G_i^0\right]^2\right)}{\sum_i\left(\alpha_i\left[G_i^0\right]^2\right)} = \frac{\Delta G^* + \sum_i\left(w_i\,\pi_i^*\right)}{\sum_i w_i} \equiv \hat{\pi} \tag{9.8}$$

where $w_i = \alpha_i[G_i^0]^2$ is a weighting factor that considers each agent i's baseline emissions and adaptation rate at which agents can change their strategies and emissions α_i. In a cap and trade system that requires global changes in emissions reductions, the total emissions of all agents meet targets as specified by the negotiated intervention (i.e., in our case, the downward sloping trajectory in each time step $\Delta G^*(t)$ that aims toward the final emissions target). The market price that meets the negotiated target is called $\hat{\pi}_i$, and it increases with the growing rate of planned emission reduction $\Delta G^*(t)$, which can be any target for collective emissions reductions. A first option for calculating planned emissions reductions would be $\Delta G^*(t) = G(t) - G^*(t)$, i.e., the collective change in emission reductions should compensate for the gap between actual emissions and target emissions in period t. A second option $\Delta G^*(t) = G^*(t-1) - G^*(t)$ implies that the collective change in emission reductions should bridge the difference of reduction targets from previous to current time steps.

In an "unfettered" market, buyers and sellers exchange goods without any joint target, and for market clearance to occur, all trades must sum to zero (i.e., emissions reductions=emissions increases, which occurs in applications like "joint compliance" or "group permitting" markets; Hamstead and BenDor 2010). This corresponds to the case without policy intervention, $\Delta G^*(t) = 0$, which implies that the price is equal to the market-clearing price $\pi = \pi^* = \sum_i w_i \pi_i^* / \sum_i w_i = \hat{\pi}$, the weighted average of all threshold prices π_i^* with weighting coefficients w_i.

Overall, the market price π is a function of the global emission reduction goals as well as the threshold prices π_i^* of all agents in a given period, where regions with higher baseline emissions G^0 have a stronger effect as given by the weighting factors w_i. If the actual market price π diverts from the collective target price $\hat{\pi}_i$, then the collective emission reduction target $\Delta G^*(t)$ is not met.

We can see in Equation 9.7 that if the actual market price π exceeds a country's price threshold ($\pi > \pi_i^*$), then it is profitable for country i to sell emission permits. This means that the market price is higher than the value of the permits to the country itself. However, if the market price is less than the agent's threshold price ($\pi < \pi_i^*$), it becomes profitable to buy emission permits. In this case, the market price is lower than the value of the permits to the country, and the country will start buying more permits so it can continue polluting or pollute in greater quantities.

Viability Analysis

Viability Constraints for Emissions Trading

When we introduce t as an index for a given time period and calculate each region's emissions $G_i(t) = G_i^0 - R_i(t) = G_i^0[1 - r_i(t)]$ as reduced from baseline emissions, we seek the percentage emission reductions $r_i(t) \leq 1$ that continue to improve the value function V_i. These emissions reductions are partly determined by the relationship between emissions market price π (a global variable) and the specific emissions reduction of each agent r_i and the emission targets G_i^*. As a result, we will find a set of reduction curves, where all agents have their own set of boundary conditions. If the agent moves beyond these boundary conditions, the agent begins to lose value V_i or misses the emission reduction targets G_i^*. We derive these curves in Appendix 9.2.

As we did in our viability analyses in Chapter 7 and the fisheries example in Chapter 8, we develop three viability conditions of emission reduction:

1. *Marginal economic viability:* The threshold at which agent i has no incentive to increase emissions is given by the marginal value (partial derivative) $v_{r_i} = \partial V_i / \partial r_i \geq 0$. Thus, value V_i increases or is kept constant with respect to growing emission reduction r_i, a condition that has also been used to determine the threshold price π_i^*. We call this the *marginal reduction* (MR) emissions price.

2. *Absolute economic viability:* For each agent, economic viability implies that value losses are avoided $V_i \geq 0$, which is a viability condition that we will define with a new variable: the *economic threshold* emissions price $\tilde{\pi}_i$.

3. *Environmental viability:* The combined emission reductions r_i and threshold emission price $\pi \geq \hat{\pi}$ together ensure that the *actual* global emissions G do not exceed the world's emission *target* $G^* \geq G$.

Together, these viability constraints define important boundary conditions for emissions reductions and emission prices. To maintain the first and third viability conditions, the market price of emission reduction should be high enough (above the individual and collective threshold prices π_i^* and $\hat{\pi}$) to induce further emission reduction, but not too high (below the economic threshold price $\tilde{\pi}_i$) that agents experience value loss. The case where all conditions are met is

$$\pi_i^* < \hat{\pi} < \pi < \tilde{\pi}_i$$

In the middle range between the marginal reduction (MR) π_i^* and economic threshold prices $\tilde{\pi}_i$, both economic viability conditions are valid for the individual agents, where the collective threshold $\hat{\pi}$ is supposed to be between π_i^* and π. The viability graph (price vs. emission reduction; Figure 9.4) presents the MR and economic threshold prices

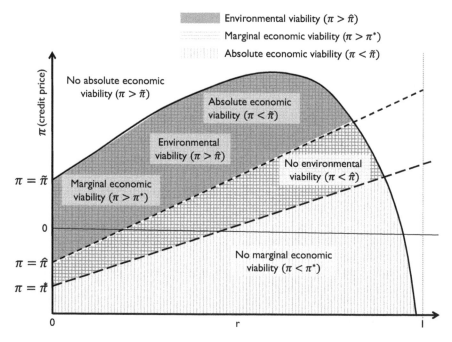

FIGURE 9.4

Zones of viability in an emissions trading scheme based on threshold conditions for emissions price π as a function of total emission reduction r and given emission target G^*. Shown are typical functions the viability conditions, including marginal economic viability ($\pi = \pi_i^*$ with boundary $v_n = 0$) absolute economic viability ($\pi = \tilde{\pi}_i$ with boundary $V_i = 0$), and the price threshold $\pi = \hat{\pi}$ which represents the environmental viability condition.

as a function of emission reduction for each agent. The following cases are of interest for each agent i:

1. $\pi_i^* < \pi < \tilde{\pi}_i$: This guarantees that agent i generates *positive* value and has an interest in further *reducing* its own emissions.

2. $\pi < \pi_i^* < \tilde{\pi}_i$: Agent i has *positive* value and benefits from *increasing* its own emissions.

3. $\pi_i^* < \tilde{\pi}_i < \pi$: Agent i is in the *negative* value range and benefits from *decreasing* its own emissions.

4. $\tilde{\pi}_i < \pi < \pi_i^*$: Agent i is in the *negative* value range and benefits from *increasing* its own emissions.

5. $\tilde{\pi}_i < \pi_i^* < \pi$: Agent i is in the *negative* value range and benefits from *decreasing* its own emissions.

Testing the Viability Constraints

Since these conditions apply differently for each agent, it is the task of the market manager (or other decision makers) to set the target emissions G_i^* low enough to keep the market price high for a sufficient number of agents. Whether this is possible also depends on the initial conditions at time $t=0$ and whether the situation for $r_i=0$ is within these price boundaries. Take, for example, an agent whose economy generates value loss; the absolute economic viability condition $V_i \geq 0$ can never be met[6], because emission reductions make it even more negative. Differently stated, if value is already negative without emission trading, then positive value cannot be achieved for a positive price, and the economic threshold price $\tilde{\pi}_i$ becomes negative.

While we have established two economic viability conditions defining threshold conditions on price, we have one environmental viability condition given by Equation 9.8. If the actual price meets this price threshold $\pi = \hat{\pi}$ the collective emission reductions of all regions together are able to meet the global emission target G^*. However, if for $\pi < \hat{\pi}$, the emissions target is missed, while for $\pi > \hat{\pi}$ it is exceeded.

The emissions target G^* is set according to several criteria, including damage limitation, risk reduction, and sustainability. While we discussed how the emission targets could be defined earlier, it is clear that any environmental viability condition must be based on a temperature goal that avoids catastrophic problems, such as limiting average future temperature increase to 2°C. This can be translated into a total emissions level G^* (and associated emission pathways $G^*(t)$) that would be compatible with the temperature goal, assuming that the causal mechanisms between emissions and temperature rise are sufficiently understood. An environmental viability condition represents a third boundary condition that restrains emissions reduction and affects agents' total value. Changes to G^* shift the shape of the viability curves connecting π and r_i for the economic (total value) viability condition. Similar to the two economic price thresholds, we can consider the price ranges that meet all, some, or no viability conditions.

In Figure 9.4, we demonstrate this viability analysis for a single hypothetical agent, observing the boundary conditions (total value, marginal value, and environmental sustainability) and observing where they intersect.

---

[6] The agent's value may be increased as climate change-induced damage D(G) is reduced, but not by the agent's actions alone.

Modeling Emissions Trading Scenarios

As we discussed at the start of this chapter, while an emissions trading model could represent the actions of all individual countries, here we consider the interactions of countries as they are grouped into 11 world regions (Figure 9.5). These regions follow the regional breakdown defined by Leimbach and Toth (2003). We do this partly for simplicity, but also under the assumption that nations within these regions have some similarity to each other and will set up some sort of self-governing system to help allocate the costs and benefits of permit purchases or sales. Therefore, the results are depicted in the following figures where the following acronyms are used for the 11 regions: Sub-Saharan Africa (AFR); China, Mongolia, Vietnam, Cambodia, and Laos (CPA); Eastern Europe (EEU); Former Soviet Union (FSU); Latin America and the Caribbean (LAM); Middle East and North Africa (MEA); North America (NAM); Pacific OECD (POECD [a.k.a. PAO]: Japan, Australia, New Zealand); Other Pacific Asia (PAS); South Asia (SAS); and Western Europe (WEU).

We draw on data for each region, including regional GDP and emissions, collected by Scheffran and Leimbach (2006). The data and scenarios used in exploring this model build on previous efforts in developing the Integrated assessment of CLImate Protection Strategies (ICLIPS) climate impact model (Leimbach and Toth 2003). The ICLIPS project arose out of the Potsdam Institute for Climate Impact Research (Germany) and focused on seeking GHG emissions pathways that prevent both climate change impacts and

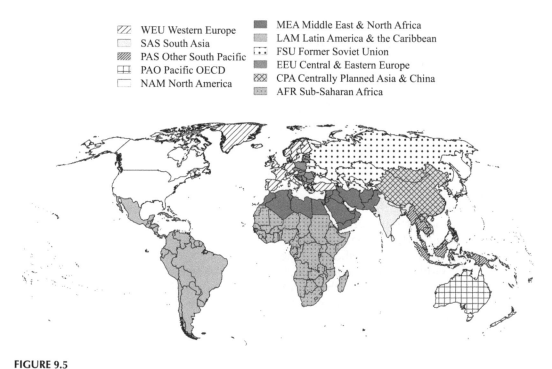

FIGURE 9.5

Map of 11 regions involved in our hypothetical emissions trading scheme. Regions adapted from Gritsevskyi and Schrattenholzer (2003) and Scheffran and Leimbach (2006).

intolerable mitigation costs (Toth 2003). ICLIPS itself was a multistep, dynamic extension of previous efforts developed by Scheffran (2004).

Our simulations start parameterizing initial emissions G_i^0 (where $r_i = 0$), which determine the value function $V_i(0)$ for all agents (regions). During subsequent time steps, emission reductions $r_i(t)$ are the main control variables of each region to determine the emission price π and the value functions $V_i(t)$. For each time step t, their changes are selected according to the decision rules $\Delta r_i(t) = \alpha_i v_{r_i}(t)$, which then determine the emission reduction $r_i(t+1) = r_i(t) + \Delta r_i(t)$. We further assume option 1 (on page 259) for calculating yearly, collective emissions reductions $\Delta G^*(t) = G(t) - G^*(t)$.

Results

Case 1. Allocations Proportionate to an Agent's Baseline Emissions

The *left column* of Figure 9.6 shows the results of the first simulation, where we can see total emissions and market price, regional emissions (total and per capita), regional values, reduction rates, marginal threshold prices and economic threshold prices for the 11 world regions over a 20-year program period.

The global emissions target is met as a result of the dynamic interaction, and the market price increases to $16.11 per metric ton of carbon at the end of the 20-year period. Nine regions stay between 6.3% Western Europe [WEU] and 33.5% Eastern Europe [EEU] emission reductions. However, China and neighboring countries (CPA; 79%) and the Former Soviet Union (FSU; 90%) hitting maximum allowed by the model) show exceptionally high emission reductions, likely because these regions highly benefit from selling emission rights as a source of income. This behavior is matched by observed threshold prices, which are very high for North America (NAM) and Western Europe (WEU) who both maintain high marginal benefits from emissions and high abatement costs for emissions reduction. This creates strong incentives to buy emission rights, which lead to fairly low reductions (9.6% and 6.3%, respectively). The opposite situation occurs in the Eastern Europe (EEU), Former Soviet Union (FSU), and China (CPA) regions, which benefit from selling emissions rights. In particular, FSU has fairly low costs, high initial emissions, low population, and low GNP, making it very easy to make large cuts in emissions.

While two regions increase value during the course of the emissions trading program (China [CPA] and the Former Soviet Union [FSU]), all others decline as a result of emission reductions (Figure 9.6B). CPA and FSU are the two regions that cut the most emissions (Figure 9.6C), even though they have very different marginal threshold prices throughout the simulation. We can see the value loss in Western Europe (WEU) and North America (NAM), leading the regions to leave their space of economic viability. Notice the relative lack of change in per-capita emissions for North America in Figure 9.6D.

Case 2: Allocations Proportionate to an Agent's Population (Rule 2)

The *right column* in Figure 9.6 depicts a simulation of the trading system based on an initial per-capita emissions allocation. The global emissions target is met as a result of the dynamic interaction and the final market price decreases to $13.53 per metric ton of carbon

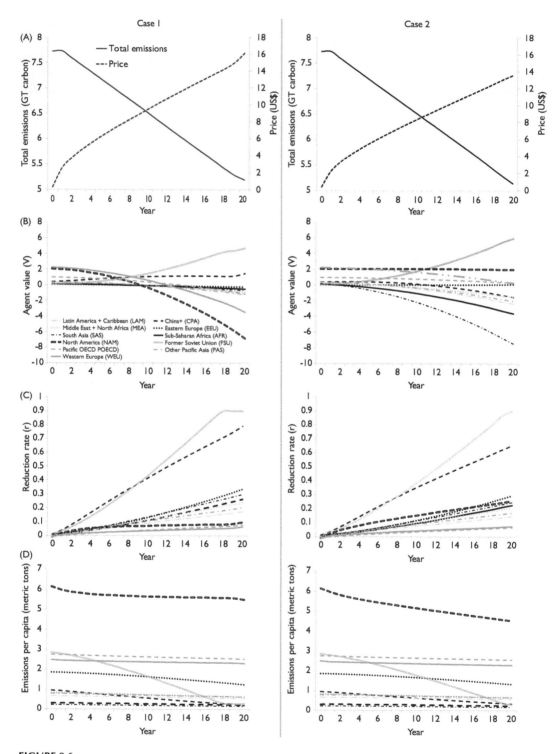

FIGURE 9.6

Baseline simulations for a global emission target of 5 GT in 20 years. Using allocation principles based on initial emission ratios (Case 1) and constant per-capita emissions (Case 2), we show comparisons of the (A) emissions and market price, (B) agent values, (C) emission reduction rates, and (D) per-capita emissions.

at the end of the period. Similar to Case 1, nine regions remain between 6.6% (Pacific OECD [POECD]) and 29.3% (Eastern Europe [EEU]) emission reductions, while China (65.1%) and the Former Soviet Union (90%) continue to show exceptionally high emission reductions. In Figure 9.6D, we see that per-capita emissions dynamics are similar for Cases 1 and 2, even though initial allocations are very different.

Agent values associated with all regions—other than the Former Soviet Union—decline as a result of emission reductions (Figure 9.6B). The Former Soviet Union appears to benefit greatly from selling emission rights as a source of income. Cases 1 and 2 differ largely in the regions that "win" or "lose" under the trading arrangement; in Case 1, we see rapid declines in the value of North America and Western Europe during increases in value in China and the Former Soviet Union while in Case 2, we see value loss in South Asia (SAS) and Sub-Saharan Africa (AFR) as the Former Soviet Union continues to gain value, partly due to low population and low marginal threshold price.

This is a prime demonstration of the controversy accompanying the initial allocation of permits. The model's extreme sensitivity to the initial allocation scheme of permits shows the wide swings that regions can have in terms of value capture within a trading scheme. Overall, we see that in the emissions trading market, a few regions become either major buyers or sellers of emissions permits, while most regions only take a small role in the market exchange.

Summary

In this chapter, we have introduced a model to better understand the collective action problem in climate policy, the associated conflicts between economic and environmental goals in an emissions trading program, and the likelihood that actors will leave such a program. Although we have spent much of this book using the VIABLE framework to set up models of agent actions, their impacts on the environment, and the feedbacks between the two, this chapter is focused on modeling the interaction of agents through a negotiated solution to anthropogenic climate impacts. In modeling the market interactions of emission trading among multiple agents, we analyze how negotiated agreements on emission reductions are implemented through a market price of emissions. This price drives the individual emission reductions of all agents, thereby defining a mechanism to overcome the collective action problem in a market of individual goal seekers.

We can begin to better understand new points of conflict that may arise during an emissions trading agreement. This model highlights many of these potential conflict points, including initial allocations, target pathways, and regions with value loss. Given the many *a priori* negotiations necessary in forming an emissions market—where emissions allocations would need to be agreed upon, as would trading platforms and arrangements to measure and enforce emissions reduction agreements and trades—we can start to identify areas where agreements could begin to break down. In this chapter, the process by which the market allocates the right to emit is actually an emergent outcome of individual agents' goal seeking processes. However, this allocation process comes at the cost that some agents experience value losses as a result of trading. The regions that lose value in the face of this type of solution move outside of their viability space, and may attempt to exit the emissions trading arrangement, if no compensation can be found. We can imagine that these regions will need additional negotiated assistance to prevent their departure and eventual failure of the agreement.

In the next chapter, we will explore how to extend the VIABLE framework to function in a spatially heterogeneous landscape, focusing on a conflict over bioenergy crop introduction that has the potential to precipitate land use and water conflicts.

Appendix 9.1: Derivation of Value Change with Respect to Emissions Reduction v_{r_i} and Threshold Price π_i^*

To find v_{r_i}, we take the derivative of Equation 9.3, yielding

$$\frac{\partial V_i}{\partial r_i} = v_{r_i} = -u_i G_i^0 + 2 d_i G_i^0 \sum_j \left[G_j^0 (1 - r_j) \right] - 2 c_i r_i G_i^0 + \pi G_i^0 - \left\{ \pi' \left[G_i - G_i^* \right] \right\} \tag{9.9}$$

where π' is the partial derivative of emission price with regard to r_i. If we set Equation 9.9 to zero, $v_{r_i} = 0$, we find the condition when the value function is indifferent to emission changes (which can be a maximum or minimum). Solving for π therefore gives us each agent's threshold emission price π_i^* for value indifference, which increases with net productivity u_i and and unit abatement c^i decreases with damage per emission unit d_i:

$$\pi = u_i - 2 d_i G + 2 c_i r_i + \left\{ \pi' \left[(1 - r_i) - \frac{G_i^*}{G_i^0} \right] \right\} = \pi_i^* \tag{9.10}$$

For a small contribution of π' (and a small value for the term it is multiplied with), this term can be neglected, which simplifies to the approximate threshold price $\pi_i^* = u_i - 2 d_i G + 2 c_i r_i$ and the approximate derivative with respect to r_i, $(\pi_i^*)' = 2 d_i G_i^0 + 2 c_i$.

Using the threshold prices, we obtain the approximate market price according to the political market clearing condition (Equation 9.8),

$$\sum_i \Delta R_i(t) = \Delta r_i(t) G_i^0 = \sum_i \alpha_i \left(\pi(t) - \pi_i^*(t) \right) \left[G_i^0 \right]^2 = \Delta G^*(t)$$

using $w_i = \alpha_i [G_i^0]^2$ as a weighting factor for each agent i:

$$\pi = \frac{\Delta G^*(t) + \sum_i \pi_i^*(t) \alpha_i [G_i^0]^2}{\sum_i \alpha_i \left[G_i^0 \right]^2} = \frac{\Delta G^*(t) + \sum_i w_i \pi_i^*(t)}{\sum_i w_i} \tag{9.11}$$

This leads to the derivative with regard to r_i:

$$\pi' = \frac{\sum_j w_j \left(\pi_j^*\right)'}{\sum_j w_j} = \frac{2\left(w_i c_i + G_i^0 \sum_j w_j d_j\right)}{\sum_j w_j} \tag{9.12}$$

Thus, by combining Equations 9.10 and 9.12, the corrected threshold price becomes the final threshold price, which now depends on the individual emissions target $G_i^*(t)$ in each time period t:

$$\pi_i^* = u_i - 2d_i G + 2c_i r_i + \left\{ \frac{2\left(w_i c_i + G_i^0 \sum_j w_j d_j\right)}{\sum_j w_j} \left[(1 - r_i) - G_i^*/G_i^0\right] \right\} \tag{9.13}$$

Using this corrected threshold price leads to the overall corrected price function π (Equation 9.11). Having found this description of the threshold price, we can also see that algebraically, Equation 9.9 collapses to

$$v_r = \frac{\partial V_i}{\partial r_i} = \left(\pi - \pi_i^*\right) G_i^0 \tag{9.14}$$

Appendix 9.2: Derivation of Viability Conditions

Marginal Economic Viability Condition

As long as the marginal value of emission reduction is positive, additional reductions are beneficial and associated with positive value gains:

$$v_{r_i} = \frac{\partial V_i}{\partial r_i} = \left(\pi - \pi_i^*\right) G_i^0 > 0$$

Resolving this for price π defines a marginal reduction (MR) threshold price for agent i as a function of emission reductions r_j of all agents

$$\pi > u_i - 2d_i \sum_j \left[(1 - r_j) G_j^0\right] + 2c_i r_i + \left\{ \frac{2\left(w_i c_i + G_i^0 \sum_j w_j d_j\right)}{\sum_j w_j} \left[(1 - r_i) - G_i^*/G_i^0\right] \right\} \equiv \pi_i^*$$

This implies that with increasing emission reduction r_i and emission targets G_i^*, the threshold price π_i^* decreases, which implies that more agents reduce emissions for the same actual price π. This MR threshold price should be below the market price π to maintain positive marginal value. For a given market price, this defines an upper limit for reduction below which marginal value of reduction is still positive: $r_i < r_i^*$.

Absolute Economic Viability Condition

Emission reduction should maintain positive value and avoid absolute value losses:

$$V_i(t) = U_i - C_i - D_i - C_i^r - \Pi_i = u_i G_i(t) - d_i [G(t)]^2 - c_i G_i^0 r_i^2 - \pi (G_i - G_i^*) > 0$$

Resolving this for price π defines an economic threshold price for agent i as a function of emission reductions r_j of all agents for $G_i > G_i^*$:

$$\pi < \frac{u_i(1-r_i) - d_i \dfrac{\left(\sum_j \left[(1-r_j)G_j^0\right]\right)^2}{G_i^0} - c_i r_i^2}{1 - r_i - G_i^* / G_i^0} \equiv \tilde{\pi}_i$$

If value is negative without emissions trading, emissions reductions can compensate for this only by selling emission rights as a source of income. Thus, the viability condition $\pi < \tilde{\pi}_i$ holds as long as the emissions target is met ($G_i = (1-r_i)G_i^0 < G_i^*$); otherwise, this requirement reverts to $\pi > \tilde{\pi}_i$. At $G_i = G_i^*$, the economic threshold price $\tilde{\pi}$ switches through infinity between positive and negative values. This implies that with increasing emission reduction r_i, the economic threshold price declines. However, this threshold price should be above the market price π to maintain positive agent value. To avoid negative value, the market price should not be too low for the seller and not too high for the buyer of emission permits. For a given market price, this defines a range of admissible reductions, which depends on the solution of the quadratic equation (due to the quadratic terms in r_i).

Environmental Viability Condition

We assume that the environmental viability condition involves meeting the target created through the initial negotiations that established the trading program. We also assume that parties begin with a temperature limit, above which environmental conditions deteriorate or change negatively. Climate models are frequently used to study the link between temperature and emissions rates, thereby fostering an emissions target that meets this environmental viability condition. As we showed in Equation 9.8, our assumption is that the environmental viability condition will be met if we abide by the global emissions target, whereby we can calculate that given the global emissions limit G^*, the trading system will yield a market price:

$$\pi > \frac{\Delta G^* + \sum_i w_i \pi_i^*}{\sum_i w_i} = \hat{\pi}$$

This can be presented where price is a function of the change rate of global emission targets ΔG^* and all emission reductions r_i.

Questions for Consideration

1. Atmospheric CO_2 concentration exceeded 400 ppm in 2015 from about 280 ppm in preindustrial times. Temperatures have risen about 1°C (1.8°F) since preindustrial times and 0.87°C relative to the 1951–1980 average, and may continue to rise 4°C relative to the 1986–2005 average by end of this century (IPCC 2014). *How much have CO_2 concentrations, temperatures, and sea-levels changed since 2015?* For updated numbers, see NASA's website: http://climate.nasa.gov/vital-signs/global-temperature.

2. Check out MIT's interactive GHG simulator: http://scripts.mit.edu/~jsterman/climate/master/. *Does playing with this simulator affect how you think about the global emissions target that we set in our model? How so? In your mind, what is the key insight from this simulation?*

3. The Nature Conservancy has created a Coastal Resilience Mapping Portal: http://maps.coastalresilience.org/. Try looking at an area near you, or an area you care about. When you explore this mapping tool, notice the features for studying the geographic extent of sea-level rise inundation. Play around a bit with different sea-level rise scenarios. *Based on this experience, how do you think we could more accurately define "emissions damages" in our model? Do you think we have underestimated or overestimated damages based on our metric?*

4. Although we treat it as homogenous in this chapter, the value function that we define for each agent could vary for each region. *How could agents' value function vary? What would make them independent of each other?*

5. Imagine that the next step in designing an emissions trading program is to find mechanisms that allocate initial emission permits from global levels down to individual humans or households. *Do you think that we should allocate emissions limits such that individuals can take part in these trading systems? Or should trading systems operate at a higher level? Say, involving large firms that we buy things from, who would pass on the costs to us indirectly? What are the advantages and disadvantages of each approach? Where do you stand on this and why?*

6. Interested in adding on to our model? Here are some ideas:

   - The benefits generated from economic output Q can be measured by a utility function $U(Q)$, which indicates the positive values generated by production associated with emissions. We focus only on the utility generated from production pathways associated with GHG emissions G. Unfortunately, we do not consider other pathways based on zero emissions (e.g., with renewables) are not considered here. This reflects the fact that with emissions G declining toward zero, the associated utility $U(G)$ generated from emissions also declines as it is increasingly replaced by carbon-free production. This is a shortcoming of our model. *How could we improve the model to incorporate a shift from GHG-inducing production to zero-emissions production?*

   - We have defined a_i as the "reactivity" of agents with respect to transacting permits. We could instead define two parameters, a_i^+ and a_i^-, indicating agent reactivity with respect to *buying* or *selling* permits. That is, the agent may be able to sell permits and reduce emissions more easily than they can buy permits and increase emissions or vice versa (see Scheffran 2004).

- The damages in terms of utility losses $D=d_G G^\delta$ induced by emissions and resulting climate change can be similarly calculated, where exponent δ indicates a potential nonlinear relationship, with increasing marginal damage if $\delta > 1$ and decreasing marginal damage if $\delta < 1$. We use $\delta=2$ as a standard case, much in the same way we treat cost as a quadratic function of emissions reduction. Therefore, we take emission-related damage as $D=d_G G^2$. While these functions are often linked to temperature (Scheffran 2008), we use damage as a function of emissions as a proxy for perceived damage affecting value, taking time discounting into account. We could add more sophisticated damage shape functions, including those described by IPCC (2001, Chapter 19) or by the National Academies of Sciences (2017).

- We assume that the negotiated emissions target $G^*(t)$ remains the same trajectory as originally negotiated (i.e., the trajectory is not renegotiated along the way, whereby G^* would be reduced midstream as the program became increasingly stringent). Over time, we can imagine a scenario where G^* (the final emissions target) is reduced as the emissions trading program gets increasingly stringent.

- We can immediately imagine the benefits of including many more explicit interactions among agents and how this might affect their trading behaviors. For example, we may see structured deals between individual agents that trade information or technology as a means of avoiding certain types of emissions trades. We can think of these as efforts to invest in technology and improvements for other agents, thereby reducing their emissions through "extramarket" investments. This is a phenomenon sometimes seen in other environmental markets, particularly in water quality trading (e.g., see Hamstead and BenDor 2010).

- We could further develop this model by considering scenarios where there is a more sophisticated trade-off among the factors that influence agent behavior. For example, we could represent a situation where increasing costs force certain agents to invest in increased emissions reduction technologies. While this might lower an agents' value in the short term, it could increase long-term value if the agent can sell this technology to other agents as a means for avoiding emissions trades and value impacts from forced emissions reductions (i.e., wind turbine technology exported from Scandinavian nations). This is an indication that agent behavior is adaptive and contributes to finding solutions to problems. This model restructuring could simply follow the concept of adding investment pathways and have agents optimize between them (i.e., buy permits, reduce emissions, become more efficient, sell efficiency technology).

- A final interesting avenue to explore could look at complex emissions target pathways, such as those suggested in the "Under 2 MOU" proposal (http://under2mou.org/), which prescribes to reduce GHG emissions 80–95% below 1990 levels by 2050 or achieve a per capita annual emissions target of less than 2 metric tons by 2050. In the face of continuing COP meetings staged around the world, this model expansion could provide strong insight into a framework for better understanding the increasing complexity of negotiated solutions to the seemingly intractable climate conflict.

Try out some of these changes; you can start by going to this book's website—http://todd.
bendor.org/viable —and tinker around with the model. What other ideas do you have to
modify the model? We invite you to contact the authors (see http://todd.bendor.org/

contact) to forward along your modifications so that we can add them to this website so others can learn from your work!

Additional Resources

Interested in emissions and other types of environmental trading? For more information on emissions trading, we refer readers to overview texts such as Kosobud, Schreder, and Biggs (2000) and Tietenberg (2006).

Author Todd BenDor's primary area of study concerns whether environmental markets, specifically "ecosystem service markets" and "offset markets" such as water quality trading and wetland and stream mitigation, improve environmental outcomes. We refer readers to his work on a broad set of market mechanisms, including BenDor and Doyle (2010), Hamstead and BenDor (2010), and Woodruff and BenDor (2015). There, readers will find a plethora of references to other interesting aspects of environmental market mechanisms.

A key concept in our model is that total emissions $G(t)$ and target emissions $G^*(t)$ are both *flow* variables, representing the amount of gases emitted *per year*. This is in comparison to other models presented in this book, where the environmental system variables are modeled as both stocks (e.g., fish populations) as well as flows (e.g., biomass growth rates). Alternatively, we could sum the total cumulative emissions from year-to-year to track total CO_2 released over time or use carbon concentration in the atmosphere as an accumulated stock variable, although we do not do that here. For a broader discussion of environmental trading and quantification as stock or flows, see BenDor and Woodruff (2014).

Interested in metrics related to emissions and emissions modeling? Exploring metrics and ratios in emissions trading. If we define output Q, energy use E, emissions G, population N, and investments in economic production C, we can use a table to help us keep track of important ratios (per-unit metrics) that are commonly used in emissions trading discussions (following the notation in Scheffran [1994, 2004]). As we showed in our explanation of the VIABLE framework in Chapter 7 and in Chapter 8's discussion of fisheries modeling, we can use these terms to develop a modeling framework to establish the goals and values driving regions' pollution levels of regions (see the following box).

• $g_e = G/E$: Emission intensity of energy, the amount of emissions produced per unit of energy use	• $c_g = C/G$: Investment intensity of emissions (investment in economic production per emission unit)
• $q_e = Q/E$: Energy productivity (inverse of energy intensity e_q)	• $g_c = G/C$: Emissions per unit investment in economic production (emissions per cost unit)
• $q_n = Q/N$: Output per capita	• $c_e = C/E$: Investment in production per unit of energy (production intensity of energy)
• $q_g = Q/G$: Emission productivity (inverse of g_q)	• $c_q = C/Q$: Investment required per unit of production (production unit cost)
• $u_q = U/Q$: Utility per production unit Q. For given u_q we have utility $U = u_q Q$.	
• $d_g = D/G$: Damage per emission unit[7]	
• $\pi_g = \Pi/G$: Trading price per emission unit (we will simplify this as $\pi \equiv \pi_g$)	

---

[7] Every agent contributes to the climate damage experienced by all other agents, a fact that represents an important coupling factor. For incremental changes, actors only take their own contribution into consideration, leading to the collective action problem.

Having defined the basic emissions and economic variables and their ratios, we can define numerous useful functional relationships, such as economic output $Q = q_g G = q_g g_c C = q_u U$, emissions $G = g_q Q = g_c C = D/d_g$, and so on. Perhaps the most famous of these relationships is the Kaya formula $G = g_e e_q q_n N$, which has been used to create and identify different emissions-related variables (IPCC 2014). This formula connects the emission intensity of energy $g_e = G/E$, energy intensity $e_q = E/Q$, and economic output per capita $q_n = Q/N$ (Waggoner and Ausubel 2002). To avoid misunderstanding, the ratio factors are only constant for linear relationships between the respective variables. They become more complex when there are nonlinear relationships between variables, which complicate calculations of partial derivatives. For example, they can be functions of the variables themselves (e.g., representing learning, escalation, or satisfying behavior of individual actors).

The abatement costs, or the costs to reduce (mitigate) emissions, are specified separately as reduction costs $F = c_R R = c_R r G^0 = c r^\gamma G^0$, where c_R is the unit abatement cost, and c is a factor in the nonlinear functional relationship. This assumes that costs are proportionate to the magnitude G^0 to which reduction is applied, but that the effect of percentage reduction r is nonlinear.

Likewise, damage from emissions can be quantified as $D = d_G G^\delta$, where the exponent δ indicates the degree of nonlinearity. Therefore, we can take emission-related damage as $D = d_g G = d_G G^\delta$, taking $\delta = 2$ as we did in the abatement cost case. For nonlinear relationships (exponent $\neq 1$) we use another representation, in this case d_G as a proportionality factor. If $\delta = 1$, then $d_g = d_G$.

References

Antes, Ralf, Bernd Hansjürgens, and Peter Letmathe, eds. 2006. *Emissions Trading and Business*. New York: Springer Physica-Verlag.

Anthoff, David, Cameron Hepburn, and Richard S. J. Tol. 2009. "Equity Weighting and the Marginal Damage Costs of Climate Change." *Ecological Economics* 68 (3): 836–49.

Arrhenius, Svante. 1896. "On the Influence of Carbonic Acid in the Air upon the Temperature of the Ground." *London, Edinburgh, and Dublin Philosophical Magazine and Journal of Science (Fifth Series)* 41 (April): 237–75.

Aubin, Jean-Pierre, Telma Bernado, and Patrick Saint-Pierre. 2005. "A Viability Approach to Global Climate Change Issues." In *The Coupling of Climate and Economic Dynamics*, edited by Alain Haurie and Laurent Viguier, Vol. 22, 113–43. Advances in Global Change Research. Amsterdam, Netherlands: Springer.

Aubin, Jean-Pierre, Luxi Chen, and Marie-Hélène Durand. 2012. "Dynamical Allocation Method of Emission Rights of Pollutants by Viability Constraints under Tychastic Uncertainty." *Environmental Modeling & Assessment* 17 (1–2): 7–18.

Avi-Yonah, Reuven S., and David M. Uhlmann. 2009. "Combating Global Climate Change: Why a Carbon Tax Is a Better Response to Global Warming than Cap and Trade." *Stanford Environmental Law Journal* 28 (3): 3–49.

Barnett, Jon. 2003. "Security and Climate Change." *Global Environmental Change-Human and Policy Dimensions* 13 (1): 7–17.

Barnett, Jon, and W. Neil Adger. 2007. "Climate Change, Human Security and Violent Conflict." *Political Geography* 26 (6): 639–55.

Beatley, Timothy. 2009. *Planning for Coastal Resilience: Best Practices for Calamitous Times*. Washington, DC: Island Press.

BenDor, Todd, and M. W. Doyle. 2010. "Planning for Ecosystem Service Markets." *Journal of the American Planning Association* 76 (1): 59–72.

BenDor, Todd, and Sierra Woodruff. 2014. "Moving Targets and Biodiversity Offsets for Endangered Species Habitat: Is Lesser Prairie Chicken Habitat a Stock or Flow?" *Sustainability* 6 (3): 1250–59.

Boyd, J., D. Burtraw, A. McConnell, V. Krupnick, R. Newell, K. Palmer, J. Sanchirico, et al. 2003. "Trading Cases: Five Examples of the Use of Markets in Environmental and Resource Management." In *The RFF Reader in Environmental and Resource Policy*, edited by Wallace E. Oates, Vol. 37, 56–65. Washington, DC: Resources for the Future.

Brassel, K.-H., O. Edenhofer, M. Mhring, and K. G. Troitzsch. 2000. "Modelling Greening Investors." In *Tools and Techniques for Social Science Simulation*, edited by Ramzi Suleiman, 317–343. Heidelberg, Germany: Springer-Verlag Telos.

Bruckner, Thomas, Georg Hooss, Hans-Martin Fussel, and Klaus Hasselmann. 2003. "Climate System Modeling in the Framework of the Tolerable Windows Approach: The ICLIPS Climate Model." *Climactic Change* 56: 119–37.

Bruckner, Thomas, G. Petschel-Held, F. L. Tóth, H. M. Füssel, C. Helm, Marian Leimbach, and H. J. Schellnhuber. 1999. "Climate Change Decision-Support and the Tolerable Windows Approach." *Environmental Modeling and Assessment* 4 (4): 217–34.

Campbell, Kurt M., Jay Gulledge, John R. McNeill, John Podesta, Peter Ogden, Leon Fuerth, R. James Woolsey, et al. 2007. *The Age of Consequences: The Foreign Policy and National Security Implications of Global Climate Change*. Washington, DC: Center for Strategic and International Studies.

CARB. 2011. "Assembly Bill 32 Overview." Sacramento, CA: California Air Resources Board. www.arb.ca.gov/cc/ab32/ab32.htm

CARB. 2018. "Cap and Trade Program." Sacramento, CA: California Air Resources Board. www.arb.ca.gov/cc/capandtrade/capandtrade.htm

Carmin, JoAnn, Isabelle Anguelovski, and Debra Roberts. 2012. "Urban Climate Adaptation in the Global South: Planning in an Emerging Policy Domain." *Journal of Planning Education and Research* 32 (1): 18–32.

Carson, R. 1962. *Silent Spring*. New York: Houghton Mifflin Harcourt.

CEJAC. 2017. *The California Environmental Justice Advisory Committee's Declaration in Support of Carbon Pricing Reform in California*. Sacramento, CA: California Environmental Justice Advisory Committee. www.arb.ca.gov/cc/ejac/meetings/02142017/20170215ca-ej-declaration-on-carbon-pricing-reform-approved.pdf

Chen, Yihsu, Andrew L. Liu, and Benjamin F. Hobbs. 2011. "Economic and Emissions Implications of Load-Based, Source-Based, and First-Seller Emissions Trading Programs under California AB32." *Operations Research* 59 (3): 696–712.

Cleetus, Rachel. 2011. "Finding Common Ground in the Debate between Carbon Tax and Cap-and-Trade Policies." *Bulletin of the Atomic Scientists* 67 (1): 19–27.

Crocker, T. D. 1966. "The Structuring of Atmospheric Pollution Control Systems." In *The Economics of Air Pollution*, edited by H. Wolozin, 61–86. New York: W.W. Norton & Co.

Dales, Jon Harkness. 1968. *Pollution, Property and Prices*. Toronto, ON: University of Toronto Press.

Edenhofer, Ottmar, Nico Bauer, and Elmar Kriegler. 2005. "The Impact of Technological Change on Climate Protection and Welfare: Insights from the Model MIND." *Ecological Economics* 54: 277–92.

EDGAR. 2017. "CO_2 Time Series 1990–2015 per Region/Country." Brussels, Belgium: European Commission Emissions Database for Global Atmospheric Research (EDGAR). CO_2 Time Series 1990–2015 per Region/Country. October 30, 2017. http://edgar.jrc.ec.europa.eu/overview.php?v=CO2ts1990-2015

EIA. 2015. *International Energy Statistics*. Washington, DC: U.S. Energy Information Administration. www.eia.gov/beta/international/data/browser/

Elkins, Paul, and Terry Baker. 2001. "Carbon Taxes and Carbon Emissions Trading." *Journal of Economic Surveys* 15 (3): 325–76.

Ellerman, A. Denny, and Barbara K. Buchner. 2007. "The European Union Emissions Trading Scheme: Origins, Allocation, and Early Results." *Review of Environmental Economics and Policy* 1 (1): 66–87.

Ellerman, A. Denny, Frank J. Convery, and Christian de Perthuis. 2010. *Pricing Carbon: The European Emissions Trading Scheme.* Cambridge, England: Cambridge University Press.

Finus, M. 2001. *Game Theory and International Environmental Cooperation.* Northampton, MA: Edward Elgar.

Fisher, A. C. 1981. *Resource and Environmental Economics.* Cambridge, England: Cambridge University Press.

Fourier, Joseph. 1824. "Remarques Générales Sur Les Températures Du Globe Terrestre Et Des Espaces Planétaires." *Annales de Chimie et de Physique* 27: 136–67.

Gritsevskyi, Andrii, and Leo Schrattenholzer. 2003. "Costs of Reducing Carbon Emissions: An Integrated Modeling Framework Approach." *Climatic Change* 56 (1–2): 167–84.

Grundig, F., H. Ward, and E. R. Zorick. 2001. "Modeling Global Climate Negotiations." In *International Relations and Global Climate Change,* edited by U. Luterbacher and D. F. Sprinz. Cambridge, MA: MIT Press.

Hamrick, Kelley, and Gallant Gallant. 2017. *Unlocking Potential: State of the Voluntary Carbon Markets 2017.* Washington, DC: Ecosystem Marketplace. http://forest-trends.org/releases/p/sovcm2017

Hamstead, Zoe, and Todd BenDor. 2010. "Over-Compliance in Water Quality Trading Programs: Findings from a Qualitative Case Study in North Carolina." *Environment and Planning C* 28: 1–17.

Harlow, Robert L. 1974. "Conflict Reduction in Environmental Policy." *Journal of Conflict Resolution* 18 (3): 536.

Haya, Barbara, Aaron Strong, Emily Grubert, and Danny Cullenward. 2016. "Carbon Offsets in California: Science in the Policy Development Process." In *Communicating Climate-Change and Natural Hazard Risk and Cultivating Resilience,* edited by J. Drake, Y. Kontar, J. Eichelberger, T. Rupp, and K. Taylor, 241–54. Advances in Natural and Technological Hazards Research. Cham, Switzerland: Springer.

Heal, Geoffrey, and Bengt Kristom. 2002. "Uncertainty and Climate Change." *Environmental and Resource Economics* 22: 3–39.

Hopkins, Lewis D. 2001. *Urban Development: The Logic of Making Plans.* Washington, DC: Island Press.

Hsiang, S. M., M. Burke, and E. Miguel. 2013. "Quantifying the Influence of Climate on Human Conflict." *Science* 341 (1212): 1235367.

Ide, Tobias, P. Michael Link, Jürgen Scheffran, and Janpeter Schilling. 2016. "The Climate-Conflict Nexus: Pathways, Regional Links, and Case Studies." In *Handbook on Sustainability Transition and Sustainable Peace,* edited by Hans Günter Brauch, Úrsula Oswald Spring, John Grin, Jürgen Scheffran, 285–304. Dordrecht, Netherlands: Springer.

Ide, Tobias, and Jürgen Scheffran. 2014. "On Climate, Conflict and Cumulation: Suggestions for Integrative Cumulation of Knowledge in the Research on Climate Change and Violent Conflict." *Global Change, Peace & Security* 26(3): 263–79.

IPCC. 2001. "Climate Change 2001: Climate Change Impacts, Adaptation and Vulnerability." Geneva, Switzerland: Intergovernmental Panel on Climate Change. www.ipcc.ch/ipccreports/tar/wg2/index.php?idp=678

IPCC. 2014. "Climate Change 2014: Impacts, Adaptation, and Vulnerability." Geneva, Switzerland: Intergovernmental Panel on Climate Change. www.ipcc.ch/report/ar5/wg2/

Ipsen, Dirk, Roland Rösch, and Jürgen Scheffran. 2001. "Cooperation in Global Climate Policy: Potentialities and Limitations." *Energy Policy* 29 (4): 315–26.

Jaeger, Carlo C., and Julia Jaeger. 2011. "Three Views of Two Degrees." *Regional Environmental Change* 11 (1): 15–26.

Jensen, Jesper, and Tobias N. Rasmussen. 2000. "Allocation of CO_2 Emissions Permits: A General Equilibrium Analysis of Policy Instruments." *Journal of Environmental Economics and Management* 40 (2): 111–36.

Johnston, Derek M., Stephan E. Sefcik, and Naomi S. Soderstrom. 2008. "The Value Relevance of Greenhouse Gas Emissions Allowances: An Exploratory Study in the Related United States SO2 Market." *European Accounting Review* 17 (4): 747–64.

Karl, Thomas R., and Kevin E. Trenberth. 2003. "Modern Global Climate Change." *Science* 302 (5651): 1719–23.

Kemfert, Claudia. 2001. *International Games of Climate Change Policies: The Economic Effectiveness of Partial Coalition Games (Working Paper)*. West Layfayette, IN: Purdue University Global Trade Analysis Project. www.gtap.agecon.purdue.edu/resources/download/243.pdf

Khanna, Neha. 2001. "Analyzing the Economic Cost of the Kyoto Protocol." *Ecological Economics* 38 (1): 59–69.

Kosobud, Richard F., Douglas L. Schreder, and Holly M. Biggs, eds. 2000. *Emissions Trading: Environmental Policy's New Approach*. New York: John Wiley and Sons.

Kossoy, Alexandre, and Pierre Guigon. 2012. *State and Trends of the Carbon Market 2012*. New York: World Bank. http://hdl.handle.net/10986/13336

Krabs, W., and S. Pickl. 2003. *Analysis, Controllability and Optimization of Time-Discrete Systems and Dynamical Games*. Heidelberg, Germany: Springer.

Leimbach, Marian. 2003. "Equity and Carbon Emissions Trading: A Model Analysis." *Energy Policy* 31 (10): 1033–44.

Leimbach, Marian, and F. L. Toth. 2003. "Economic Development and Emission Control over the Long Term: The ICLIPS Aggregated Economic Model." *Climatic Change* 56 (1–2): 139–65.

Lenton, Tim M., H. Held, E. Kriegler, J. W. Hall, W. Lucht, S. Rahmstorf, and H. J. Schellnhuber. 2008. "Tipping Elements in the Earth's Climate System." *Proceedings of the National Academy of Sciences* 105 (6): 1786–93.

Lippke, Bruce, and John Perez-Garcia. 2008. "Will Either Cap and Trade or a Carbon Emissions Tax Be Effective in Monetizing Carbon as an Ecosystem Service." *Forest Ecology and Management* 256 (12): 2160–65.

Metcalf, Gillbert E., and David Weisbach. 2009. "The Design of a Carbon Tax." *Harvard Environmental Law Review* 33: 499.

Metz, B., O. R. Davidson, P. R. Bosch, R. Dave, and L. A. Meyer, eds. 2007. *Climate Change 2007: Mitigation of Climate Change* (IPCC Fourth Assessment Report). New York: Cambridge University Press.

Morthorst, P. E. 2003. "National Environmental Targets and International Emission Reduction Instruments." *Energy Policy* 31 (1): 73–83.

Moss, Scott. 2002. "Agent Based Modelling for Integrated Assessment." *Integrated Assessment* 3 (1): 63–77.

National Academies of Sciences. 2017. *Valuing Climate Damages: Updating Estimation of the Social Cost of Carbon Dioxide*. Washington, DC: National Academies Press.

National Assessment Synthesis Team. 2001. *Climate Change Impacts on the United States: The Potential Consequences of Climate Variability and Change*. Cambridge, England: Cambridge University Press.

Newman, Peter, Timothy Beatley, and Heather Boyer. 2009. *Resilient Cities: Responding to Peak Oil and Climate Change*. Washington, DC: Island Press.

Nordås, Ragnhild, and Nils Petter Gleditsch. 2007. "Climate Change and Conflict." *Political Geography* 26 (6): 627–38.

Olson, Mancur. 1971. *The Logic of Collective Action: Public Goods and the Theory of Groups* (Revised Edition). Cambridge, MA: Harvard University Press.

Ott, Konrad, Gernot Klepper, Stephan Lingner, Achim Schäfer, Jürgen Scheffran, and Detlef Sprinz. 2003. *Reasoning Goals of Climate Protection - Specification of Article 2 UNF- CCC (Research Report 202 41 252)*. Berlin, Germany: Federal Ministry of the Environment, Nature, Conservation and Nuclear Safety (Europäische Akademie GmbH, Bad Neuenahr-Ahrweiler).

Parmesan, Camille, and Gary Yohe. 2003. "A Globally Coherent Fingerprint of Climate Change Impacts across Natural Systems." *Nature* 421: 37–42.

Petschel-Held, G., H. J. Schellnhuber, T. Bruckner, F. L. Toth, and K. Hasselmann. 1999. "The Tolerable Windows Approach: Theoretical and Methodological Foundations." *Climatic Change* 41 (3/4): 303–31.

Philibert, Cedric, and Jonathan Pershing. 2001. "Considering the Options: Climate Targets for All Countries." *Climate Policy* 1 (2): 211–27.

Pickl, Stefan. 2001. "Convex Games and Feasible Sets in Control Theory." *Mathematical Methods of Operations Research : ZOR* 53 (1): 51–66.

Pittock, Jamie. 2011. "National Climate Change Policies and Sustainable Water Management: Conflicts and Synergies." *Ecology and Society* 16 (2): 25.

Pizer, William A., and Xiliang Zhang. 2018. *China's New National Carbon Market (NI WP 18-01).* Durham, NC: Duke University, Nicholas Institute for Environmental Policy Solutions.

REN21. 2017. *Renewables 2017 Global Status Report.* Paris, France: REN21, United Nations Environment Programme. www.ren21.net/status-of-renewables/global-status-report/

Reuveny, Rafael. 2007. "Climate Change-Induced Migration and Violent Conflict." *Political Geography* 26 (6): 656–73.

Reuveny, Rafael. 2008. "Ecomigration and Violent Conflict: Case Studies and Public Policy Implications." *Human Ecology* 36 (1): 1–13.

Rico, Renee. 1995. "The U.S. Allowance Trading System for Sulfur Dioxide: An Update on Market Experience." *Environmental and Resource Economics* 5 (2): 115–29.

Sathaye, Jayant, and P. R. Shukla. 2013. "Methods and Models for Costing Carbon Mitigation." *Annual Review of Environment and Resources* 38: 137–68.

Sauquet, Alexandre. 2014. "Exploring the Nature of Inter-Country Interactions in the Process of Ratifying International Environmental Agreements: The Case of the Kyoto Protocol." *Public Choice* 159 (1–2): 141–58.

Scheffran, Jürgen. 1994. "Modelling International Security Problems in the Framework of the SCX Model." In *Mathematische Methoden in Der Sicherheitspolitik,* edited by J. Hermanns, B. v. Stengel, and A. Tolk, 103–20. Neubiberg, Germany: IASFOR.

Scheffran, Jürgen. 2002. "Economic Growth, Emission Reduction and the Choice of Energy Technology in a Dynamic-Game Framework." In *Operations Research Proceedings 2001: Selected Papers of the International Conference on Operations Research (OR 2001),* edited by P. Chamoni, 329–36. Duisburg, Germany: Springer.

Scheffran, Jürgen. 2004. "Interaction in Climate Games: The Case of Emissions Trading." In *Entscheidungstheorie Und -Praxis in Industrieller Produktion Und Umweltforschung [Decision Theory and Practice in Industrial Production and Environmental Research],* edited by J. Geldermann and M. Treitz, 1–18. Aachen, Germany: Shaker.

Scheffran, Jürgen. 2006. "The Formation of Adaptive Coalitions." In *Advances in Dynamic Games,* edited by A. Haurie, S. Muto, L.A. Petrosjan, 163–178. Boston: Birkhäuser.

Scheffran, Jürgen. 2008. "Preventing Dangerous Climate Change." In *Global Warming and Climate Change,* edited by Velma I. Grover, Vol. 2, 449–82. Boca Raton, FL: CRC Press.

Scheffran, Jürgen. 2016. "Der Vertrag von Paris: Klima Am Wendepunkt?" *WeltTrends* 24 (February): 4–9.

Scheffran, Jürgen, and Antonella Battaglini. 2011. "Climate and Conflicts: The Security Risks of Global Warming." *Regional Environmental Change* 11: 27–39.

Scheffran, Jürgen, Michael Brzoska, H. G. Brauch, P. M. Link, and J. Schilling, eds. 2012. *Climate Change, Human Security and Violent Conflict: Challenges for Societal Stability.* Berlin, Germany: Springer.

Scheffran, Jürgen, Michael Brzoska, Jasmin Kominek, P. Michael Link, and Janpeter Schilling. 2012. "Climate Change and Violent Conflict." *Science* 336 (6083): 869–71.

Scheffran, Jürgen, and Marian Leimbach. 2006. "Policy-Business Interaction in Emission Trading between Multiple Regions." In *Emissions Trading and Business,* edited by Ralf Antes, Bernd Hansjürgens, and Peter Letmathe. New York: Springer Physica-Verlag.

Scheffran, Jürgen, and Stefan Pickl. 2000. "Control and Game-Theoretic Assessment of Climate Change: Options for Joint Implementation." *Annals of Operations Research* 97 (1): 203–12.

Seifert, Jan, Marliese Uhrig-Homburg, and Michael Wagner. 2008. "Dynamic Behavior of CO_2 Spot Prices." *Journal of Environmental Economics and Management* 56 (2): 180–94.

Springer, Urs. 2003. "The Market for Tradable GHG Permits under the Kyoto Protocol: A Survey of Model Studies." *Energy Economics* 25 (5): 527–51.

Springer, Urs, and Matthew Varilek. 2004. "Estimating the Price of Tradable Permits for Greenhouse Gas Emissions in 2008–12." *Energy Policy* 32 (5): 611–21.

Stefan, Bachu. 2000. "Sequestration of CO2 in Geological Media: Criteria and Approach for Site Selection in Response to Climate Change." *Energy Conversion and Management* 41 (9): 953–70.

Svirezhev, Yu M., W. von Bloh, and H. J. Schellnhuber. 1999. "'Emission Game': Some Applications of the Theory of Games to the Problem of CO2 Emission." *Environmental Modeling and Assessment* 4 (4): 235–42.

The World Bank. 2016. *State and Trends of Carbon Pricing.* New York: The World Bank.

Tietenberg, Thomas H. 2006. *Emissions Trading: Principles and Practice.* Washington, DC: Resources for the Future.

Toth, Ferenc L. 2003. "Integrated Assessment of Climate Protection Strategies." *Climatic Change* 56 (1–2): 1–5.

Tyndall, John. 1869. *Heat Considered as a Mode of Motion.* New York: D. Appleton.

UNFCCC. 2015. *Adoption of the Paris Agreement (Conference of Parties-21).* Paris, France: United Nations Framework Convention on Climate Change.

UNFCCC. 2015. "Lima Call for Climate Action Puts World on Track to Paris 2015." Bonn, Germany: United Nations Framework Convention on Climate Change. http://newsroom.unfccc.int/lima/lima-call-for-climate-action-puts-world-on-track-to-paris-2015

United Nations. 1992. "United Nations Framework Convention on Climate Change." 1992. http://unfccc.int/resource/docs/convkp/conveng.pdf

Waggoner, P. E., and J. H. Ausubel. 2002. "A Framework for Sustainability Science: A Renovated IPAT Identity." *Proceedings of the National Academy of Sciences* 99 (12): 7860–65.

WBGU. 1995. "Scenario for the Derivation of Global CO2 Reduction Targets and Implementation Strategies: Statement on the Occasion of the First Conference of the Parties to the Framework Convention on Climate Change in Berlin." Berlin, Germany: German Advisory Council on Global Change. www.wbgu.de/fileadmin/user_upload/wbgu.de/templates/dateien/veroef-fentlichungen/sondergutachten/sn1995/wbgu_sn1995_engl.pdf

WBGU. 2008. "World in Transition - Climate Change as a Security Risk." Berlin, Germany: German Advisory Council on Global Change. www.wbgu.de/fileadmin/user_upload/wbgu.de/tem-plates/dateien/veroeffentlichungen/hauptgutachten/jg2007/wbgu_jg2007_engl.pdf

WBGU. 2009. *Solving the Climate Dilemma: The Budget Approach (Special Report).* Berlin, Germany: German Advisory Council on Global Change. www.wbgu.de/en/special-reports/sr-2009-budget-approach

Weber, Michael, Volker Barth, and Klaus Hasselmann. 2005. "A Multi-Actor Dynamic Integrated Assessment Model (MADIAM) of Induced Technological Change and Sustainable Economic Growth." *Ecological Economics,* Technological Change and the Environment, 54 (2): 306–27.

Woodruff, Sierra C., and Todd K. BenDor. 2015. "Is Information Enough? The Effects of Watershed Approaches and Planning on Targeting Ecosystem Restoration Sites." *Ecological Restoration* 33 (4): 378–87.

10

Modeling Bioenergy and Land Use Conflict

When we try to pick out anything by itself, we find it hitched to everything else in the Universe.

—*John Muir (1911, p. 110), My First Summer in the Sierra*
American naturalist, author, environmental philosopher, and "Father of the National Parks"

Introduction

Responding to climate change and sustaining ecosystem services have been identified as two of the greatest challenges facing society today (Millennium Ecosystem Assessment 2005b; Pittock 2011). Countries have many ways to respond to climate change. In the last chapter, we discussed market-based policy for offsetting emissions. In this chapter, we will discuss another pathway toward emissions reduction that involves adopting new energy generation techniques that mitigate greenhouse gas (GHG) emissions. Muir's quote is foreshadowing; some pathways toward emissions reductions bring along with them a host of other issues, including the potential for land use conflict, environmental damage, and broad economic impacts.

One growing energy source is the use of "bioenergy" crops—plants grown and harvested to produce biofuels, such as ethanol, methanol, or biodiesel fuels, or a combustion to directly generate heat or electricity. Bioenergy crops are also often called "biomass" crops, referring to the rapid production of large amounts of biological material (specifically, lignocellulose found in dried plant matter). Biofuels have been heralded as one of the most promising strategies for reducing our dependence on fossil fuels and lowering CO_2 emissions (Farrell et al. 2006; Ragauskas et al. 2006). They have also been touted for their ability to support local agriculture and developing economies (Goldemberg 2007). Bioenergy is intended to be climate-neutral, since any carbon emitted from burning biomass fuel was originally sequestered from the atmosphere by the biomass plants, thereby creating a kind of "carbon loop."

However, while farming, bioenergy crops has been proposed as a solution to domestic energy security and global climate problems, the process of producing biofuels through large-scale cultivation of bioenergy crops has the potential to exacerbate a variety of fuel, water, and land use conflicts across the United States and other countries. Moreover, some of these technologies for processing bioenergy crops are highly water-dependent and have significant impacts on biodiversity, water quality, habitat, and general ecosystem functions. Fortunately, policy makers now have a wide range of options in managing the relationship between energy, land use, and water, since there is a large variation in the environmental "footprint" of different energy technologies.

In this chapter, we use the *Values and Investments for Agent-Based interaction and Learning in Environmental systems* (VIABLE) modeling platform to examine patterns of agricultural land use change and potential resulting conflict, which occur when introducing bioenergy sources. We will focus on the State of Illinois, USA, which is a major producer of ethanol, and the area where two promising high-yield perennial grasses, miscanthus (*Miscanthus giganteus*) and switchgrass (*Panicum virgatum*), are expected to play major future roles as energy crops in the Midwest United States.

We are particularly interested in retrospectively studying a situation in the Town of Royal, a small town in central Illinois that was slated to begin construction of an ethanol plant in the late 2000s. Although construction was mothballed due to major financial barriers, the proposal created extensive conflict within the community, with many concerned about large-scale water use and land use changes, and others hopeful that the plant would create badly needed jobs in the region. Therefore, while constructing the model, we collect data that reconstruct the landscape, financial, and market structures during this time period.

The model that we develop, which is based on our previous work (Scheffran and BenDor 2009), will examine the dynamics of farmers as they individually make decisions to plant, harvest, and process bioenergy and nonbioenergy crops. By simulating these decisions, this model helps us to assess the viability of different mixes of biomass crops that can be successfully harvested in a given area over the long term. Finding this mix means determining the spatial pattern of land that farmers jointly allocate to traditional agriculture and bioenergy crops.

This chapter will demonstrate a spatial extension of the VIABLE framework as we will use geographic information systems (GIS) data to spatialize aspects of this system, such as crop yields, agricultural land availability, and agricultural costs, to simulate farmer value based on their selected mix of crops. However, we will diverge from our last two application chapters in not looking directly at *viability corridors* (i.e., land use patterns that meet ecological and economic constraints for maintaining farmer value). Therefore, we will not implement a viability analysis as we have earlier (Aubin and Saint-Pierre 2007; Aubin, Bayen, and Saint-Pierre 2011). Instead, we focus our attention on understanding the land use patterns that result from agent[1] interactions, market demand, and an important policy intervention: government investment in subsidies for biofuels (Scheffran and BenDor 2009). One of our goals then is to assess the conditions under which agents are likely able and willing to produce biofuels, and the implications of land use change, crop production (including for food), and water demand.

We begin this chapter by introducing basic aspects of bioenergy, biomass crops, and bioenergy policy. We then discuss conflict potential due to bioenergy introduction and follow this with a look at previous modeling efforts in this area. Next, we will introduce a VIABLE model of farmer decision making, first describing a single farmer agent and then spatializing the model to account for multiple agents interacting over space and time. We will explore several scenarios, including simulations with and without bioenergy crop introduction, and bioenergy crops with subsidies. Finally, we conclude by summarizing our findings and their implications, as well as discussing ways to improve the model and apply it to real-world conflicts.

---

[1] As in previous chapters, we use the phrases "stakeholder," "actor," "party," and "agent," interchangeably.

What Is Bioenergy?

Bioenergy crops usually comprise a variety of perennial grasses and trees and are often produced using conventional agricultural practices (see Figure 10.1). The potential impacts of energy from homegrown bioenergy sources are expected to make major contributions to the domestic energy supply and contribute to rural economic development.

The Potential Impacts of Bioenergy

Bioenergy crops can contribute to a variety of energy uses, including electricity production through biomass incineration and refinement into biogas and biofuels, such as ethanol and biodiesel (Rosillo-Calle, Hemstock, and De Groot 2006; Figure 10.2). On the whole, renewable energy accounted for 19% of global energy consumption in 2012, with over half of all renewable energy production—representing over 9% of total energy demand—coming from traditional biomass (REN21 2015). This comprises nearly 50% of renewable energy. In comparison, demand for nuclear accounts for nearly a third of that of biomass (2.6% of total energy usage).

In 2006, there were 110 ethanol biorefineries in operation and 73 under construction in the United States (Renewable Fuels Association 2006). By 2014, this number had grown to 2010 ethanol biorefineries across 28 states in the United States alone, producing an estimated 57 billion liters of ethanol annually (Renewable Fuels Association 2014; Figure 10.3). Compare this to figures from 2005, where *global* biofuel production hovered around 107 million liters per day, which itself represented a doubling of 2001 production levels (Worldwatch 2006).

Considerable research funding has been used in recent years to search for more efficient and cost-effective methods of converting biomass crops into ethanol fuel (Farrell et al. 2006; Khanna, Scheffran, and Zilberman 2010; Scheffran 2010b). For example, the Worldwatch Institute (2006) projected that biofuel consumption will grow until 2030 to supply 37% of the transportation fuel demand in the United States, and up to 75% if mandated increases in national automobile efficiency standards over that period are realized (NHTSA 2011). The US Department of Energy has also recently called for replacing 30% of America's gasoline consumption with biofuels by the year 2030 (US DOE 2011).

Ecological Impacts and Net Energy Value

The ecological footprint of biomass crops is a function of several factors that affect the ability of biofuels to grow in sufficient quantities to meet legislated benchmarks and successfully replace fossil fuels. This reality has led some biofuels, like switchgrass, to be developed as a high-yield monoculture, which can weaken soil function and require greater fertilizer, pesticide, and water inputs than native species crops grown in polycultures (Groom, Gray, and Townsend 2008).

While bioenergy is intended to be carbon-neutral, additional GHGs are inevitably produced during the bioenergy life cycle; for example, fossil fuels are used during the production of fertilizer. To sustainably implement bioenergy systems, the environmental, economic, and societal impacts of such systems must be assessed, as must be the entire life cycle surrounding material and energy flows entering and leaving these systems (Scheffran 2010a).

To do this, researchers and professionals have created a number of metrics for measuring biofuel environmental impacts. One of these environmental impact metrics concerns the level of GHGs emitted during the biofuel production and usage process (Figure 10.4).

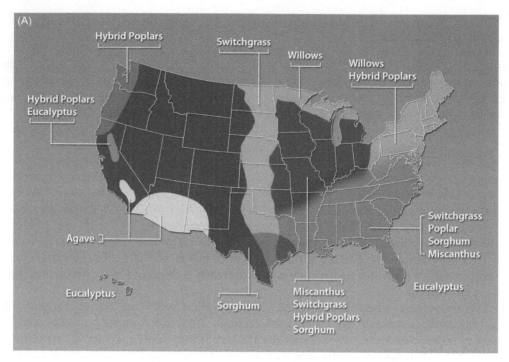

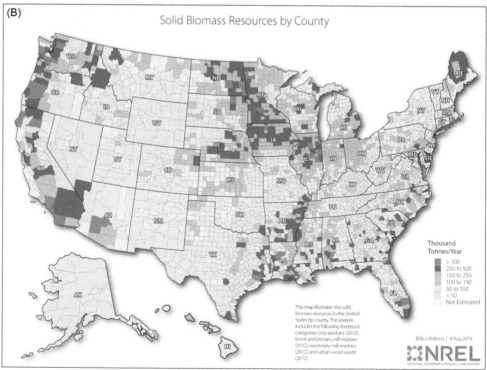

FIGURE 10.1
(A) Geographic distribution of potential biomass crops and (B) available biofuel resources in the United States.
Sources: (a) Oak Ridge National Laboratory Biomass Program, US DOE (2006, 2015) and (b) Milbrandt (2005)
and NREL (2014).

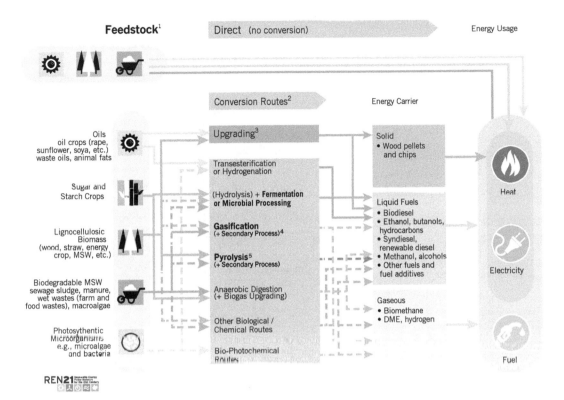

FIGURE 10.2

General bioenergy conversion pathways. [1]Solid lines represent commercial pathways and dotted lines represent emerging bioenergy routes. [2]Parts of each feedstock, e.g., crop residues, could also be used in other routes. [3]Each route also gives coproducts. [4]Biomass "upgrading" includes any one of the densification processes (pelletization, pyrolysis, torrefaction, etc.). Anaerobic digestion processes release methane and CO_2, and removal of CO_2 provides methane, the major component of natural gas; the upgraded gas is called biomethane. [5]There could be other thermal processing routes such as hydrothermal, liquefaction, etc. DME = dimethyl ether. Source: Renewable Energy Policy Network for the 21st Century, REN21 (2015, Figure 6).

Another metric involves assessing the *energy balance* of a biofuel crop, which is measured as the ratio of the biofuel energy output to the energy input required to produce and use fuel (called the energy output-to-input ratio; slightly different than *net energy*, defined in Figure 10.4's caption).

Switchgrass and Miscanthus as Bioenergy Crops

As stated earlier, we are interested in exploring two important bioenergy sources that have been proposed to augment the use of corn in ethanol production, switchgrass and miscanthus. Switchgrass (*Panicum virgatum*), which is also known as tall panic grass, Wobsqua grass, wild redtop, or thatchgrass, is a warm season grass that has historically been a dominant species of the central North American prairie (McLaughlin and Kszos 2005). Switchgrass has been determined to be a strong candidate crop for bioenergy production, based on its resilience in poor soil and climate conditions, rapid growth characteristics, and low fertilization and herbicide requirements. In research trials, the biomass productivity of switchgrass ranges from 9.9 to 23 tons/ha (averaging 13.4 tons/ha) in research trials (McLaughlin and Kszos 2005).

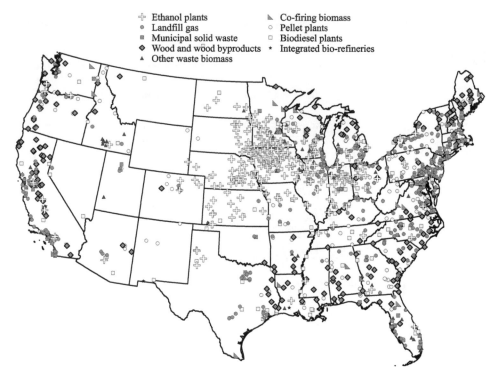

FIGURE 10.3
Biofuels facilities by type across the contiguous United States. Source: Map generated from data collected by NREL (2017).

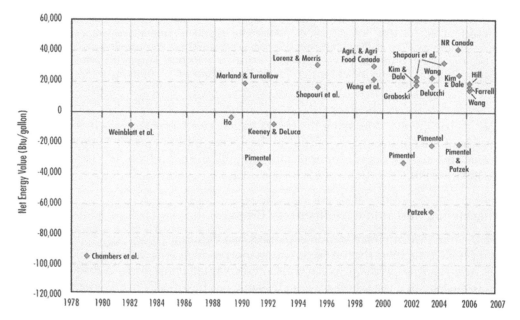

FIGURE 10.4
A summary of studies on the net energy value of ethanol. *Net energy* is defined as Btu content in a volume of ethanol minus the fossil fuel energy used to produce that volume of ethanol. Source: Argonne National Laboratory; Wang (2007).

While switchgrass is an established domestic plant throughout the midwestern United States, miscanthus (*Miscanthus giganteus*; Figure 10.5) is a perennial grass from east Asia that is genetically similar to sugar cane (Clifton-Brown, Long, and Jorgensen 2001). The utility of miscanthus for energy production has been explored in extensive test trials (Heaton, Voigt, and Long 2006), which indicate harvestable miscanthus yields ranging from 10 to 40 tons/ha throughout Europe.

Farming and Bioenergy Policy

For decades, US federal farm policy has been specifically designed to affect the markets of different crops, with a large amount of federal support directed at maintaining price supports for the production of commodity row crops, including soybeans and corn (Atwell, Schulte, and Westphal 2011). These subsidies have had widespread impacts on land use at broad scales, both from a commodity and conservation perspective. Programs such as the US Department of Agriculture's Wetland Reserve Program (WRP; Parks and Kramer 1995; Stroman and Kreuter 2016) and Conservation Reserve Program (CRP; Reichelderfer and Boggess 1988; Secchi and Babcock 2015) have been instrumental in conserving millions of acres of former farmland for conservation purposes.

An important challenge for responsible energy policy is to achieve an economically and environmentally sustainable transition toward cleaner and more secure forms of energy, while reducing the vulnerability of the energy supply to future disruptions, disasters, and conflicts within the energy sector (Dorian, Franssen, and Simbeck 2006; Fiddaman 2007). Growing concerns about climate change and energy security have increased interest in developing renewable energy sources for meeting electricity, heating, and fuel needs in the United States. Domestic energy sources that displace fossil fuels can not only improve environmental quality and mitigate climate change, but can also contribute to energy security by diminishing dependence on foreign fuel sources (NRDC 2004; Worldwatch 2006).

FIGURE 10.5
Miscanthus crops have been cultivated in Illinois and yields have increased dramatically over the last three decades. Author Jürgen Scheffran's wife, Marianne, is shown for scale.

The Energy Policy Act of 2005 (P.L. 109–58) established concrete targets for renewable fuels ("renewable fuels standards") and subsidized ethanol production with a tax credit of $0.51 per gallon (Farrell et al. 2006). Following this, congress passed the Energy Independence and Security Act of 2007 (P.L. 110–40), expanding renewable fuel standards and requiring the annual use of 9 billion US gallons (34,000,000 m^3) of biofuels in 2008, with a goal of 36 billion gallons (140,000,000 m^3) annually by 2022, where no more than 15 billion gallons (57,000,000 m^3) is produced from corn-based ethanol, and no less than 16 billion (60,000,000 m^3) from cellulosic biofuels (Schnepf and Yacobucci 2013).

Similarly, several states have played an important role in supporting ethanol production through the introduction of standards for renewable energies (primarily renewable portfolio standards that require utilities to produce some of their energy from renewable sources). For example, the California Low-Carbon Fuel Standard requires a reduction of at least 10% in carbon intensity of transportation fuels that, according to Farrell and Sperling (2007), has led to a higher level of ethanol usage in petroleum products than would have occurred otherwise. As of 2012, 30 states (and the District of Columbia) have implemented renewable portfolio standards (EIA 2012), many of which mandate vehicles to use certain levels of renewable fuels (Schnepf 2005).

Bioenergy in Illinois

The State of Illinois is part of the American "Corn Belt," a region with excellent soil and temperate climate that stretches across 13 states in the north-central US (Atwell, Schulte, and Westphal 2011). The state is one of the leading producers of ethanol in the United States and is commonly referenced as having some of the most productive farmland in North America. While tall grass prairie formerly comprised much of the region, the landscape is now dominated by intensive, row-crop agricultural production as the state is currently the source of significant corn and soybean production. This predominance of row-crop agriculture in Illinois can be seen in Figure 10.6.

The potential for conflicts due to biofuels introduction in Illinois abound, including water availability and water-related ecological conflicts, conflicts with coal energy producers, and environmental conflicts over new monocrop production.

Bioenergy Crops and Conflict

The considerable potential for bioenergy crop growth and investment raises the issue of whether the large-scale land use, farming, and infrastructure changes necessary for production of bioenergy on this scale may generate or exacerbate environmental and economic conflict. Recent work has also looked at the impacts of biofuel production as a driver of conflicts at multiple scales (Vidosh, Praksh, and Alok 2011; Venghaus and Selbmann 2014). For example, Atwell, Schulte, and Westphal (2011) found that in Iowa, stakeholders have expressed concern about the impact of these rapidly developing markets on regional and state-wide economic, social, and environmental sustainability. We can group these potential conflicts into categories that include conflicts over food production, degradation of water quality and availability, biodiversity and habitat damage, changes to the agricultural economic system (e.g., Scheffran 2010a), and disputes over subsidies or energy production.

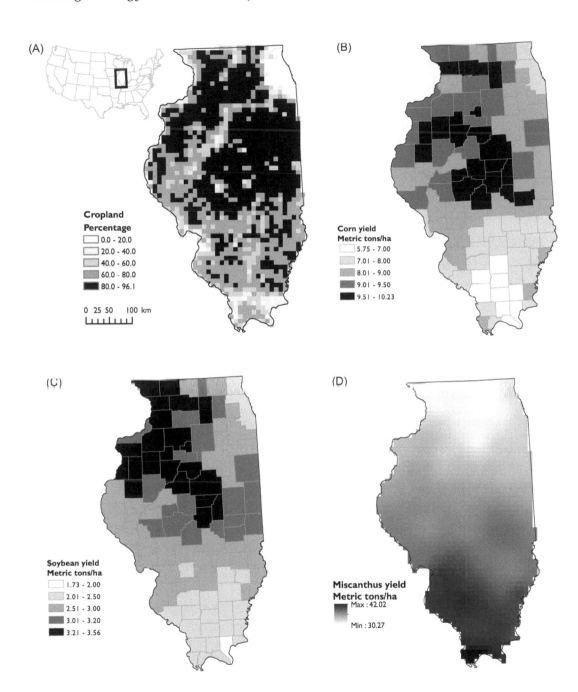

FIGURE 10.6
(A) Agricultural land distribution (2007) in Illinois (Source: NASS 2006; see data discussion in this chapter's additional resources section). Illinois (B) corn and (C) soybean yields at the county level (1997–2001 average (Source: University of Illinois Farm Decision Outreach Council; farmdoc 2014). Potential (D) miscanthus and (E) switchgrass yields (Sources: Dhungana 2007; Khanna, Dhungana, Clifton-Brown 2008). Although the yield distributions look very similar, note the different scales and slight differences in the yield distributions, particularly in Central Illinois. (F) Illinois agricultural regions as determined by the University of Illinois Farm Decision Outreach Council (farmdoc 2007). Adapted from Scheffran and BenDor (2009).

(Continued)

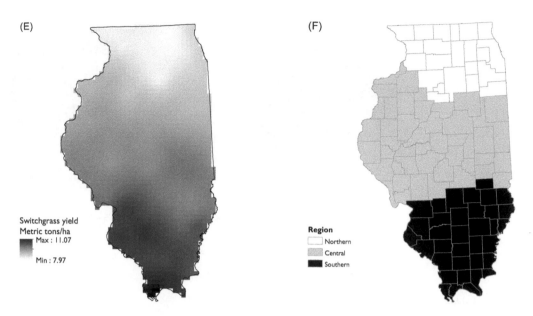

FIGURE 10.6 (CONTINUED)
(A) Agricultural land distribution (2007) in Illinois (Source: NASS 2006; see data discussion in this chapter's additional resources section). Illinois (B) corn and (C) soybean yields at the county level (1997–2001 average (Source: University of Illinois Farm Decision Outreach Council; farmdoc 2014). Potential (D) miscanthus and (E) switchgrass yields (Sources: Dhungana 2007; Khanna, Dhungana, Clifton-Brown 2008). Although the yield distributions look very similar, note the different scales and slight differences in the yield distributions, particularly in Central Illinois. (F) Illinois agricultural regions as determined by the University of Illinois Farm Decision Outreach Council (farmdoc 2007). Adapted from Scheffran and BenDor (2009).

Conflicts over Food and Water

Perhaps the most visible concern is that land requirements place bioenergy crops in competition with needs for conventional agriculture of food and nonfood commodities (i.e., food, animal feed, and seed; Phalan et al. 2011). For example, biofuel production is now being promoted in countries such as India and China, where water and food shortages may place energy creation in conflict with other domestic water and food demands (de Fraiture, Giordano, and Liao. 2008).

Likewise, conflicts can occur over water availability. Although there has been widespread promotion of biofuel and forestry expansion, decision makers often fail to understand or predict potential conflicts between energy generation, water resources, and ecosystems, leaving these conflicts to go unrecognized and unmanaged (Pittock 2011). For example, given that water is fully allocated in many regions (Falkenmark and Molden 2008) and demand for water by other users is increasing dramatically, even modest increases in water demand for energy production may have enormous effects on human users and the environment (Pittock 2011). In these regions, expected growth in energy consumption and population over the coming decades will exacerbate water shortages.

Heavy water use resulting from wide-scale biofuels cultivation and refinement may also threaten biodiversity and create water pollution and management problems (Berndes 2002; National Research Council 2007), which are likely to create new types of conflict (Scheffran and Summerfield 2009). Currently, 20–50% of the land area in a majority of terrestrial biomes has been converted to food production for a growing global population (Millennium Ecosystem Assessment 2005a). The sheer spatial scale of potential biofuel crop production has a similar potential to significantly affect local biodiversity through pollution, soil degradation, and climate impacts from their cultivation, transportation, refining, and burning (Cook, Beyea, and Keeler 1991; Worldwatch 2006).

Perhaps the biggest impact on biodiversity could occur from the likely expansion of agricultural lands for biofuels into sensitive areas, similar to the expansion of grazing areas into South American rainforests. Large-scale deforestation for the purpose of planting energy crops could exacerbate, rather than diminish, the effects of global warming (Reinhardt, Rettenmaier, and Gärtner 2007). This type of land use change could also dramatically degrade the habitats of many species while devastating the ecosystem services produced by complex ecological systems (Groom, Gray, and Townsend 2008; Gasparatos, Stromberg, and Takeuchi 2011). Historical antecedents for this concern can be seen internationally as the expansion of palm oil plantations has damaged natural habitats throughout Malaysia (Chiew 2009), and domestically, as expansion of corn ethanol production has threatened lands enrolled in the US Department of Agriculture's CRP and WRP (Marshall 2007).

Spikes in the price of corn ethanol, which refiners blend with gasoline to meet renewable fuel mandates, have led to concerns that increased ethanol fuel use may increase gasoline price volatility. This had already led to substantial disputes in the US Congress (e.g., Rascoe 2011; Goode 2013). After a 2012 drought in the United States' Midwest ravaged corn crops, commodity prices were pushed so high that ethanol became too expensive to produce and nearly 10% of the nation's ethanol plants halted production (Eligon and Wald 2013).

Finally, biofuel *subsidies* are frequently the center of intense controversy (e.g., Grafton, Kompas, and Van Long 2010; NSAC 2011). Why does this happen? First, biofuel introduction can cause tensions with stakeholders who have invested in the current energy system with which bioenergy seeks to compete. Additionally, the cost of energy crop cultivation relative to its demand is a key component in determining the competitiveness of bioenergy systems. Successful, widespread bioenergy cultivation requires that these crops be economically competitive with alternative renewable and nonrenewable energy systems. Bioenergy crops must provide farmers with an income that is comparable to what they could earn producing conventional crops on the same land. Along these lines, most biofuel research has aggressively focused on creating biofuels at costs comparable to petroleum, which assumes that large-scale agribusinesses similar to that of corn will be created for energy crops.

Corn is an apt model; more than 90% of biofuels produced and used in the United States come from corn, which requires among the highest fertilizer and pesticide inputs of any US crop (NASS 2007) and the highest water and energy inputs per acre of any biofuel crop (National Research Council 2007). Growing monocultures of many types of crops (biofuel or not) in the manner that corn is currently grown will also require heavy energy inputs and high pesticide, fertilizer, and water use (Groom, Gray, and Townsend 2008; Figure 10.7). These inputs come with a high financial cost, as well as high social, environmental, and economic costs, which are not often internalized in agricultural output prices. For instance, the high nitrogen inputs from crops, particularly corn, in the Mississippi watershed are implicated in the expansion of the hypoxic zone in the Gulf of Mexico (Mitsch 1999; National Research Council 2007).

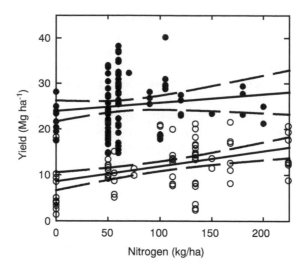

FIGURE 10.7
Impact of nitrogen inputs on miscanthus and switchgrass yields. The linear relationship between nitrogen fertilizer inputs and yields can incentivize overfertilization by farmers, which can cause dramatic downstream water quality and habitat impacts. Source: Heaton, Voigt, and Long (2004). Used with Permission.

Without significant demand for bioenergy crops, little production would occur. Likewise, without production, comparatively little infrastructure would be built. This chicken-and-egg scenario around bioenergy crop production has directly led to the creation of biofuel subsidies, which has contentiously pitted farming coalitions and environmental groups against other parts of the energy industry and limited government advocates (Mosnier et al. 2012). In this chapter's model simulations and additional resources, we will devote time to thinking about how to represent biofuel subsidies (e.g., Grafton, Kompas, and Van Long 2010; NSAC 2011).

Previous Efforts to Model Agricultural Dynamics

Improving policies that link energy production and its environmental impacts will need to include engagement of senior political leaders, iterative policy development, multiagency and stakeholder processes, and stronger accountability and enforcement measures (Pittock 2011).

Combined water quantity–quality modeling, like the type done by de Moraes et al. (2010) can yield important insights into the impacts of new land use patterns. However, in order for this type of modeling to be functional, we first need to have an idea of what land use patterns will look like in the future. Although some work has looked at ways of reducing the environmental impacts of biofuel crops (e.g., through site-suitability analysis; Förster et al. 2008), we also need to understand where farmers are likely to grow biofuel crops by their own accord, and how policies (e.g., water quality or use restrictions) are likely to affect the decisions of individual farmers.

Agent-based modeling (ABM) has been used to study the individual- and village-level social interactions leading to the system-level social and ecological dynamics of agricultural conflicts (Castella, Trung, and Boissau 2005; Barnaud, Bousquet, and Trébuil 2008;

Ng et al. 2011; Abdollahian, Yang, and Nelson 2013). Systems approaches, implemented at the farm scale, allow us to assess individual trade-offs and the interactions between agents (farmers) and the biophysical, social, economic, and political environments of their farms (Cabrera, Breuer, and Hildebrand 2008).

Previous work in this area began by representing farmers as agents who hold heterogeneous amounts of land and capital (animals, labor, technology) and interact with other farmer agents, whether it be through land markets (e.g., Balmann 2000), or through other factors, like exchanges of water or technological innovation (Berger 2001). This heterogeneity is an important factor in farm differentiation, a factor that has long interested agricultural economists, who have developed numerous farm typologies (Landais 1998; Barnaud, Bousquet, and Trébuil 2008).

Perennial crops, which do not have to be replanted every year, can often reduce soil erosion substantially.

"Social capital," is concisely defined by Stave (2010) as "the social bonds and norms that enable and regulate the interactions of people in communities." Others have defined social capital along the lines of the levels of social relationships, norms, and institutions present (Barnaud, Bousquet, and Trébuil 2008). We discuss social capital at greater length in Chapter 4.

Using the *Companion Modeling* ABM technique discussed at length in Chapters 6 and 7, Barnaud, Bousquet, and Trébuil (2008) studied a heterogeneous community of small-scale farmers in Northern Thailand to understand the allocation of rural credit as a means of equitably expanding the use of *perennial crops* in an erosion-prone watershed. The authors found that informal money exchanges between farmers reduced bankruptcy risks, thus exemplifying the role of *social capital* in ensuring system stability. In a similar study, Castella, Trung, and Boissau (2005) modeled the spatial and dynamic interactions among Vietnamese farmers' strategies and decision rules, institutions that affected resource usage and access, and changes in their environment (biophysical, social, and economic). Their study helped illuminate the causal relationships between land tenure policy (i.e., policies pertaining to land ownership rights) in the lowlands and high rates of hillside deforestation.

A variety of work has also employed ABM for addressing water management issues in agriculture. Several studies have even used modeling to assess the implications of changing land uses and landscape patterns due to biofuel introduction (Graham, English, and Noon 2000; Keeney and Hertel 2009; Lapola et al. 2010; Ng et al. 2011), including our previous work in this area upon which this chapter is based (Scheffran and BenDor 2009).

Additionally, there are several good examples of collaborative, stakeholder-led solutions to water management problems, including studies by Schlüter and Pahl-Wostl (2007), who used water management models in the Amudarya River basin in Central Asia. Additionally, Becu et al. (2003) and Castella et al. (2007) modeled individual farmer decisions over water in Northern Thailand and Vietnam, respectively, while Holtz and Pahl-Wostl (2012) modeled farmer decisions and reactions to pumping quotas and other policies in Spain. Although our work is similar to these past efforts, instead of focusing directly on water management, we will focus on the effects of ethanol plant construction and government subsidies in affecting farmer behavior and crop prices during the introduction of biofuels.

Model Design and Structure

Although agents in this model will not actually move in the landscape, environmental factors that vary across the landscape create non-uniform environments for growing and cultivating crops. Therefore, we will use geographically contextual input into our dynamic bioenergy model, built on the VIABLE platform that we presented in Chapter 7. While the placement of each farmer within the landscape is important, for the moment, the position of farmers relative to their neighbors does not affect their behavior (which we recommend as a topic to explore in a model expansion). Farmers harvest and sell crops in a common market that extends well beyond their neighborhood within the landscape. As a result, like the previous models that we have explored, the agents in this model will interact through a central market mechanism, whereby bioenergy crops are sold and converted to fuel.

Accordingly, we represent farmers as agents that are responsive to market signals that depend heavily on the ability and willingness of other farmers within the state to plant and cultivate biofuel crops. This means that farmers interact as a network of agents, whose aggregate decisions can significantly alter the locations of biofuel introduction and the chances of biofuel crop success. These decisions are beholden to systemic delays and complex feedbacks. Given this complexity, we will explore how the individual decisions of farmer agents can impact the behavior of agriculture in the state as a whole.

ABMs of land use change are similar to other types of ABMs, and usually include four different features: (1) agents as decision-making units (in this case, farmers), (2) the spatial representation of the study environment, (3) rules that describe how agents interact (e.g., land markets, movement), and (4) methods that describe interactions between agents and their environment (Holtz and Pahl-Wostl 2012). Like other ABMs, we will continue to see a tension between the complexity used to represent agent behaviors and interactions, the approaches of parameter specification, empirical data collection, and model verification and validation.

The Farmer Agent Model

We begin our modeling process in this instance by constructing a simple model to represent the ability and willingness of an individual farmer (cell) to grow, harvest, and profit from farming a mix of chosen crops. Like our previous VIABLE models, this "cellular" model, which represents the behavioral structure of a single agent, will focus on the important determinants of the dynamics of harvesting. These dynamics depend on prioritization of land devoted to individual crops, investment costs associated with cultivation, and farmer profit, all of which are based on the mix of crops grown and their market prices.

The harvest $h_i^k = r_i^k A_i^k B_i^k f^k$ of each of the four simulated crops (as in previous chapters, this is given by the index k) is a function of the available arable land area A_i^k (hectares), the biomass yield per hectare B_i^k, the fraction of biomass produced that is actually harvested f^k (we assume this to be 90% [i.e., 0.9] for all crops), and the *priority* r^k given to crop k, which is the fraction of the farmer's land that will be planted with a given crop $\left(\sum_k r^k = 1\right)$. Priority is a key variable that determines the extent to which farmers decide to cultivate multiple crops. As we will see, the primary source of dynamics within this model comes from shifting farmer decisions about the type and extent of their crop mix.

BOX 10.1   VARIABLE DEFINITIONS

All are given for each agent $i = 1,...,n$ unless otherwise specified

- V_i = farmer agent value, which is the farmer's net profit, calculated as revenue minus investments
- h_i^k = The harvest of each crop k by farmer i (metric tons)
- h^k = The total harvest (supply) of a crop k across all farmers $i = 1,...,n$ (metric tons)
- p^k = Market price of crop k (US$)
- C_i = Investment costs for each farmer i (US$)
- c_i^k = The total harvest unit cost of crop k (US$ per metric tons of harvest)
- A_i^k = Available arable land area of crop k for farmer i (hectares)
- B_i^k = Biomass yield per hectare of crop k for farmer i (metric tons per hectare)
- f^k = Fraction of biomass type k produced that is actually harvested (we assume this to be 90% [i.e., 0.9] for all crops)
- r_i^k = Priority given to crop k by farmer i which is the fraction of the farmer's land $0 \le r_i^k \le 1$ that will be planted with a given crop $\left(\sum_k r_i^k = 1\right)$
- α_i = Adaptation rate at which farmer agents i can change their crop mix
- v_i^k = marginal change in profit caused by a marginal change of priority for that crop, given by the partial derivative $v_i^k = \partial V_i / \partial r_i^k$
- D^k = monetary demand for crop k (US$)
- S_i = Subsidies received by a farmer from biofuels subsidies and carbon credits (US$)
- s^k = Unit subsidy (US$ per metric ton of crop produced)
- λ = Conversion factor to account for the net biofuel output a biorefinery produces from a metric ton of biomass, which weights the price per ton of corn to create the price of perennial grasses per metric ton $(p^{SG} = p^{Misc} = \lambda p^{Corn})$

To calculate farmer revenue and profit, we need to estimate the crop market prices p^k and investment costs C_i for each farmer i. To do this, we could use a whole variety of price functions that agricultural economists have derived or use data for specific agricultural markets. For simplicity, we will use the price function derived in Scheffran (2006), which calculates market prices at the market equilibrium of demand and supply balance as the ratio $p^k = D^k / h^k$ between the money that all consumers together are willing to spend for that crop (monetary demand D^k) and the total harvest h^k (supply) over all farmers $i = 1,...,n$. While this price function simplifies the model substantially, it is still important to note that we are modeling price endogenously, which most other modeling studies leave as exogenous (Holtz and Pahl-Wostl 2012).

Thus, market prices for each crop are dependent on the total harvest supply and require information from all agents. Now that we have a way of calculating the price of a unit of crop income, we can calculate the *value function* V_i for each individual farmer i, which is the farmer's net profit, as calculated by taking the revenue for each farmer i from selling the harvest h_i^k of each crop at a market price p^k, and subtracting the investments (costs) C_i needed to cultivate all of the crops, which is a function of the per-hectare unit cost c_k^i of harvesting for each crop.

Additionally, we must consider any subsides $S_i = s^k h_i^k$ received by the farmer from biofuels subsidies and carbon credits, both of which are assumed to be proportionate to harvest (s^k is a unit subsidy, US\$ per metric ton). Therefore, the value function can be given as

$$V_i = \sum_k p^k h_i^k - C_i + S_i$$

Farmers begin to accrue capital when their revenue exceeds their costs (positive profit), while capital depletion occurs when costs exceed revenues (negative profit). We represent investment costs as crop-specific, per-hectare unit costs that are accrued over the entire land area available. These costs are then multiplied by the amount of land in the given crop to find the total investment costs for each crop k.

Since the individual farmer is largely at the whim of market prices, their main tool for enhancing their own profit potential is the mix and extent of crops they produce. To find the right mix of crops, we use a decision rule where farmer agents change their priority for a crop based on several factors. These factors include the *adaptation rate* α_i at which farmer agents i can change their crop mix. Similar to models presented in previous chapters, α_i is a way of taking into account a farmer's skill, adaptability, or technical resources to rapidly make land use decisions (Atwell, Schulte, and Westphal 2011). Another factor influencing crop priority is the marginal change in profit caused by a marginal change of priority for that crop, given by the partial derivative $v_i^k = \partial V_i / \partial r_i^k$. This closely mimics the fisheries model in Chapter 8.

If we take the priority-weighted average of this change in crop profit $\sum_l r_i v_i^l$, then we ensure that the priorities represent a fraction $0 \le r_i^k \le 1$, and that they all add up to one $\left(\sum_k r_i^k = 1\right)$. This can be normalized between zero and one by dividing by $\sum_l v_i^l$, thereby smoothing agents' reactions to possible rapid changes in profit margins. Finally, we assume that α_i is the same for each crop k. This yields the farmers' decision rule, which seeks change in a logistic fashion (we use the index l to denote summation across all crops because we are already using k outside the summations):

$$\frac{dr_i^k}{dt} = \alpha_i r_i^k (v_i^k - \textstyle\sum_l r_i^l v_i^l) \tag{10.1}$$

As we explored previously in the fisheries model in Chapter 8, it is possible to create a whole class of techniques for representing how agents seek to maximize or increase their value based on their choice of allocation priorities. The most accurate descriptive decision rule, however, would be derived from stakeholder and expert input based on the exact system being modeled. We should note that although we are using a decision rule that negates the need to model farmers' spatial relationship relative to other farmers, we could easily alter this to account for their neighbor's behavior. For example, we could represent a type of *satisficing*, as risk-averse farmers followed their more risk-tolerant neighbors instead of attempting to optimize their crop mix (see Chapters 7 and 8 for more on satisficing).

As this evolutionary game plays out among competing crops, farmer agents iteratively shift their crop priority toward the direction of growing profits, following the derivative of the profit function with respect to that crop priority (see Appendix 10.1 for more on this). Thus, farmers test how quickly small changes in the mix of planted crops can affect their profit on a yearly basis. If one crop yields a high rate of gain, it behooves the farmer to increase the priority of that crop, while lowering the priority of other, less-profitable crops.

This approach leads to a changing distribution of crops over time. However, the speed of changes to crop priorities depends on the adaptation rate α_i. We estimate the adaptation rate as 0.1/year for all farmers, although this value can be adjusted on either an individual or regional basis. For example, farmers in one region may be able to adapt their equipment more easily or cheaply, or, depending on their experience, some farmers may have faster or slower adaptation rates to changing agricultural demand.

Over time, in a manner similar to fisher investments in Chapter 8, we are likely to see farmers invest their entire land area in the most profitable crop. In Illinois, and the rest of the US Midwest, this has historically been either corn or soybeans. However, as more biofuels are introduced into the market and government incentives begin to make cultivation of alternate crops viable, this may change.

Spatial Extension of the Viable Model Framework

We will extend our agent based conflict modeling techniques by simulating the behavior and decisions of farmers throughout Illinois in a *spatially explicit* manner. As we will show, this behavior heavily depends on the farmers' *location*, which determines their ability to cultivate certain types of crops, their yields, and their access to bioenergy processing facilities (e.g., ethanol plants). For example, the production of biomass is more likely to be profitable, and therefore successful, in areas closer to production and demand centers (e.g., ethanol biorefineries). As a result, the profitability of cultivating specific crops varies significantly throughout space and the cost of transporting biomass from production regions to ethanol plants must be considered in any model of land use competition and conflict (Kang et al. 2010).

As we will demonstrate, using spatially relevant inputs is another way of introducing heterogeneity into ABM. Although this appears to complicate the model, remember that using spatial data is just another way of adding parameters to each agent, just like the differences in GDP and emissions gave heterogeneity to regions in the emissions model we presented in Chapter 9. We will simulate the cultivation of four crops, including corn and soybeans, which are used for both traditional (food and animal feed) and bioenergy purposes, and miscanthus and switchgrass, which are harvested exclusively for bioenergy production.

To do this, we will rely on NetLogo's GIS capability (first established in Version 4.0 [2007]), which allows us to import our spatially explicit data as a series of standard, raster GIS maps (ESRI[2] ASCII format). NetLogo transforms GIS *raster* data (i.e., pixel data representing individual units on a two-dimensional plane) into the NetLogo *view*—the NetLogo "space" of sorts—allowing raster cell values to populate patch variables in NetLogo. This logic is demonstrated in Figure 10.8.

---

[2] Environmental Systems Research Institute.

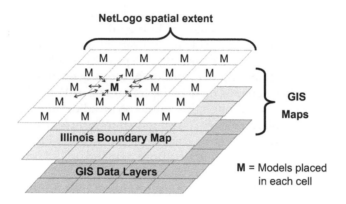

FIGURE 10.8
Interactions between cell-based agents (farms). The reality of our model, however, is that agents do not interact locally (i.e., we do not simulate spread of technology or information between neighboring farmers), but rather interact globally. Therefore, there are interactions between agents through the dynamic feedback of the spatially distributed market.

Constraining the Model Spatially

For simplicity, we constrain this study to a closed system, focusing on the market for crops in Illinois where the supply and demand across state boundaries is outside the scope of the conflict being studied. As a result, we assume that supply is entirely directed at meeting a given demand within Illinois alone. Our choice to limit the spatial boundaries of the model to the State of Illinois reflects a common approach to simplifying models (e.g., Atwell, Schulte, and Westphal 2011). This choice also underscores the reality that many investment and policy decisions that affect agricultural land use change are made at the state, local, and individual farmer levels.

We must reiterate that it remains essential to gain stakeholder buy-in and improve the modeling process through stakeholder-led collaborative modeling efforts (van den Belt 2004). This fact becomes even more pronounced in spatial-dynamic modeling as the role of data and agent heterogeneity becomes more complex (BenDor and Kaza 2012).

Spatial Resolution

An important factor to consider during data collection is the spatial resolution of the model; we can think of this as the unit of analysis for geographically explicit modeling. Spatial data resolution is typically defined by the relative size of the units observed. For example, parcel-level, county-level, or state-level data denotes how fine-grained spatial data is collected.

Ideally, we would actually have a map with outlines of individual farms, or better yet, a map showing all of the individual parcels of land that are under the direction of individual farmer decision makers. Then, we could model the land use effects of decisions made by individual farmer agents. In a real-world implementation of this process, participatory processes involving farmers and other stakeholders throughout the state could aid in this data collection process.

However, not only do these maps not exist (or would be very difficult to compile for such a large area), but much of our data exist at low spatial resolutions, which would require us to aggregate farm-level data and decrease our resolution, regardless. Given this wide variation of data resolution, we select a fairly aggregated (large) unit of analysis for our model that corresponds to the size of a "township," which is a 6 by 6 mile land area that is the primary unit of the Public Land Survey System, the common method of subdividing and describing land in the midwestern and western United States (National Atlas 2011). Each of these townships will, for our purposes, function as an individual farmer unit (agent), and our model will simulate the joint behavior of the $n = 1,568$ township-sized agents (arrayed over a 37×65 grid). See this chapter's additional resources section for more information about the creation of the input maps of crop yields, farmer investments, and location-specific agricultural cost profiles.

Initializing the Model to Equilibrium

We first establish initial conditions for the model in a landscape with only corn and soybean crops, assuming zero initial priorities for miscanthus and switchgrass. We initialize the priorities for both crops at 0.5 for all, and let the simulation run long enough that a near-equilibrium distribution between the two crops was achieved. After this initialization period, the ending fraction of corn and soybeans in each cell determines their initial priority r^{lt} for future model simulations (Figure 10.9). Prices for corn and soybeans equilibrate at $116.00 and $264.53 per ton, equivalent to $2.95 and $7.20 per bushel, respectively. This fits well during the 2000–2006 period that we are aiming to replicate, where Illinois corn and soybean prices ranged between $1.91 and $3.35, and $4.55 and $7.51 per bushel, respectively (NASS 2007).

Figure 10.9 shows that in a two-crop landscape, farmers often establish equilibrium conditions after specializing in a single crop ($r^k = 1$), thereby exercising competitive advantages based on their location and environmental conditions (again, remember that we are not considering crop rotation in this model). When running the model, profits after 50 years are particularly high in central and northern Illinois and reach negative numbers only in largely nonagricultural parts of southern, northeastern, and western Illinois (see Figure 10.6).

Scenario Analysis for New Biofuel Crops

Now, we can generate different scenarios that we expect will generate major impacts on the rates and patterns of biofuel crop introduction.

Scenario 1: Simulating the Introduction of a New Ethanol Refinery

How can we understand the impact of a new ethanol plant on the landscape? To do this, we will simulate an actual proposed case where a new ethanol plant was planned for construction in the town of Royal, IL (Champaign County; Figure 10.10) in 2007. Construction of this plant was originally slated to begin in the Spring of 2007, and the plant was expected to come online by January 2009 (Cook 2007). In June 2008, however, the plant's developers halted construction due to adverse economic conditions, including financial markets that made it difficult for commercial loans, high input costs due to the rising costs of corn (corn

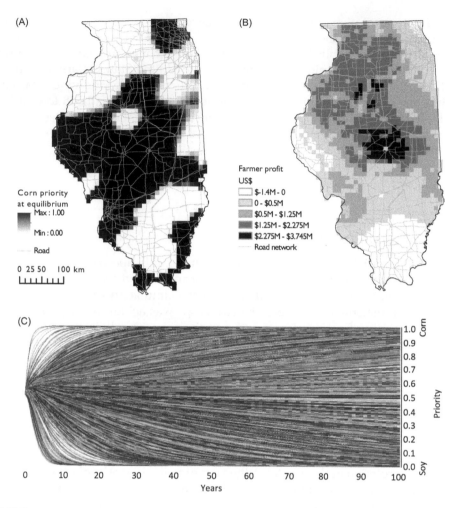

FIGURE 10.9
The initial distribution of corn and soybean priorities across Illinois (without switchgrass or miscanthus introduction) and associated profits after 50 years. (A) The spatial arrangement of corn. Soybeans experience a complimentary priority to corn ($r^{soybean} = 1 - r^{corn}$). The relative absence of gray signifies specialization by farmers in corn or soybeans. (B) Annual profit per cell across Illinois after 50 years. (C) Patterns of corn and soybean priorities, where each line represents the path of one cell (farmer agent) through time ($n = 1568$). The diverging pattern suggests that most agents tend to specialize in corn or soybeans, rather than cultivating a mix of the two. This divergence is represented spatially in Panel A. Adapted from Scheffran and BenDor (2009).

prices increased to over \$4.25 a bushel in 2008), high oil prices, and continuing uncertainty over whether the next president would support ethanol (Mitchell 2008).

Although economic barriers scrapped this plan, much of the town remained hopeful that an ethanol plant will be constructed in the future, providing much needed jobs in an area dominated by large-scale agriculture. Therefore, we retrospectively simulate the impacts of the plant's construction as a way of understanding the effect of possible future plant construction and the subsequent reaction of farmer agents to new infrastructure in the landscape.

To set up this model, we can first introduce transportation costs to the nearest ethanol refinery as large-scale crop movement and refinement begins (Figure 10.10; ethanol plants outside of Illinois are not considered here). These costs are subtracted from the profit

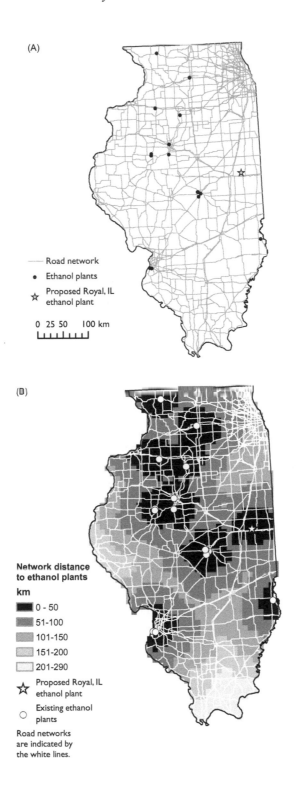

FIGURE 10.10
(A) Road network and ethanol plant map, including proposed Royal, IL plant. (B) Network distance to ethanol plants, including new Royal, IL plants. Sources: Renewable Fuels Association (2014) and Cook (2007); Adapted from Scheffran and BenDor (2009).

function for each farmer and are calculated as the product of the volume of deliveries and a constant per ton-mile charge, plus the return trip cost. To calculate the transportation costs from the farm to the ethanol processing facility, we use the approach developed by Khanna, Dhungana, and Clifton-Brown (2008), who estimate the cost of transportation of switchgrass or miscanthus per metric ton-km as \$(1.12+0.07d), where d is the travel distance in kilometers between the on-farm storage area and the power plant. This was originally stated in 1983 US\$ and is adjusted to 2007 US\$ by multiplying by 2.09. To calculate d, we use a GIS technique known as a "cost–distance" analysis to calculate the travel distance along the state's highways and major roads, from each cell to the closest ethanol plant, and have included in the cost calculation an additional GIS layer imported into NetLogo (see this book's website: http://todd.bendor.org/viable). We will begin by simulating a scenario where we introduce ethanol production (and associated transportation costs) into a landscape where there is no demand for miscanthus and switchgrass.

Scenario 2: Simulating Switchgrass and Miscanthus Introduction

To begin exploring the effects of government policies (or the results of collaborative negotiations on biofuel investments), we must first test the effects of an exogenously introduced demand for biofuel crops without any sort of policy intervention or subsidies. This effectively simulates the gradual development of a new demand stream, perhaps through new automobile or residential fuel technologies created by private industries or through incentives or investment by the federal government (e.g., Obama Administration 2011). We do this by initializing the total monetary biofuel demand D at zero and increase it linearly at a rate of \$100 million a year for the 50-year run of the model, until monetary demand totals \$5 billion per year—a value that is somewhat lower than the current total monetary demand for soybeans and corn in the state (roughly \$6.8 billion per year).

We allocated the total biofuel demand among the four different sources of biofuels. We assume that all four crops compete for this new, "exogenous" biofuel demand: corn, switchgrass, and miscanthus for ethanol production and soybeans for biodiesel production. While we estimate projections of switchgrass and miscanthus demand, we draw on actual values of future corn and soybean demand from the National Agricultural Statistics Service (NASS). For simplicity, we assume that biodiesel receives 10% of the total monetary demand and ethanol the rest. The prices of the four crops are now driven by this growing demand, which leads to increasing supplies and shifting land priorities, depending on expected profits for each of the crops.

Corn is harvested for both food and ethanol production, thereby competing with miscanthus and switchgrass in the growing market for bioenergy crops. Here, we see the genesis of the controversy that exists around the interaction between biofuel markets and their effects on the price function for corn. Corn's price is affected by the increasing supply of miscanthus and switchgrass, each of which is multiplied by a factor that measures the conversion rate of a metric ton of miscanthus into ethanol, compared with the gallons per ton of corn. We have noted throughout this book that conflict modeling can easily take into account the role of evolving technological capability of agents (Dudley 2008) by making assumptions about how harvesting and production efficiencies will change in the future. In this case, agricultural technologies are important in influencing what types of crop growth and land use practices are possible and economical. New technological pathways associated with emerging types of bioenergy production are influencing new land cover and care patterns, which in turn can either enhance or erode ecosystem services (Atwell, Schulte, and Westphal 2011).

Here, we will implement an evolutionary technology scheme by theorizing that the conversion efficiencies from miscanthus and switchgrass into ethanol will change over time, largely due to better biochemical-conversion processes. We assume that for current technology, corn has better conversion efficiency into ethanol than perennial grasses, and that there is a linear increase in this conversion efficiency for both, at a higher rate for cellulosic biomass compared with corn. This is due to anticipated advances in research and development in second-generation cellulosic ethanol (for more details on estimates of crop-to-biofuel conversion factors, see Wu, Wang, and Huo 2006).

The price of the perennial grasses per-metric ton is coupled with the price per ton of corn, weighted by the conversion factor to account for the net biofuel output a biorefinery produces from a ton of biomass. We assume that the conversion efficiencies and prices of miscanthus and switchgrass are equal.

Scenario 3: Introduction of a Biofuel Subsidy

Finally, we introduce a policy intervention jointly with the new demand for biofuels by adding a $50/metric ton subsidy for bioenergy crops. While we could alter the exact form of this subsidy to represent a multitude of different proposed subsidies or tax credits (Steenblik 2007), we take a simple alternative and add the subsidy into the value function as a per-unit harvest payment. This will allow us to better understand the relationship between subsidy level, spatial pattern, and price.

Simulation Results

Scenario 1: No Demand for Miscanthus or Switchgrass

We begin by simulating a scenario where we introduce ethanol production into the system by considering the transportation costs to ship corn and soybeans to plants throughout the state. We first simulate these transportation costs under the scenario where there is no demand for miscanthus and switchgrass. As shown in Figure 10.11, adding transportation costs do not appear to lead to broad changes in the distribution of corn and soybeans. The priority (preference) for corn increases slightly in areas surrounding ethanol plants in the northern part of the state, while dropping significantly in areas that are at great distances from ethanol plants.

Scenario 2: Introduction of Miscanthus and Switchgrass

As Figure 10.12 shows, the introduction of miscanthus and switchgrass does not lead to major changes in the spatial patterns of corn and soybeans, but rather lowers the dominance of corn and soybean in the landscape, particularly in southeastern and southern Illinois due to the high yields of miscanthus in the area. Accordingly, most landscape changes are due to increases in miscanthus cultivation, which commands priorities ($p^{miscanthus}$) of 5%–10% in most areas (maximum 33%), whereas switchgrass struggles in the market, yielding 2%–3% in most areas (maximum 4%–5%) due to the lower yield compared with miscanthus.

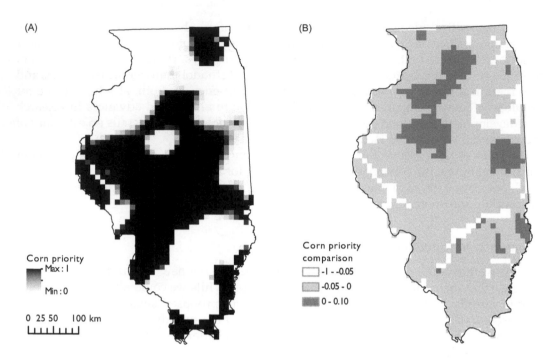

FIGURE 10.11
(A) Corn priority (at $t = 2057$) with integrated travel costs and (B) a comparison of corn priority with and without travel costs ($r_{costs}^{corn} - r_{nocosts}^{corn}$). Adapted from Scheffran and BenDor (2009).

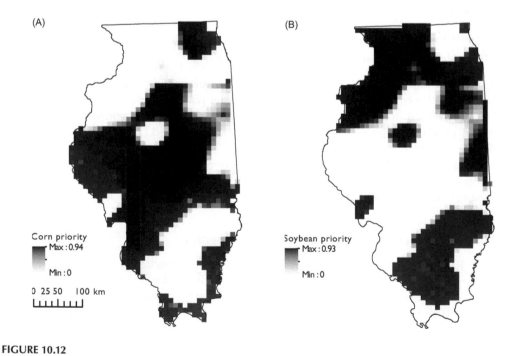

FIGURE 10.12
Spatial distribution after miscanthus and switchgrass introduction of (A) corn, (B) soybeans, (C) miscanthus, and (D) switchgrass priorities. Adapted from Scheffran and BenDor (2009).

(*Continued*)

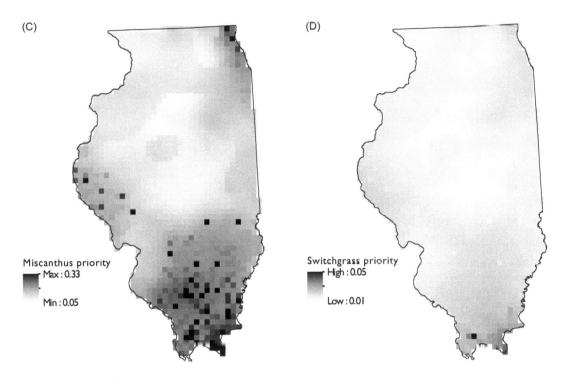

FIGURE 10.12 (CONTINUED)
Spatial distribution after miscanthus and switchgrass introduction of (A) corn, (B) soybeans, (C) miscanthus, and (D) switchgrass priorities. Adapted from Scheffran and BenDor (2009).

Altogether, miscanthus harvest is more than ten times the switchgrass harvest throughout the state, concentrated mainly in the southern and western parts of the state. Although the priorities of perennial grasses appear to be small compared with the food crops, their share of the increasing demand contributes to significantly growing profits of an average of more than $4 million per township-size cell and up to $9 million in some cells, as shown in Figures 10.13 and 10.14.

We see an important insight from this experiment when we look more closely at the market prices after the introduction of miscanthus and switchgrass; prices initially rise for all crops as a result of the growing demand for biofuels (Figure 10.13). Due to the rising conversion efficiency of switchgrass and miscanthus, their price per metric ton exceeds the corn price after 35 years. Corn reaches a maximum price after about 40 years and levels off, an indication that supply has kept up with growing demand.

Since both perennial grasses are assumed to have the same biomass-to-ethanol conversion efficiency, the price per ton of miscanthus remains equal to the price of switchgrass, despite the significant differences in total harvest. Profit margins are low for switchgrass due to the high unit harvesting cost and lower yield relative to miscanthus. Due to the dual role of soybeans as a food crop and as a crop for biodiesel production, the initial high price continues to increase in the face of growing demand.

As we see in Figure 10.12, the considerable increase in miscanthus harvest and the higher prices garnered for biofuels do not trigger a significant widespread reduction in corn and soy cultivation. This is largely due to high miscanthus yields, which require less land than corn to satisfy the same demand.

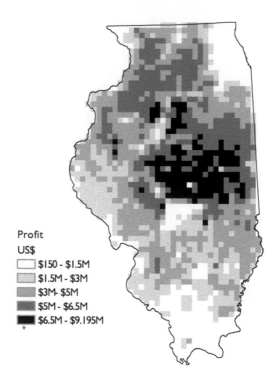

FIGURE 10.13

Spatial distribution of farmer profits after miscanthus and switchgrass introduction (*t* = 2057, assuming introduction in 2007). Adapted from Scheffran and BenDor (2009).

Scenario 3: Biofuel Subsidies

In our final scenario test, we introduce a subsidy that is proportionate to the harvest of switchgrass and miscanthus. We can say that a subsidy of $50/metric ton is "successful," in that we see increases in profits and further shifts in priorities toward the miscanthus and switchgrass bioenergy crops.

Although total miscanthus harvest increases from 16 to 19 million tons, the distribution of crops within the state does not change substantially. However, the impacts of the subsidy on market prices and total harvests are rather small.

Summary

In this chapter, we have explored agricultural land use change that may occur during the widespread implementation of renewable bioenergy crops in the State of Illinois. We began with an introduction to biofuels issues, history, and policy. We discussed biofuel-induced conflict and some of the many efforts to create agricultural simulation models. We then created our farmer agent model and extended it spatially.

Using GIS data on crop yields, agricultural land availability, and agricultural costs, we simulated the profitability of farmers based on their selected mix of crops. This mix

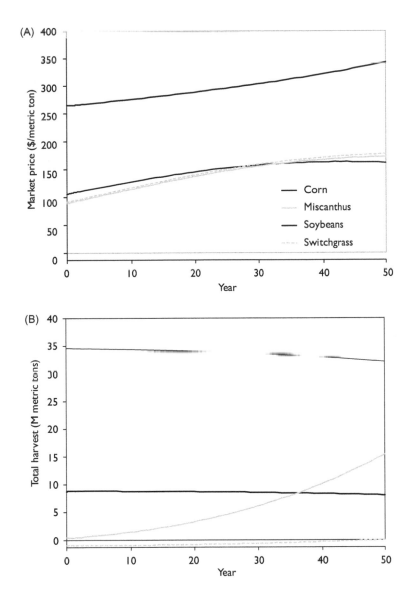

FIGURE 10.14

The dynamics of (A) crop market prices and (B) total yearly harvests. Switchgrass and miscanthus prices are artificially offset for visualization but are otherwise identical. Adapted from Scheffran and BenDor (2009).

generated revenues based on crop prices and costs associated with cultivating certain crops within different regions in Illinois. Crop prices were determined by the relative supply from all other farmers simulated in the model. Like our model of fisheries, farmers could then optimize their profit potential by changing their crop mix (investment allocation) on a yearly basis in order to take advantage of more profitable crops. This model allows us to identify where (and when) it is profitable for farmers to switch from conventional agriculture to bioenergy crop production. As a result, it also allows us to explore the implications of this large-scale land use change.

In the context of a biofuels-related viability analysis, Groom, Gray, and Townsend (2008) argue that among the central goals of any biofuel policy must be efforts to minimize risks to biodiversity, ecosystems, and the climate. Likewise, similar to our analyses of fisheries (Chapter 8) and emissions trading (Chapter 9), sustainable criteria and solutions to large-scale conflicts require definitions of sustainable pathways for all actors involved (Scheffran 2010a). While we have shown that impact mitigation can be partially accomplished through efforts to increase the efficiency of biofuel production and decreasing the land area needed to grow sufficient quantities of the feedstock, additional innovative solutions will be required to reduce the total negative environmental and economic impacts of biofuels.

Like our previous models, the model presented in this chapter greatly simplifies many of the conflicts and decision-making behaviors of farmers and other stakeholders. For example, farmers' decisions are not necessarily based on purely rational economic evaluations. Other factors that influence their willingness to adopt biofuel crops include community social norms, legal knowledge, networks and regional institutions, and information from government agencies, nonprofit organizations, and agricultural and conservation groups. All of these factors can play a key role in mediating the influence of macro-scale markets, technologies, and policies on farmers' land use decisions (Atwell, Schulte, and Westphal 2011).

However, what is important is that this model establishes a methodological framework for examining the availability, feasibility, and economics of bioenergy sources in areas where conflicts may prevent or complicate the introduction or use of biofuels. Two of the innovative facets of this model include its explicit treatment of land competition between crops, as well as its inclusion of the dynamics of land availability. This model framework has been expanded and applied to other regional contexts such as China (Shu, Schneider, and Scheffran 2015; Shu et al. 2017).

Several interesting conclusions come out of this chapter's exercise. First, simulations indicate that even small priorities devoted to miscanthus and switchgrass can increase net profits significantly. This is likely due to large yields from these energy crops, particularly in areas that have traditionally been poor for cultivating corn and soybeans. This leads us to additional questions about the potential effects of miscanthus and switchgrass (or other energy crops) on decisions of farmers to idle land from crop production due to its "marginal" ability to grow conventional crops. This may be a conflict that could unintentionally emerge in the rush to introduce biofuels, pitting environmentalists against the biofuel subsidies for which they may have lobbied.

Second, the lag experienced between supply and increasing demand initially drives up prices for energy crops, after which price increases begin to slow down as farmers allocate more land to these crops. This model demonstrates that planning for this pattern could be important to prevent market volatility and associated conflicts, during the initial stages of biomass crop introduction. This price surge—and ensuing food cost-related fallout—may be exacerbated by costs experienced by farmers transitioning from conventional to biomass crops, which we do not consider in this model, but could also be included in the yearly harvest cost calculations.

Biofuels are emerging alternatives to petroleum-based fuels and may serve as an important piece of the carbon-emissions-reduction puzzle. However, certain feedstock production practices can cause harm to land, soil, water, and climate. The VIABLE model created in this chapter, if used with stakeholder coalitions, could act as a starting point for assessing areas where biofuel expansion may create negative consequences for —and conflict over—ecosystems, food security, and human health (Groom, Gray, and Townsend 2008). Furthermore, it could be used to help direct policies to foster biofuels with low risk to biodiversity. This approach demonstrates the importance of demand factors and external assistance (subsidies) for creating bioenergy markets as well as the potential value (profit) increases to farmers as a result of the widespread adoption of biomass crops.

Appendix 10.1: Agent Model of Biofuel Investment and Harvesting

To avoid errors in the logistic function around priority change for each crop, we will assume base (initial) priorities for miscanthus and switchgrass of 5% for all farmers. We will also assume that price p^k is given as demand D^k divided by supply (harvest h^k), $p^k = D^k/h^k$. The value function V of agents is calculated as profit minus cost: $V_i = \sum_k p^k h_i^k - C_i = \sum_k \left(\frac{D^k}{h^k} h_i^k - c_i^k h_i^k \right)$. When we add in per-unit subsidies s^k for each crop, we see that this becomes

$$V_i = \sum_k \left(\frac{D^k}{h^k} h_i^k - c_i^k h_i^k + s^k h_i^k \right).$$

As we calculated in Chapter 8 in the case of fisheries, the change in crop priority Δr^k (ignoring index i for each spatial agent to simplify notation) can be taken as a function of priority, the rate of change of profit V with respect to changes in allocation priorities r for crop k ($v_r^k = \partial V / \partial r^k$):

$$\Delta r^k(t) = \alpha r^k \left(v_r^k - \sum_l r^l v_r^l \right)$$

Here, v_r^k can be taken as

$$v_r^k = \frac{\partial V_i}{\partial r_i^k} = \frac{D^k}{(h^k)^2} A_i^k B_i^k f^k \left(h^k - h_i^k \right) - c_i^k A_i^k B_i^k f^k + s^k A_i^k B_i^k \tag{10.2}$$

where for each crop, D^k is total demand, h^k is total harvest from all farmers, A_i^k is the area in hectares available for each farmer, B_i^k is the per-hectare biomass yield of each farmer, f^k is the fraction of biomass produced that is actually harvested (we assume this to be 90% [i.e., 0.9] for all crops), and c_i^k is the total harvest unit cost.

Questions for Consideration

1. Some researchers have suggested that by more efficiently using biomass to produce energy, it may be possible to minimize land competition and reduce associated conflicts, including problematic interactions between food and fuel (Patil et al. 2008), and biofuel resource abuse (McGranahan 1986). *Of the conflicts that we have discussed, which do you think would be alleviated by bioenergy efficiency increases? Which would not? Why?*

2. In the model we build in this chapter, farmer agents do not interact with their neighbors. An explicit spatial representation would only really be relevant if spatial patterns of farmer movement or knowledge transfer were of interest. *Under what circumstances should we try to model direct, spatially explicit agent interaction? What sorts of questions would we be asking?*

3. Farmers' decisions are not necessarily based on purely rational economic evaluations. *What are the other factors that influence their willingness to adopt biofuel crops? How do these factors mediate the influence of macroscale markets, technologies, and policies on farmers' land use decisions?*

4. When we discussed land use conflicts, we noted that biofuel crops can also compete with other land uses, such as housing and other urban uses. *Why would this be?* Hint: look for literature on the 'bid-rent' curve, which views a city sort of like a balloon, and agricultural land costs almost like outside air pressure keeping the balloon from expanding. A good place to start is with the seminal work of urban planner and economist, William Alonso (1960).

5. There is a significant *literature on* the impacts that biofuel crops can also have on biodiversity and ecosystem services. Read the discussion of wildlife and bioenergy trade-offs by David Stoms et al. (2012). *What are the trade-offs, exactly? How does this work affect your understanding of the impacts of biofuel crops?*

6. While our price function simplifies the model substantially, it is still a way of modeling price endogenously, which is rare among many other studies. *If we wanted to take a more detailed look into this conflict, how could we use historical data and/or participatory input to more accurately represent prices experienced by real-world farmers?*

7. If one crop yields a high rate of gain, it behooves the farmer to increase the priority of that crop, while lowering the priority of other, less-profitable crops. For the sake of brevity, we have ignored the costs of switching between crops—a type of transaction cost in seeking higher profit. *What is an analogous example of this type of switching costs? How could we integrate switching costs into the model? Would these costs remain the same year-to-year or could these costs change over time?*

8. While we could explore any number of conflicts driven by proposed biofuel introduction and development, in this chapter we have focused on a simulation exercises that represents biofuel subsidies. *How could we improve our representation of subsidies? How might subsidies become dynamic? How could subsidies be made smarter to decrease their potential for inducing conflict?*

9. Few biofuel policies include provisions to protect biodiversity and ecosystem health, with the notable exception of the US Advanced Clean Fuels Act of 2007 (Groom, Gray, and Townsend 2008). *Are there any others? Any from other nations? How do these policies balance biofuel production and subsequent biodiversity impacts?*

10. Interested in adding on to our model? Here are some ideas:

    - Our choice to limit the spatial boundaries of the model to the State of Illinois reflects a common approach to simplifying models (e.g., Atwell et al. 2011). A more comprehensive version of this model could, and should, include interstate and international trade.

    - Perhaps the biggest impact on biodiversity is the likely expansion of agricultural lands for biofuels into sensitive areas, similar to the expansion of grazing areas into South American rainforests. While we do not consider this in the model presented in this chapter, it would be relatively easy to add mechanisms for expanding the planted area beyond the existing agricultural area in Illinois.

    - Similar to models presented in previous chapters, we defined α_i (adaptation rate or crop switching rate) as a way of taking into account a farmer's skill, adaptability, or technical resources to rapidly make land use decisions (Atwell, Schulte, and Westphal 2011). Interested in making this more sophisticated? Emerging technologies can strongly influence farmers' land use decisions. A good place to start learning about this is through is the work of Holtz and Pahl-Wostl (2012), who use a sophisticated model incorporating technical options and farmer skills.

- As biofuel crop prices increase, we may see large-scale transfers of land from the National Resource Conservation Service's Conservation Reserve Program (CRP) to full-time biofuel production. In theory, by allowing farmers to produce commercial crops on retired lands (set aside for ecological preservation purposes), regulators could reduce governmental budgetary burdens while increasing farmer income. Although we do not explicitly model this behavior, this is a major, unintended "side effect" of biomass crop introduction that could have enormous impacts on carbon sequestration, sedimentation, water use and run-off, biodiversity, and other factors (Perlack et al. 2005; Scheffran 2010b). A model extension could help to predict where CRP land could be induced to come back into biofuel production.

- How do the costs experienced by farmers transitioning from conventional to biomass crops get passed on to consumers through crop market prices? Would biofuel-induced increases in crop prices further surge due to these transitions? We do not consider this behavior in our model, but this could be included in the yearly harvest cost calculations.

- In the question for consideration, we asked *why* you might model farmers interacting with their neighbors. Here, we ask, *how* would you model farmers interacting with their neighbors? Model extensions here could simulate information transfers, material transfers, or modified, "satisficing" decision rules.

- Other factors that could augment our model include analysis of
  1. more complex farming dynamics like crop rotation;
  2. additional biofuel species;
  3. weather uncertainties;
  4. best management practices that farmers could use to minimize fertilizer, pesticide, and energy inputs (e.g., conservation tillage);
  5. calculation of the ecological footprint of biofuels (i.e., evaluating the entire life cycle of biofuel production, use, and waste disposal; Blaschek, Ezeji, and Scheffran 2010); and
  6. calculation of the GHG emissions of biofuels over their life cycle to determine how much carbon is sequestered by given crops (Scheffran 2010a).
  7. Finally, this model could be used to analyze how farmers can implement biofuel polycultures, where multiple crop priorities exist together, and $0 < r^k < 1$. Polycultures may aid in reducing soil depletion and improve habitat and other environmental impacts (Groom, Gray, and Townsend 2008).

Try out some of these changes; you can start by going to this book's website—http://todd. bendor. org/viable —and tinker around with the model. What other ideas do you have to modify the model? We invite you to contact the authors (see http://todd.bendor.org/ contact) to forward along your modifications so that we can add them to this website so others can learn from your work!

Additional Resources

How did we create the initial farmland map? For the time period we were originally looking at (2007), we started by creating a land use map using the US Department of Agriculture (USDA) NASS (NASS 2006) Cropland Data Layer. The NASS aggregated this data from satellite imagery (Jensen 2000) into 13 standardized categories with an emphasis on agricultural land cover. NASS used broad land use categories to define land that is not under cultivation, such as nonagricultural, pasture/rangeland, waste, wooded, and farmstead lands. In this case, classification decisions were based on extensive field observations collected during the annual NASS June Agricultural Survey (NASS 2006).

Due to the discrepancy of the ground resolution of this base agricultural land use map (30 by 30 meters) and the resolution selected for this project (6 by 6 miles), the land cells on the base map were aggregated using the ESRI ArcToolbox GIS software (ESRI 2006), a software package designed to manipulate and analyze geospatial data.

How did we create the yield maps?

Corn and soybeans. To approximate actual yield values, we use data on soybean and corn production from the Illinois Crop Yields Historical NASS Database, which lists wheat, corn, and soybean yields for each county in Illinois between 1972 and 2007 (Farmdoc 2014b). One problematic issue in geospatial analysis occurs when data exist at a lower resolution (i.e., larger spatial scale) than the unit of analysis (township). Since the crop yield data is only available at the US county level, we need a way to accurately improve the resolution of the data. In geostatistics, a straightforward way of doing this involves using techniques that spatially "interpolate" known county data as a way of creating a continuous map of soybean and corn yields throughout Illinois. The method of spatial interpolation that we used is known as "Kriging," which estimates yields at a given point in space by creating a function that links geographic location to observations of yields at nearby, known locations (Cressie 1993). To be specific, we used "ordinary Kriging" on county centroids and applied a spherical interpolation model with average standard errors of 140 and 440 kg/ha for soybean and corn yields, respectively. This continuous map (which looks almost like a topographical map) can then be spatially aggregated to the township resolution.

Switchgrass and miscanthus. As miscanthus and switchgrass have not been planted extensively anywhere in the United States, we drew on research by Dhungana (2007) to estimate potential yields as a function of soil quality, climate, and other environmental conditions (Figure 10.7). Just like the corn and soybean yield data, our estimates on switchgrass and miscanthus yields from Dhungana (2007) were only available at the county level and need to be interpolated to the higher township resolution. Khanna et al. (2005) estimated that switchgrass yields in Illinois likely have a geographic production pattern that is similar to miscanthus and usually yields roughly 26% as much total biomass (although this is not always true for the entire state).

This data collection and adjustment process yields four maps showing expected yields for miscanthus, corn, switchgrass, and soybean crops throughout the 1,568 townships within the State of Illinois. We should note that small fraction of biomass growth is not actually harvested; most crops require that "stubble" be left on the field to aid in the growth of next season's crop. This is why when we calculate the biomass yield per hectare B_i^k, we account for the fraction of biomass produced that is actually harvested f^k (we assume this to be 90% [i.e., 0.9] for all crops).

How did we think about crop value and farmer investment costs? We collected data on the costs (investments) that farmers incurred harvesting crops from the Illinois Farm Business Farm Management Association through the University of Illinois Farm Decision Outreach Council (farmdoc 2014a, 2014b), which maintains cost records for corn and soybeans back to 2001. Farmer investments, which include direct costs (e.g., fertilizer, pesticides, seed, storage, drying, crop insurance), power costs (e.g., machine use/lease/depreciation, utilities, fuel), and overhead costs (e.g., labor, building repair/rent/depreciation, insurance) can vary spatially, with the data being separated into northern, central, and southern regions within Illinois (Figure 10.6).

A 6-year average of costs for each region was taken using 2001–2006 cost data, while high and low productivity areas were averaged in central Illinois. For miscanthus and switchgrass, we again used research literature to inform the harvest unit costs, which were estimated by John and Watson (2007) and were aggregated to the farm "gate price" (net value upon leaving the farm, which contrasts retail prices later paid by consumers) of miscanthus and switchgrass production. Data on cost variations for these crops was not available throughout the state.

The value of production of corn and soybeans was estimated from the Illinois Crop Yields Historical NASS Database (Sherrick 2005; NASS 2007; farmdoc 2014a, 2007), from which the average "value of production," the average amount of soy and corn sold multiplied by the average selling price, was calculated for 2000–2006 (again, we were concentrating on a situation that occurred in 2007). During this period, Illinois produced an average ~$2.6 billion worth of soybeans and ~$4.1 billion worth of corn per year. Since no miscanthus or switchgrass was produced for profit in the state (this is still mostly true today), the initial value for both crop yields was assumed to be $0.

How does this chapter's spatial modeling compare in NetLogo vs. classic system dynamics (SD) software? Figure 10.15 depicts this model as if it was constructed solely to model miscanthus introduction. If this was purely a system dynamics model, we could build the entire model as four identical pieces that were each parameterized to be

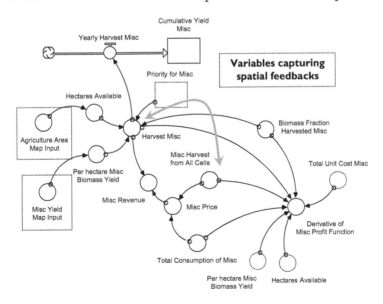

FIGURE 10.15
System dynamics representation of the model for miscanthus, including spatial connection between agents through a central market.

crop-specific. Alternatively, we could use functions for replicating the model (e.g., the "subscripting feature in Vensim or "arrays" in STELLA). However, this model could not be spatialized, unless we somehow were able to describe the spatial interactions using additional subscripts (this laborious process is discussed in Ford 1999), or we used another software system that could spatialize SD models (e.g., BenDor and Metcalf 2006).

Instead, in this chapter, we have capitalized on a major advantage of using NetLogo: the ability to import and export standard spatial data (allowing for pre- and postprocessing in GIS). Without this capability, we would need to use a set of scripts to create unique identifiers for each farm cell, link those identifiers to our spatial data to create a nonspatial database, and import that database into NetLogo (see a description of this laborious procedure in Scheffran and BenDor 2009). After running the model, we would then need to reverse that process to analyze and visualize output from the model. Instead, all of this can be done internally in NetLogo through the use of its GIS extension (Wilensky and Rand 2015), which is included in the model by listing the command, **extensions [GIS]** at the top of the model code.

Interested in the trajectory of bioenergy growth? In his 2006 State of the Union Address, former US President George W. Bush announced the goal of increasing domestic production capacity to 133 billion liters of renewable fuels by 2017, compared with 18.2 billion liters of ethanol produced for 2006 (Renewable Fuels Association 2006). In 2011, the Obama Administration announced a private–public partnership, through which the US Departments of Agriculture, Energy and the Navy will invest up to $510 million to produce biofuels to power military and commercial transportation between 2011 and 2014 (part of the Administration's "Blueprint for a Secure Energy Future"; Obama Administration 2011). Other countries have established various policies to support biofuels (for an overview, see Khanna, Scheffran, and Zilberman 2010; Scheffran 2010b).

One pioneering assessment of the US biomass potential (popularly known as "the Billion-Ton" study, referencing the total production of dry plant material, "lignocellulose") determined that the two largest sources of biomass alone, forested and agricultural areas, are capable of producing more than 1.1 billion tons of biomass per year—enough to replace more than one-third of the US demand for transportation fuels (Perlack et al. 2005; US DOE 2011). The full resource potential could be available by the middle of this century with a large-scale bioenergy industry and numerous biorefineries. This potential growth in bioenergy infrastructure could lead to a seven-fold increase in the annual US biomass production, while allowing the United States to produce enough food for its own use and for export (see Figure 10.2). This level of infrastructure improvement, coupled with anticipated government subsidies and negative public reactions to recent oil price shocks, may lead to a strong growth in biofuel usage.

However, other projections are not so optimistic; Miranowski and Rosburg (2013, p. 85) conclude, that "...global, market-based expansion of biofuel production will be limited in the absence of high oil prices [$140/barrel] or government incentives and mandates." Moreover, to replace this much gasoline consumption with ethanol (as it is produced today), the United States would need to produce approximately 62 billion gallons of ethanol a year, which, by some estimates, would require devoting 60 million acres of farmland to ethanol-generating bioenergy crops (Brainard 2007), or around 6.5% of the total US agricultural land area (US Census 2012). However, with all possible efficiency gains (higher yields, more efficient energy conversion, lower fuel consumption in cars), federal goals could be achieved on one-sixth of the land area needed under current conditions of ethanol production, making it possible to produce far more biomass on the same land or use less land and achieve an equivalent energy output (National Commission on Energy Policy 2004).

Interested in net-energy values of bioenergy fuels? The net values of the energy balance metric are likely to vary substantially, and depend on plant selection, growth, and harvesting methods, as well as transportation and conversion processes. For example, studies have shown that the energy output-to-input ratio (i.e., amount of energy produced based on amount of fossil-fuel energy used for production process) for corn ethanol ranges from 0.8:1 to 1.45:1, while gasoline achieves 0.8:1 at best. These values remain close to the values determined by a plethora of studies dating back to the early 2000s (Pimentel 2003; Kim and Dale 2004; Sheehan et al. 2004; Brinkman et al. 2005; Farrell et al. 2006; Hill et al. 2006).

> *Cellulosic ethanol* technologies produce ethanol from the lignocellulose in dry plant matter, rather than the starch matter in feedbacks such as corn (Somma, Lobkowicz, and Deason 2010).

With advanced production techniques and process optimization methods, this energy balance could be considerably improved. Significant hope for future petroleum replacement with biofuels has followed recent research into *cellulosic ethanol* technologies, which may push ethanol energy output-to-input ratio as high as 10:1 (Powlson, Richie, and Shield. 2005; Schmer et al. 2008; Solomon 2007). However, so far, the underlying assumptions have not materialized.

Improving these output-to-input ratios could have dramatic effects on GHG emissions; relative to fossil fuels, studies have found that the production and combustion of ethanol and biodiesel reduce GHG emissions by 12% and 41%, respectively (Hill et al. 2006). More recent studies for corn ethanol estimate an even higher life cycle GHG reduction of 24% relative to gasoline emissions (Wang et al. 2011). When you consider that much of the CO_2 that biofuel crops remove from the atmosphere through photosynthesis is actually stored in the soils and sediments that are not burned as biomass fuel, the full life cycle of production, deconstruction, and fermentation processes to create biofuels is actually even more "carbon negative" (Post and Kwon 2000).

Interested in switchgrass and miscanthus yields? In 2004–2005 miscanthus test trials in Illinois, dry matter per unit area was significantly greater than for switchgrass, particularly in the center of the state where dry biomass production peaked at 60.8 tons/ha. Miscanthus yields of 48.5 tons/ha were observed in the southern parts of the state and yields of 38.1 tones/ha in the north (Heaton, Voigt, and Long 2006). Earlier trials with miscanthus demonstrated little nitrogen contribution to runoff water and low levels of water use compared with corn or soybeans (Beale and Long 1997; Beale, Morison, and Long 1999). The crop can photosynthesize well at low temperatures and attain high yields with little nitrogen input. Like switchgrass, miscanthus has been shown to be effective at carbon sequestration and soil quality improvement.

McLaughlin and Kszos (2005), Perlack et al. (2005), and Cook (2007) believe that the increases in switchgrass yields will likely continue, with innovative breeding efforts to increase yields by 60% to 20 tons/ha by 2030.

Interested in potential biofuel-induced conflict? Illinois is an interesting place to study biofuel introduction. Brower et al. (1993) long ago concluded that "homegrown biomass energy could create jobs in Illinois, keep energy dollars in state, reduce air pollution and soil erosion, and provide many other environmental benefits, all at competitive costs." Although it may be environmentally suitable, the attractiveness of bioenergy crops may be lower in central Illinois, since the region produces corn and soybean yields that are much higher than in southern Illinois. This may lead to significant land competition within the region.

A study by Bournakis et al. (2005) analyzed the economic impacts of 1% annual increases in the fraction of electricity generated from renewable resources, assuming that levels reach at least 16% in 2020. This study noted that Illinois has considerable wind energy, biomass, and biowaste resources to meet these targets, which would require construction of renewable energy facilities capable of delivering about 28 terawatt hours (TWh) in 2020. Bioenergy crops also have the potential to displace coal in power plants and thereby reduce carbon emissions as well as sequester carbon in the soil in Illinois (Dhungana 2007). The industries responsible for the conversion of corn into ethanol carry considerable economic and political weight within Illinois and are likely to experience rapid future growth. In particular, local markets play an important role in triggering this process (John and Watson 2007).

References

Abdollahian, Mark, Zining Yang, and Hal Nelson. 2013. "Techno-Social Energy Infrastructure Siting: Sustainable Energy Modeling Programming (SEMPro)." *Journal of Artificial Societies and Social Simulation* 16 (3): 6.

Alonso, William. 1960. "A Theory of the Urban Land Market." *Papers in Regional Science* 6 (1): 149–57.

Atwell, Ryan C., Lisa A. Schulte, and Lynne M. Westphal. 2011. "Tweak, Adapt, or Transform: Policy Scenarios in Response to Emerging Bioenergy Markets in the U.S. Corn Belt." *Ecology and Society* 16 (1): 10.

Aubin, Jean-Pierre, and Patrick Saint-Pierre. 2007. "An Introduction to Viability Theory and Management of Renewable Resources." In *Advanced Methods for Decision Making and Risk Management in Sustainability Science*, edited by Jürgen P. Kropp and Jürgen Scheffran, 56–95. New York: Nova Science Publishers.

Aubin, Jean-Pierre, Alexandre M. Bayen, and Patrick Saint-Pierre. 2011. *Viability Theory: New Horizons* (2nd Edition). Berlin, Germany: Springer-Verlag.

Balmann, A. 2000. "Modeling Land Use with Multi-Agent Systems: Perspectives for the Analysis of Agricultural Policies." In *Microbehavior and Macroresults: Proceedings of the Tenth Biennial Conference of the International Institute of Fisheries Economics and Trade* (July 10–14, 2000), edited by R. S. Johnston. Corvallis, OR : International Institute of Fisheries Economics and Trade.

Barnaud, Cécile, François Bousquet, and Guy Trébuil. 2008. "Multi-Agent Simulations to Explore Rules for Rural Credit in a Highland Farming Community of Northern Thailand." *Ecological Economics* 66 (4): 615–27.

Beale, C. V., and S. P. Long. 1997. "Seasonal Dynamics of Nutrient Accumulation and Partitioning in the Perennial C-4-Grasses Miscanthus x Giganteus and Spartina Cynosuroides." *Biomass & Bioenergy* 12: 419–28.

Beale, C. V., J. I. L. Morison, and S. P. Long. 1999. "Water Use Efficiency of C-4 Perennial Grasses in a Temperate Climate." *Agricultural and Forest Meteorology* 96: 103–15.

Becu, Nicolas, P. Perez, A. Walker, O. Barreteau, and C. Le Page. 2003. "Agent Based Simulation of a Small Catchment Water Management in Northern Thailand: Description of the CATCHSCAPE Model." *Ecological Modelling* 170: 319–31.

Belt, Marjan van den. 2004. *Mediated Modeling: A System Dynamics Approach to Environmental Consensus Building*. Washington, DC: Island Press.

BenDor, Todd, and Nikhil Kaza. 2012. "A Theory of Spatial System Archetypes." *System Dynamics Review* 28 (2): 109–30.

BenDor, Todd, and Sara Metcalf. 2006. "The Spatial Dynamics of Invasive Species Spread." *System Dynamics Review* 22 (1): 27–50.

Berger, T. 2001. "Agent-Based Spatial Models Applied to Agriculture: A Simulation Tool for Technology Diffusion, Resource Use Changes and Policy Analysis." *Agricultural Economics* 25 (2–3): 245–60.

Berndes, G. 2002. "Bioenergy and Water–Implications of Large-Scale Bioenergy Production for Water Use and Supply." *Global Environmental Change* 12: 253–71.

Blaschek, Hans P., Ezeji Thaddeus C., Scheffran, Jürgen , eds. (2010) *Biofuels from Agricultural Wastes and Byproducts*. New York: Wiley-Blackwell.

Bournakis, A. D., J. Cuttica, S. Mueller, and G. J. D. Hewings. 2005. *The Economic and Environmental Impacts of Clean Energy Development in Illinois*. Chicago: University of Illinois at Chicago and at Urbana-Champaign Energy Resources Center.

Brainard, J. 2007. "The Big Deals in Biofuels." *Chronicle of Higher Education* 53 (33): 1–18.

Brinkman, N., M. Wang, T. Weber, and T. Darlington. 2005. *Well-to-Wheels Analysis of Advanced Fuel/Vehicle Systems: A North American Study of Energy Use, Greenhouse Gas Emissions, and Criteria Pollutant Emissions*. Argonne, IL: Argonne National Lab.

Brower, M., M. Tennis, E. Denzler, and M. Kaplan. 1993. *Powering the Midwest: Renewable Electricity for the Economy and the Environment—Analysis of the Potential of Renewable Energy for Electricity Generation in the Midwest*. Boston, MA: Union of Concerned Scientists.

Cabrera, Victor, Norman Breuer, and Peter Hildebrand. 2008. "Participatory Modeling in Dairy Farm Systems: A Method for Building Consensual Environmental Sustainability Using Seasonal Climate Forecasts." *Climatic Change* 89 (3): 395–409.

Castella, Jean-Christophe, Suan Pheng Kam, Dang Dinh Quang, Peter H. Verburg, and Chu Thai Hoanh. 2007. "Combining Top-down and Bottom-up Modelling Approaches of Land Use/Cover Change to Support Public Policies: Application to Sustainable Management of Natural Resources in Northern Vietnam." *Land Use Policy* 24 (3): 531–45.

Castella, Jean-Christophe, Tran Ngoc Trung, and Stanislas Boissau. 2005. "Participatory Simulation of Land-Use Changes in the Northern Mountains of Vietnam: The Combined Use of an Agent-Based Model, a Role-Playing Game, and a Geographic Information System." *Ecology and Society* 10 (1): 27.

Chiew, Thang Hooi. 2009. *Malaysia Forestry Outlook Study (Working Paper No. APFSOS II/WP/2009/02)*. Bangkok, Thailand: UN Food and Agriculture Organization, Regional Office for Asia and the Pacific.

Clifton-Brown, J. C., S. P. Long, and U. Jorgensen. 2001. "Miscanthus Productivity." In *Miscanthus for Energy and Fiber*, edited by M. B. Jones and M. Walsh, 46–67. London, England: James and James (Science Publishers) Ltd.

Cook, Anne. 2007. "Ethanol Project in Progress—Pike County Effort Expands to Site near Royal Breaking Ground This Spring." *The (Champaign-Urbana, IL) News-Gazette*.

Cook, J. H., J. Beyea, and K. H. Keeler. 1991. "Potential Impacts of Biomass Production in the U.S. on Biological Diversity." *Annual Reviews of Energy and Environment* 16: 401–31.

Cressie, N. 1993. *Statistics for Spatial Data*. New York: Wiley.

Dhungana, B. 2007. *Economic Modeling of Soil Carbon Sequestration and Carbon Emission Reduction through Bioenergy Crops in Illinois*. PhD Thesis, Urbana: University of Illinois at Urbana-Champaign, Department of Agricultural and Consumer Economics.

Dorian, James P., Herman T. Franssen, and Dale R. Simbeck. 2006. "Global Challenges in Energy." *Energy Policy* 34 (15): 1984–91.

Dudley, Richard G. 2008. "A Basis for Understanding Fishery Management Dynamics." *System Dynamics Review* 24 (1): 1–29.

EIA. 2012. "Most States Have Renewable Portfolio Standards." Washington, DC: Energy Information Administration, Department of Energy. www.eia.gov/todayinenergy/detail.cfm?id=4850

Eligon, John, and Matthew L. Wald. 2013. "Days of Promise Fade for Ethanol. New York: The New York Times." www.nytimes.com/2013/03/17/us/17ethanol.html

ESRI. 2006. *What Is ArcGIS 9.2?* Redlands, CA: ESRI Press.

Falkenmark, M., and D. Molden. 2008. "Wake up to Realities of River Basin Closure." *Water Resources Development* 24 (2): 201–15.

farmdoc. 2007. "Per Acre Revenue and Costs for Illinois Crops." Urbana, IL: University of Illinois Farm Decision Outreach Council. 2007. http://farmdoc.illinois.edu/

farmdoc. 2014a. "Illinois Farm Management Handbook." Urbana, IL: University of Illinois Farm Decision Outreach Council. 2014. www.farmdoc.illinois.edu/manage/index.asp#handbook

farmdoc. 2014b. "Illinois Farm Management Handbook: Yield and Price Tools." Urbana, IL: University of Illinois Farm Decision Outreach Council. 2014. www.farmdoc.illinois.edu/manage/pricing/index.asp

Farrell, Alexander E., and D. Sperling. 2007. *A Low-Carbon Fuel Standard for California: Part 1: Technical Analysis*. Berkeley/Sacramento: California Energy Commission.

Farrell, Alexander E., Richard J. Plevin, Brian T. Turner, Andrew D. Jones, Michael O'Hare, and Daniel M. Kammen. 2006. "Ethanol Can Contribute to Energy and Environmental Goals." *Science* 311 (5760): 506–8.

Fiddaman, Thomas S. 2007. "Dynamics of Climate Policy." *System Dynamics Review* 23 (1): 21–34.

Ford, Andrew. 1999. *Modeling the Environment: An Introduction to System Dynamics Modeling of Environmental Systems*. Washington, DC: Island Press.

Förster, Michael, Yvonne Helms, Alfred Herberg, Antje Köppen, Kathrin Kunzmann, Dörte Radtke, Lutz Ross, et al. 2008. "A Site-Related Suitability Analysis for the Production of Biomass as a Contribution to Sustainable Regional Land-Use." *Environmental Management* 41 (4): 584–98.

Fraiture, C. de, M. Giordano, and Y. Liao. 2008. "Biofuels and Implications for Agricultural Water Use: Blue Impacts of Green Energy." *Water Policy* S1: 67–82.

Gasparatos, Alexandros, Per Stromberg, and Kazuhiko Takeuchi. 2011. "Biofuels, Ecosystem Services and Human Wellbeing: Putting Biofuels in the Ecosystem Services Narrative." *Agriculture, Ecosystems & Environment* 142 (3–4): 111–28.

Goldemberg, J. 2007. "Ethanol for a Sustainable Energy Future." *Science* 315: 808–10.

Goode, Darren. 2013. "Rising Cost of Corn Ethanol Credits Alarms Hill." Washington, DC: Politico. www.politico.com/story/2013/03/rising-cost-of-corn-ethanol-credits-alarms-hill-89043.html

Grafton, R. Quentin, Tom Kompas, and Ngo Van Long. 2010. *Biofuels Subsidies and the Green Paradox (CESifo Working Paper, No. 2960)*. Munich, Germany: Leibniz Information Centre for Economics. www.econstor.eu/bitstream/10419/30713/1/620444681.pdf

Graham, Robin L., Burton C. English, and Charles E. Noon. 2000. "A Geographic Information System-Based Modeling System for Evaluating the Cost of Delivered Energy Crop Feedstock." *Biomass and Bioenergy* 18 (4): 309–29.

Groom, Martha J., Elizabeth M. Gray, and Patricia A. Townsend. 2008. "Biofuels and Biodiversity: Principles for Creating Better Policies for Biofuel Production. *Conservation Biology* 22 (3): 602–9.

Heaton, Emily, Tom Voigt, and Stephen P. Long. 2004. "A Quantitative Review Comparing the Yields of Two Candidate C4 Perennial Biomass Crops in Relation to Nitrogen, Temperature and Water." *Biomass and Bioenergy* 27 (1): 21–30.

Emily, T. B. Voigt, and S. P. Long. 2006. *Miscanthus X Giganteus: The Results of Trials Alongside Switchgrass (Panicum Virgatum) in Illinois*. Urbana, IL: University of Illinois at Urbana-Champaign, Department of Plant Biology.

Hill, J., E. Nelson, D. Tilman, S. Polasky, and D. Tiffany. 2006. "Environmental, Economic, and Energetic Costs and Benefits of Biodiesel and Ethanol Biofuels." *Proceeding of National Academy of Sciences of the United States* 103: 11206–10.

Holtz, Georg, and Claudia Pahl-Wostl. 2012. "An Agent-Based Model of Groundwater Over-Exploitation in the Upper Guadiana, Spain." *Regional Environmental Change* 12 (1): 95–121.

Jensen, John R. 2000. *Remote Sensing of the Environment: An Earth Resource Perspective*. Upper Saddle River, NJ: Prentice Hall.

John, Stephen, and Adam Watson. 2007. *Establishing a Grass Energy Crop Market in the Decatur Area: Report of the Upper Sangamon Watershed Farm Power Project*. Decatur, IL: The Agricultural Watershed Institute. www.agwatershed.org/PDFs/Biomass_Report_Aug07.pdf

Kang, Seungmo, Hayri Önal, Yanfeng Ouyang, Jürgen Scheffran, and Ü Deniz Tursun. 2010. "Optimizing the Biofuels Infrastructure: Transportation Networks and Biorefinery Locations in Illinois." In *Handbook of Bioenergy Economics and Policy*, edited by Madhu Khanna, Jürgen Scheffran, and David Zilberman, 151–73. New York: Springer.

Keeney, Roman, and Thomas W. Hertel. 2009. "The Indirect Land Use Impacts of United States Biofuel Policies: The Importance of Acreage, Yield, and Bilateral Trade Responses." *American Journal of Agricultural Economics* 91 (4): 895–909.

Khanna, M., B. R. Dhungana, and J. C. Clifton-Brown. 2008. "Costs of Producing Miscanthus and Switchgrass for Bioenergy in Illinois." *Biomass and Bioenergy* 32 (6): 482–93.

Khanna, M., H. Onal, B. Dhungana, and M. Wander. 2005. *The Economics of Soil Carbon Sequestrating Using Biomass Crops in Illinois*. Urbana: University of Illinois at Urbana-Champaign.

Khanna, M., Jürgen Scheffran, and D. Zilberman, eds. 2010. *Handbook of Bioenergy Economics and Policy*. Berlin, Germany: Springer-Verlag.

Kim, S., and B. E. Dale. 2004. "Cumulative Energy and Global Warming Impact from the Production of Biomass for Biobased Products." *Journal of Industrial Ecology* 7 (3–4): 147–62.

Landais, E. 1998. "Modelling Farm Diversity: New Approaches to Typology Building in France." *Agricultural Systems* 58 (4): 505–27.

Lapola, David M., Ruediger Schaldach, Joseph Alcamo, Alberte Bondeau, Jennifer Koch, Christina Koelking, and Joerg A. Priess. 2010. "Indirect Land-Use Changes Can Overcome Carbon Savings from Biofuels in Brazil." *Proceedings of the National Academy of Sciences* 107 (8): 3388–93.

Marshall, L. 2007. "Thirst for Corn: What 2007 Plantings Could Mean for the Environment (Energy: Biofuels, Policy Note 2)." Washington, DC: World Resource Institute. http://pdf.wri.org/policy note_thirstforcorn.pdf

McGranahan, G. 1986. *Searching for the Biofuel Energy Crisis in Rural Java*. Madison, WI: Unpublished Ph.D. Dissertation, University of Wisconsin at Madison.

McLaughlin, Samuel B., and Lynn Adams Kszos. 2005. "Development of Switchgrass (Panicum Virgatum) as a Bioenergy Feedstock in the United States." *Biomass and Bioenergy* 28 (6): 515–35.

Milbrandt, A. 2005. *A Geographic Perspective on the Current Biomass Resource Availability in the United States (Technical Report NREL/TP-560–39181)*. Golden, CO: National Renewable Energy Laboratory.

Millennium Ecosystem Assessment. 2005a. *Ecosystems and Human Wellbeing: Biodiversity Synthesis*. Washington, DC: World Resources Institute.

Millennium Ecosystem Assessment. 2005b. *Millennium Ecosystem Assessment: Ecosystems and Human Wellbeing*. Washington, DC: Island Press.

Miranowski, John, and Alicia Rosburg. 2013. "Long-Term Biofuel Projections under Different Oil Price Scenarios." *AgBioForum* 15 (4): 79–87.

Mitchell, Tim. 2008. "Firm Pulls Plug on Royal Ethanol Plant (6/18/08)." *Champaign News-Gazette*. www.news-gazette.com/news/business/2008-06-18/firm-pulls-plug-royal-ethanol-plant.html

Mitsch, William J. 1999. "Hypoxia Solution through Wetland Restoration in America's Breadbasket." *National Wetlands Newsletter* 21 (6): 9–10, 14.

Moraes, Marcia Maria Guedes Alcoforado de, Ximing Cai, Claudia Ringler, Bruno Edson Albuquerque, Sergio P. Vieira da Rocha, and Carlos Alberto Amorim. 2010. "Joint Water Quantity-Quality Management in a Biofuel Production Area—Integrated Economic-Hydrologic Modeling Analysis." *Journal of Water Resources Planning and Management* 136 (4): 502–11.

Mosnier, Aline, Petr Havlík, Hugo Valin, Justin S. Baker, Brian C. Murray, Siyi Feng, Michael Obersteiner, et al. 2012. *The Net Global Effects of Alternative U.S. Biofuel Mandates: Fossil Fuel Displacement, Indirect Land Use Change, and the Role of Agricultural Productivity Growth*. Durham, NC: Nicholas Institue for Environmental Policy Solutions, Duke University.

Muir, John. 1911. *My First Summer in the Sierra.* Cambridge, MA: The Riverside Press.

NASS. 2006. *National Agriculture Statistics Service1:100,000-Scale 2005 Cropland Data Layer: A Crop-Specific Digital Data Layer for Illinois (2006 March 14).* Washington, DC: United States Department of Agriculture, National Agricultural Statistics Service.

NASS. 2007. *USDA-NASS Quick Stats (Corn Field and Soybeans).* Washington, DC: United States Department of Agriculture, National Agricultural Statistics Service. www.nass.usda.gov/Quick_Stats/

National Atlas. 2011. "The Public Land Survey System (PLSS)." The National Map Small Scale. Washington, DC: US Geological Survey. 2011. https://nationalmap.gov/small_scale/a_plss.html

National Commission on Energy Policy. 2004. *Ending the Energy Stalemate-A Bipartisan Strategy to Meet America's Energy Challenges.* Washington, DC: The National Commission on Energy Policy.

National Research Council. 2007. *Water Implications of Biofuels Production in the United States.* Washington, DC: National Academy Press.

Ng, Tze Ling, J. Wayland Eheart, Ximing Cai, and John B. Braden. 2011. "An Agent-Based Model of Farmer Decision-Making and Water Quality Impacts at the Watershed Scale under Markets for Carbon Allowances and a Second-Generation Biofuel Crop." *Water Resources Research* 47 (9): W09519.

NHTSA. 2011. "Summary of Fuel Economy Performance." Washington, DC: National Highway Traffic Safety Administration. November 19, 2011. www.nhtsa.gov/fuel-economy

NRDC. 2004. *Growing Energy: How Biofuels Can Help End America's Oil Dependence.* Washington, DC: Natural Resources Defense Council.

NREL. 2014. "Biomass Maps." Washington, DC: National Renewable Energy Laboratory. 2014. www.nrel.gov/gis/biomass.html

NREL. 2017. "The Biofuels Atlas." Washington, DC: National Renewable Energy Laboratory. 2017. https://maps.nrel.gov/biofuels-atlas/

NSAC. 2011. "Conflicting Congressional Action on Ethanol Subsidies." Washington, DC: National Sustainable Agriculture Coalition. 2011. http://sustainableagriculture.net/blog/ethanol-votes/

Obama Administration. 2011. "Blueprint for a Secure Energy Future." Washington, DC: Obama Administration. https://obamawhitehouse.archives.gov/sites/default/files/blueprint_secure_energy_future.pdf

Parks, Peter J., and Randall A. Kramer. 1995. "A Policy Simulation of the Wetlands Reserve Program." *Journal of Environmental Economics and Management* 28 (2): 223–40.

Patil, Vishwanath, Khanh-Quang Tran, and Hans Ragnar Giselrod. 2008. "Towards Sustainable Production of Biofuels from Microalgae." *International Journal of Molecular Sciences* 9 (7): 1188–95.

Perlack, R. D., L. L. Wright, A. Turhollow, R. L. Graham, B. Stokes, and D. C. Erbach. 2005. *Biomass as Feedstock for a Bioenergy and Bioproducts Industry: The Technical Feasibility of a Billion-Ton Annual Supply.* Oak Ridge, TN: U.S. Department of Energy and U.S. Department of Agriculture, Oak Ridge National Laboratory.

Phalan, Ben, Malvika Onial, Andrew Balmford, and Rhys E. Green. 2011. "Reconciling Food Production and Biodiversity Conservation: Land Sharing and Land Sparing Compared." *Science* 333 (6047): 1289–91.

Pimentel, D. 2003. "Ethanol Fuels: Energy Balance, Economics and Environmental Impacts Are Negative." *Natural Resources Research* 12 (2): 127–34.

Pittock, Jamie. 2011. "National Climate Change Policies and Sustainable Water Management: Conflicts and Synergies." *Ecology and Society* 16 (2): 25.

Post, W. M., and K. C. Kwon. 2000. "Soil Carbon Sequestration and Land-Use Change: Processes and Potential." *Global Change Biology* 6: 317–28.

Powlson, D. S., A. B. Richie, and I. Shield. 2005. "Biofuels and Other Approaches for Decreasing Fossil Fuel Emissions from Agriculture." *Annals of Applied Biology* 146: 193–201.

Ragauskas, Arthur J., Charlotte K. Williams, Brian H. Davison, George Britovsek, John Cairney, Charles A. Eckert, William J. Frederick, et al. 2006. "The Path Forward for Biofuels and Biomaterials." *Science* 311 (5760): 484–89.

Rascoe, Ayesha. 2011. "U.S. Lawmakers Say Ethanol Mandate May Hike Gasoline Price." Washington, DC. https://uk.reuters.com/article/usa-ethanol-lawmakers/update-4-u-s-lawmakers-say-ethanol-mandate-may-hike-gasoline-price-idUKL1N0CC2VK20130321

Reichelderfer, Katherine, and William G. Boggess. 1988. "Government Decision Making and Program Performance: The Case of the Conservation Reserve Program." *American Journal of Agricultural Economics* 70 (1): 1–11.

Reinhardt, G., N. Rettenmaier, and S. Gärtner. 2007. *Rain Forest for Biodiesel? Ecological Effects of Using Palm Oil as a Source of Energy.* Berlin/Heidelberg/New York: WWF Germany.

REN21. 2015. *Renewables 2015: Global Status Report.* Paris, France: REN21, United Nations Environment Programme. www.ren21.net/wp-content/uploads/2015/07/REN12-GSR2015_Onlinebook_low1.pdf

Renewable Fuels Association. 2006. *Ethanol Industrial Outlook 2006.*Washington, DC: Renewable Fuels Association.

Renewable Fuels Association. 2014. *Ethanol Industry Outlook 2014.* Washington, DC: Renewable Fuels Association.

Rosillo-Calle, Frank, Sarah Hemstock, and Peter De Groot. 2006. *The Biomass Assessment Handbook. Bioenergy for a Sustainable Environment.* London, England: Earthscan.

Scheffran, Jürgen. 2006. "Tools in Stakeholder Assessment and Interaction." In *Stakeholder Dialogues in Natural Resources Management and Integrated Assessments: Theory and Practice*, edited by S. Stoll Kleemann and M. Welp, 153–185. Berlin, Germany: Springer.

Scheffran, Jürgen. 2010a. "Criteria for a Sustainable Bioenergy Infrastructure and Lifecycle." In *Plant Biotechnology for Sustainable Production of Energy and Co Products*, edited by P. Mascia, J. Scheffran, and J. Widholm, 409–47. Biotechnology in Agriculture and Forestry. Berlin/Heidelberg, Germany: Springer.

Scheffran, Jürgen. 2010b. "The Global Demand for Biofuels: Technologies, Markets and Policies." In *Biomass to Biofuels: Strategies for Global Industries*, edited by A. Vertes, H. P. Blaschek, H. Yukawa, and N. Qureshi, 27–54. New York: Wiley.

Scheffran, Jürgen, and Todd BenDor. 2009. "Bioenergy and Land Use: A Spatial-Agent Dynamic Model of Energy Crop Production in Illinois." *International Journal of Environment and Pollution* 39 (1/2): 4–27.

Scheffran, Jürgen, and G. Summerfield, eds. 2009. *Swords & Ploughshares: Sustainable Biofuels and Human Security (Volume 17(2): Summer 2009).* Urbana, IL: University of Illinois, Program in Arms Control, Disarmament, and International Security.

Schlüter, Maja, and Claudia Pahl-Wostl. 2007. "Mechanisms of Resilience in Common-Pool Resource Management Systems: An Agent-Based Model of Water Use in a River Basin." *Ecology and Society* 12 (2): 4.

Schmer, M. R., K. P. Vogel, R. B. Mitchell, and R. K. Perrin. 2008. "Net Energy of Cellulosic Ethanol from Switchgrass." *Proceedings of the National Academy of Sciences* 105 (2): 464–69.

Schnepf, Randy. 2005. *Agriculture-Based Renewable Energy Production (CRS Report for Congress).* Washington, DC: Congressional Research Service.

Schnepf, Randy, and Brent D. Yacobucci. 2013. *Renewable Fuel Standard (RFS): Overview and Issues (R40155). CRS Report for Congress.* Washington, DC: Congressional Research Service.

Secchi, Sylvia, and Bruce A. Babcock. 2015. "Impact of High Corn Prices on Conservation Reserve Program Acreage." *Iowa Ag Review* 13 (2): 2.

Sheehan, J., A. Aden, K. Paustian, K. Killian, J. Bremer, M. Walsh, and R. Nelson. 2004. "Energy and Environmental Aspects of Using Corn Stover for Fuel Ethanol." *Journal of Industrial Ecology* 7 (3–4): 117–46.

Sherrick, Bruce. 2005. "Illinois Crop Yields Historical NASS Database v. 5.1." Urbana, IL: University of Illinois Farm Decision Outreach Council. 2005. www.farmdoc.illinois.edu/

Shu, Kesheng, Jürgen Scheffran, Uwe A. Schneider, Liang E. Yang, and John Elflein. 2017. "Reconciling Food and Bioenergy Feedstock Supply in Emerging Economies: Evidence from Jiangsu Province in China." *International Journal of Green Energy* 14 (6): 509–21.

Shu, Kesheng, Uwe A. Schneider, and Jürgen Scheffran. 2015. "Bioenergy and Food Supply: A Spatial-Agent Dynamic Model of Agricultural Land Use for Jiangsu Province in China." *Energies* 8 (11): 13284–307.

Solomon, S., D. Qin, M. Manning, Z. Chen, M. Marquis, K. B. Averyt, M. Tignor andet al. 2007. *Climate Change 2007: The Physical Science Basis, Contribution of Working Group I to the Fourth Assessment Report of the Intergovernmental Panel on Climate Change.* Cambridge, England: Cambridge University Press.

Somma, Dan, Hope Lobkowicz, and Jonathan P. Deason. 2010. "Growing America's Fuel: An Analysis of Corn and Cellulosic Ethanol Feasibility in the United States." *Clean Technologies and Environmental Policy* 12: 373–80.

Stave, Krystyna. 2010. "Participatory System Dynamics Modeling for Sustainable Environmental Management: Observations from Four Cases." *Sustainability* 2 (9): 2762–84.

Steenblik, R., ed. 2007. *Biofuels—At What Cost? Government Support for Ethanol and Biodiesel in Selected OECD Countries.* Geneva, Switzerland: Global Subsidies Initiative, International Institute for Sustainable Development.

Stoms, David M., Frank W. Davis, Mark W. Jenner, Theresa M. Nogeire, and Stephen R. Kaffka. 2012. "Modeling Wildlife and Other Trade-offs with Biofuel Crop Production." *Gcb Bioenergy* 4 (3): 330–41.

Stroman, Dianne, and Urs P. Kreuter. 2016. "Landowner Satisfaction with the Wetland Reserve Program in Texas: A Mixed-Methods Analysis." *Environmental Management* 57 (1): 97–108.

US Census. 2012. "2012 National Projections." Washington, DC: US Census Bureau. 2012. www.census.gov/population/projections/data/national/2012.html

US DOE. 2006. *Breaking the Biological Barriers to Cellulosic Ethanol: A Joint Research Agenda (DOE/SC/EE-0095).* Washington, DC: U.S. Department of Energy, Office of Science and Office of Energy Efficiency and Renewable Energy.

US DOE. 2011. *U.S. Billion Ton Update : Biomass Supply for a Bioenergy and Bioproducts Industry.* Washington, DC: Energy Efficiency and Renewable Energy (Office of the Biomass Program), U.S. Department of Energy.

US DOE. 2015. *Lignocellulosic Biomass for Advanced Biofuels and Bioproducts: Workshop Report (DOE/SC-0170).* Washington, DC: Office of Science, U.S. Department of Energy.

Venghaus, Sandra, and Kirsten Selbmann. 2014. "Biofuel as Social Fuel: Introducing Socio-Environmental Services as a Means to Reduce Global Inequity." *Ecological Economics* 97: 84–92.

Vidosh, Mahate, Verma Praksh, and Chaube Alok. 2011. "Impacts of Bio-Fuel Use: A Review." *International Journal of Engineering Science and Technology* 3 (5): 3776–82.

Wang, Michael Q. 2007. *Ethanol: The Complete Energy Lifecycle Picture.* Chicago, IL: Argonne National Laboratory. www1.eere.energy.gov/vehiclesandfuels/pdfs/program/ethanol_brochure_color.pdf

Wang, Michael Q., Jeongwoo Han, Zia Haq, Wallace E. Tyner, May Wu, and Amgad Elgowainy. 2011. "Energy and Greenhouse Gas Emission Effects of Corn and Cellulosic Ethanol with Technology Improvements and Land Use Changes." *Biomass and Bioenergy* 35 (5): 1885–96.

Wilensky, Uri, and William Rand. 2015. *An Introduction to Agent-Based Modeling: Modeling Natural, Social, and Engineered Complex Systems with NetLogo.* Cambridge, MA: MIT Press.

Worldwatch. 2006. *Biofuels for Transportation, Global Potential and Implications for Sustainable Agriculture and Energy in the 21st Century: Extended Summary.* Washington, DC: Worldwatch Institute.

Wu, M., Michael Q. Wang, and H. Huo. 2006. *Fuel-Cycle Assessment of Selected Bioethanol Production Pathways in the United States.* Argonne, IL: Argonne National Laboratory.

11

The Future of Modeling Environmental Conflict and Cooperation

In every investigation, in every extension of knowledge, we're involved in action. And in every action, we're involved in choice. And in every choice, we're involved in a kind of loss, the loss of what we didn't do. We find this in the simplest situations.... Meaning is always obtained at the cost of leaving things out.... In practical terms this means, of course, that our knowledge is always finite and never all encompassing.... This makes the world of ours an open world, a world without end.

—J. Robert Oppenheimer (1961), American physicist and "Father of the atomic bomb"

Reflecting on Our Goals

In this book, we have strived to accomplish two goals. First, we have demonstrated how a number of disparate fields can be tied together to approach conflict understanding and resolution from both technical and anthropological standpoints. These fields include alternative dispute resolution, law, geography, conflict and peace studies, negotiation theory, sociology, political science, system dynamics (SD), and agent-based modeling (ABM), among others. We have argued that it is now possible to tear down the disciplinary divisions that have historically siloed the professions engaged in conflict studies and their resolutions. These fields can be linked in their study of environmental conflict under the banner of complexity science, a powerful approach to studying problems that explicitly considers how forces interact to collectively create structures and behaviors that are emergent and often very difficult to predict. Adaptive systems, including those at play during conflicts, adjust to this emergent behavior, thereby linking changes in the structure of individual parts of the system to the behavior of the whole.

Second, we have sought to introduce readers to the application of SD and ABM to the analysis of multiagent interaction, conflict, and cooperation. Our work has built on a set of computational techniques in complex adaptive systems, including SD and ABM, as well as work in institution building and coalition formation. Using this theoretical underpinning, we have developed the _Values and Investments for Agent-Based interaction and Learning in Environmental systems_ (VIABLE) modeling framework. We have also provided several in-depth examples of environmental and resource conflicts around the world. In doing so, we have also addressed cooperative approaches, including coalition building and joint management approaches, which can be applied to conflict resolution efforts.

Takeaways

In Part I of this book, we explored the vast literature on environmental conflict; from the history of conflict resolution and modeling to the types of models that have been developed. In Part II, we delved into specific modeling frameworks, including SD and ABM, which are both widely used tools that constitute burgeoning fields of inquiry. We introduced the VIABLE framework, and in Part III, we applied it to archetypes of different conflicts that have occurred around the work. So, what can we take away from this book? We offer five major conclusions.

Conflict Resolution Is Becoming an Increasingly Supported and Sophisticated Field

First, we can be buoyed by the fact that conflict resolution processes are backed by more and more evidence and are, in turn, growing more and more sophisticated; our understanding of conflict, including its causes, dynamics, and resolution potential, has expanded immensely over the last several decades. This trend will continue, particularly as disputes, and the unpopularity of violent conflict, generally, become progressively more visible to an increasingly connected world. It is our hope that conflict understanding and legibility, particularly regarding the complexity of conflict, will continue to grow through the use of the VIABLE model and other agent-based techniques (e.g., D'Aquino et al. 2002; Bousquet and Le Page 2004; Gurung, Bousquet, and Trébuil 2006; Scheffran 2016; Voinov et al. 2016; Oubraham and Zaccour 2018). As conflict analysis has reframed intractable disputes as a form of investment (i.e., aggression or war requires time, money, relationships, and often bloodshed to get what you want), we have endeavored to place these investments on an equal footing with other, more collaborative strategies to gain value for all stakeholders.

Participatory Processes and Modeling Are Gaining a Foothold

A second valuable takeaway is that we view conflict resolution as inherently participatory. Throughout this book, we outline and theorize ways of capturing conflict dynamics at the agent level; it is a fundamental goal of the dispute resolution field to understand how individuals or organizations behave and adapt during a conflict. However, we cannot adequately understand behavior without fully comprehending the objectives and values of individual agents. Doing this involves an agent's input and participation. Furthermore, even if we understand agents, in order to actualize change in their behavior or gain their acceptance of the decisions of other agents, we must earn the buy-in and support of conflict participants. This suggests a vitally important interface between complexity research and conflict resolution practice—participatory engagement during modeling interventions (Gray et al. 2018).

Cross-Pollination between Modeling Methods Makes Them More Useful

Third, divisions between modeling fields regarding the level of aggregation with which models treat the world are starting to erode. For example, within SD modeling, a methodological field that has historically been very active in linking technical modeling efforts to ethnographic stakeholder interventions, many practitioners are now examining agent-based behavior to better describe situations where systems are composed of great heterogeneity (Barnaud, Bousquet, and Trébuil 2008; Railsback and Grimm 2011; BenDor and Kaza 2012). In some situations, avoiding the *ecological fallacy*, wherein individuals are only viewed and understood based on the aggregate average of their group instead of their nuanced preferences and actions, has become paramount (Parunak, Savit, and Riolo 1998).

Likewise, in agent-based systems, we see broad discussions emerging around non-intuitive behavior, feedback mechanisms, the role of institutions and policy levers, and emergence of system structure (Wiegand et al. 2003; Grimm et al. 2005; Monticino et al. 2007). The fact is that these techniques, while developed in widely disparate fields for very different reasons, are now being applied in tandem to triangulate and heighten our understanding of conflict. This is encouraging.

The VIABLE Model Can Help Us Find Alternative Pathways to Increasing Agent Value

The fourth takeaway from this book concerns our modeling framework. We argue that understanding what agents want, how they are willing to get it, and how different agents' goals and actions interact, are the core factors in determining the trajectory of conflict potential. Additionally, the extent to which agents can learn from the strategies of others adds additional complexity to the conflict situation. In our three example applications— fisheries, emissions, and, biofuels—agents interact through markets. These indirect interactions send delayed signals back to agents, who then adjust their strategies to meet their goals. Understanding agents' goals—as they define them and act on them—is key to being able to predict when conflict interactions can occur. A huge facet of alternative dispute resolution involves finding options to create value for agents in ways that they had not previously understood (Fisher and Ury 1981).

There Are Opportunities to Better Understand the Emergent Effects of Conflicts

Finally, our fifth major takeaway is that the impacts of conflict must be explored carefully, as they often stretch far beyond individual stakeholders. Instead, they can dominate the futures of those never directly involved or concerned with disputes in the first place. In our fisheries example, we saw fish populations collapse, followed by the economic livelihoods of fishers. Behind the scenes and beyond the individual conflicting agents, we can start to understand the regional economic ramifications of a fishing industry collapse, which history has shown to be altogether devastating for fishing-dependent communities (Harris 1999). In our emissions model, we saw the emergence of a solution to climate conflict; but one that likely has far-reaching implications (e.g., economic development, public health) for the societies and industries shouldering the costs of emissions reductions or purchases. Finally, in our biofuels model, we modeled sweeping changes to the landscape, which can also change the culture and trajectory of communities across the region.

There are many additional applications of the VIABLE framework to modeling environmental disputes and their potential conclusions. Several examples include water use, threatened species, and land disputes. Fights continue domestically and internationally over water use and quality. The protection of the quality of water often means restriction of the use of water, leading to multidimensional conflicts within cities, regions, and entire river basins (Delli Priscoli and Wolf 2009). In drought-plagued regions, freshwater availability is often the topic of very high-resolution disputes between municipalities and industries, where dozens of agents may lay claim to water and become involved in long-term conflict. The uncertain long-term viability of rare or threatened species is a growing topic of concern that has sparked numerous disputes in developed and developing regions alike (Beatley 2014; Reeve 2014). These disputes involve land, individual rights, access to resources, and political philosophy. The expression of "value" for stakeholders is complex and may be difficult to express in a meaningful functional form. However, these are areas that merit exploration using the VIABLE framework. We hope our readers will use it, either conceptually or formally, for worthwhile applications in their own work.

Questions for the Future

We would like to pose four different questions about issues that generally pertain to potential extensions of the work in this book and to the future of ABM in conflict contexts.

Does Modeling Matter?

The first question concerns extensions of the excellent work of Etienne Rouwette (Rouwette, Vennix, and Mullekom 2002; Rouwette and Vennix 2006; Rouwette et al. 2011), who has spent the last 15 years tackling a key question: do modeling interventions actually work? Rouwette and others that study this topic have asked whether conflict stakeholders involved in modeling interventions actually learned anything, changed their behavior, or changed their perceptions of other stakeholders or the conflict itself. In most instances, it appears that modelers have not even attempted to determine if or how their efforts actually had any effect on the conflict.

We pose a similar question: do ABM efforts give guidance and aid in resolving conflicts? Do stakeholders learn from them? And how could ABM be improved to work in conflict situations? Unfortunately, work in this area is fairly slim (Morell et al. 2010; Brown et al. 2013; Walton 2014). It is clear that the research and practitioner communities are now in need of strong research designs for monitoring modeling interventions, as well as techniques for evaluating modeling outcomes. While there has been intensive discussion about how we get real-world data into models (e.g., Janssen and Ostrom 2006; Robinson et al. 2007), there has been little effort to get data about modeling processes and their impacts. This is an area in need of significant study and improvement.

In implementing the VIABLE framework presented in this book, we must consider the possibility that a modeling intervention will not be useful to the stakeholders embroiled in conflict. If this is the case, modelers are ethically obligated to consider how modeling can more effectively be used to inform stakeholders (Prell et al. 2007).

Are We Modeling What We Mean to Model?

Following the idea that we need to better assess the outcomes of conflict modeling comes the question of ensuring that the modeling technique we choose is indeed applicable to our goals. Agent-based techniques have made the case that modeling individuals and their behaviors can improve "model fidelity," or the degree to which a model is faithful to real-world behavior, in comparison with other techniques (Osgood 2009). An important question to ask is then: when does this argument hold for models of conflict systems? Is increased model fidelity worth the increased complexity?

Surprisingly little work has assessed the trade-offs between using agent-based and more aggregate (e.g., SD) forms of modeling in general (Parunak, Savit, and Riolo 1998), and almost no work has looked at comparative ways of representing conflict situations. We have argued that modelers have an ethical obligation to use the best modeling technique for the right situation; however, research and practitioners have yet to develop tools (e.g., decision trees or other model-selection decision support systems) to help guide modelers' efforts.

How Do We Make Conflict Modeling a Common and Standard Activity?

The third issue that we believe will drive the future of conflict modeling concerns the way that we make the case for modeling interventions in the first place. If we accept modeling to be a useful and worthwhile activity, how can we convince decision makers or stakeholders to draw on modeling interventions more frequently? In some instances, authors have argued that modeling will become more commonplace if it is more accessible to the public (e.g., open-source code, better usage guidelines, and manuals). For instance, Voinov et al. (2008) argued that accessibility may come with improved model usability and simplicity. In other cases, efforts to make complex models more accessible and useful to decision makers have led to profound rethinking of the roles of modelers and stakeholders in the modeling process. For example, one project devoted to democratizing the use of complex urban growth models has created a system where users actually help to improve the model iteratively (Deal and Pallathucheril 2009). In this case, the model becomes part of the pantheon of tools accessible to stakeholders involved in urban decision making.

Nearly all aspects of our lives are becoming increasingly interconnected with the people and places around us. As the pace of technological innovation has increased, data about the world around us have become both more plentiful and more readily available. Social and computer science research has capitalized on this trend through emerging "big data" research, which seeks otherwise-hidden patterns about human behavior and decision making in data sets that have been previously unavailable or too large to easily understand (Mayer-Schönberger and Cukier 2013). This newfound data availability may soon become standard in informing conflict resolution processes and the models used therein. How will conflict modeling techniques and participatory processes need to change to allow this? How will data change negotiation processes? Will data help or complicate? Will data build or erode trust?

Paralleling efforts to make models more usable (bottom-up, demand driven), we may also consider efforts to engage modeling as a standard part of decision processes (top-down modeling requirements set by decision makers). In the future, a model could potentially become a codified part of the decision-making process, as transportation or engineering models are part of the fabric of designing the built infrastructure around us. In the conflict sphere, is this type of framework possible (e.g., see Harich 2010)? Can formalized models, such as the ones that we have created in this book, become embedded in the fabric of conflict resolution processes? In what circumstances can modeling change the course of conflict resolution processes?

How Can Modeling Prevent Conflict?

If we continue this line of reasoning to its eventual conclusion, we ask how the use of agent-based or other conflict modeling techniques can be used to prevent conflicts in the first place. Can we institutionalize the use of conflict modeling in mediation or facilitation processes? For example, during formal appeals processes in urban land use disputes (e.g., neighborhood disputes over congestion or unwanted land uses), we might improve joint fact finding through a high-level application of VIABLE modeling. But, could we employ VIABLE modeling to instead head off conflicts before they start? This would help us to better understand the goals and strategies of agents as they rely on shared resources or spaces to prevent conflicts in the first place. This type of proactive resilience (or we could even think of it as "resistance") to conflict occurrence could become a key pathway to strategies that prevent conflicts and the high costs they incur. Pathways for doing this could involve increased efforts to publicize the use of models during conflicts, along with studies about the beneficial outcomes of modeling.

We pose these questions to readers to help them think about how they want to interact with conflict modeling in the future. While answering these questions will require years of additional research, it is our hope that with this volume, we have laid the groundwork. The VIABLE modeling framework provides a powerful platform for conceptualizing and analyzing environmental conflict in many different settings. It draws on an open and transparent modeling technique (implemented in a nearly-public domain software environment, NetLogo) as a vehicle for helping users to simulate the emergent behavior arising from interactions of different types of conflict agents, who can take a range of actions, and who operate in various environments. The VIABLE framework also allows users to analytically determine the critical boundaries (viability conditions) between different types of social interactions, helping to define and predict states of conflict or cooperation. With this tool in hand, there are promising prospects not only for understanding conflicts, but also for resolving them.

Questions for Consideration

1. *What other questions were raised for you by this book? How does your work touch on the questions that we have raised?*

2. Check out the thought-provoking debate in the journal, *Geoforum,* on the future of complexity theory between Steven Matson and Femke Reitsma (Manson 2001; Manson 2003; Reitsma 2003). *What is the future of complex systems research?*

3. Peruse the numerous case studies that are described at https://participatory-modeling.org/. *Pick a conflict to learn more about.* Think about how you might apply the VIABLE framework to that conflict. *Consider: what are the agents? What information do they have? What are their value functions? How do they invest in action paths? How do they choose between different paths? How do they interact with other agents? Consider how these interactions could lead to points of conflict. How could they lead to cooperation? How could we seek additional value for agents that leads them back into a viable corridor?*

References

Barnaud, Cécile, François Bousquet, and Guy Trébuil. 2008. "Multi-Agent Simulations to Explore Rules for Rural Credit in a Highland Farming Community of Northern Thailand." *Ecological Economics* 66 (4): 615–27.

Beatley, Timothy. 2014. *Habitat Conservation Planning: Endangered Species and Urban Growth*. Austin: University of Texas Press.

BenDor, Todd, and Nikhil Kaza. 2012. "A Theory of Spatial System Archetypes." *System Dynamics Review* 28 (2): 109–30.

Bousquet, François, and C. Le Page. 2004. "Multi-Agent Simulations and Ecosystem Management: A Review." *Ecological Modelling* 176 (3–4): 313–32.

Brown, C. Hendricks, David C. Mohr, Carlos G. Gallo, Christopher Mader, Lawrence Palinkas, Gina Wingood, Guillermo Prado, et al. 2013. "A Computational Future for Preventing HIV in Minority Communities: How Advanced Technology Can Improve Implementation of Effective Programs." *Journal of Acquired Immune Deficiency Syndromes* 63 (1): S72–S84.

D'Aquino, Patrick, Olivier Barreteau, Michel Etienne, Stanislas Boissau, Sigrid Aubert, François Bousquet, Christophe Le Page, and William S. Daré. 2002. "The Role Playing Games in an ABM Participatory Modeling Process: Outcomes from Five Different Experiments Carried out in the Last Five Years." In *Integrated Assessment and Decision Support, 1st Biennial iEMSs Meeting*, edited by A. E. Rizzoli and A. J. Jakeman, 275–80. Lugano, Switzerland: International Environmental Modelling and Software Society.

Deal, Brian, and Varkki Pallathucheril. 2009. "A Use-Driven Approach to Large-Scale Urban Modelling and Planning Support." In *Planning Support Systems: Best Practice and New Methods*, edited by Stan Geertman and John Stillwell, 29–51. Amsterdam, Netherlands: Springer.

Delli Priscoli, J., and A. T. Wolf. 2009. *Managing and Transforming Water Conflicts*. Cambridge, England: Cambridge University Press.

Fisher, R., and William Ury. 1981. *Getting to Yes: Negotiating Agreements without Giving In*. New York: Penguin Books.

Gray, Steven, Alexey Voinov, Michael Paolisso, Rebecca Jordan, Todd BenDor, Pierre Bommel, Pierre Glynn, et al. 2018. "Purpose, Processes, Partnerships, and Products. Four Ps to Advance Participatory Socio-Environmental Modeling." *Ecological Applications* 28 (1): 46–61.

Grimm, Volker, Eloy Revilla, Uta Berger, Florian Jeltsch, Wolf M. Mooij, Steven F. Railsback, Hans-Hermann Thulke, et al. 2005. "Pattern-Oriented Modeling of Agent-Based Complex Systems: Lessons from Ecology." *Science* 310 (5750): 987–91.

Gurung, Tayan Raj, François Bousquet, and Guy Trébuil. 2006. "Companion Modeling, Conflict Resolution, and Institution Building: Sharing Irrigation Water in the Lingmuteychu Watershed, Bhutan." *Ecology and Society* 11 (2): 36.

Harich, Jack. 2010. "Change Resistance as the Crux of the Environmental Sustainability Problem." *System Dynamics Review* 26 (1): 35–72.

Harris, Michael. 1999. *Lament for an Ocean: The Collapse of the Atlantic Cod Fishery, A True Crime Story*. Toronto, ON: McClelland & Stewart.

Janssen, Marco A., and Elinor Ostrom. 2006. "Empirically Based, Agent-Based Models." *Ecology and Society* 11 (2): 37.

Manson, Steven M. 2001. "Simplifying Complexity: A Review of Complexity Theory." *Geoforum* 32 (3): 405–14.

Manson, Steven M. 2003. "Epistemological Possibilities and Imperatives of Complexity Research: A Reply to Reitsma." *Geoforum* 34 (1): 17–20.

Mayer-Schönberger, Viktor, and Kenneth Cukier. 2013. *Big Data: A Revolution That Will Transform How We Live, Work, and Think*. New York: Eamon Dolan/Houghton Mifflin Harcourt.

Monticino, Michael, Miguel Acevedo, Baird Callicott, Travis Cogdill, and Christopher Lindquist. 2007. "Coupled Human and Natural Systems: A Multi-Agent-Based Approach." *Environmental Modelling & Software* 22 (5): 656–63.

Morell, Jonathan A., Rainer Hilscher, Stephen Magura, and Jay Ford. 2010. "Integrating Evaluation and Agent-Based Modeling: Rationale and an Example for Adopting Evidence-Based Practices." *Journal of MultiDisciplinary Evaluation* 6 (14): 32–57.

Oppenheimer, J. Robert. 1961. "Reflections on Science and Culture." *Colorado Quarterly* 10 (2): 101–11.

Osgood, Nathaniel. 2009. "Lightening the Performance Burden of Individual-Based Models through Dimensional Analysis and Scale Modeling." *System Dynamics Review* 25 (2): 101–34.

Oubraham, Aïchouche, and Georges Zaccour. 2018. "A Survey of Applications of Viability Theory to the Sustainable Exploitation of Renewable Resources." *Ecological Economics* 145: 346–67.

Parunak, H. Van Dyke, Robert Savit, and Rick Riolo. 1998. "Agent-Based Modeling vs. Equation-Based Modeling: A Case Study and Users' Guide." In *Multi-Agent Systems and Agent-Based Simulation*, edited by Jaime Sichman, Rosaria Conte, and Nigel Gilbert, Vol. 1534, 277–83. Berlin/Heidelberg, Germany: Springer.

Prell, Christina, Klaus Hubacek, Mark S. Reed, Claire Quinn, Nanlin Jin, Joe Holden, Tim Burt, et al. 2007. "If You Have a Hammer Everything Looks like a Nail: Traditional versus Participatory Model Building." *Interdisciplinary Science Reviews* 32 (3): 263–82.

Railsback, Steven F., and Volker Grimm. 2011. *Agent-Based and Individual-Based Modeling: A Practical Introduction*. Princeton, NJ: Princeton University Press.

Reeve, Rosalind. 2014. *Policing International Trade in Endangered Species: The CITES Treaty and Compliance*. New York: Routledge.

Reitsma, Femke. 2003. "A Response to Simplifying Complexity." *Geoforum* 34 (1): 13–16.

Robinson, Derek T., Daniel G. Brown, Dawn C. Parker, Pepijn Schreinemachers, Marco A. Janssen, Marco Huigen, Heidi Wittmer, et al. 2007. "Comparison of Empirical Methods for Building Agent-Based Models in Land Use Science." *Journal of Land Use Science* 2 (1): 31–55.

Rouwette, Etiënne A. J. A., and Jac A. M. Vennix. 2006. "System Dynamics and Organizational Interventions." *Systems Research and Behavioral Science* 23 (4): 451–66.

Rouwette, Etiënne A. J. A., Jac A. M. Vennix, and Theo van Mullekom. 2002. "Group Model Building Effectiveness: A Review of Assessment Studies." *System Dynamics Review* 18 (1): 5–45.

Rouwette, Etiënne A. J. A., Hubert Korzilius, Jac A. M. Vennix, and Eric Jacobs. 2011. "Modeling as Persuasion: The Impact of Group Model Building on Attitudes and Behavior." *System Dynamics Review* 27 (1): 1–21.

Scheffran, Jürgen. 2016. "From a Climate of Complexity to Sustainable Peace: Viability Transformations and Adaptive Governance in the Anthropocene." In *Handbook on Sustainability Transition and Sustainable Peace*, 305–46. Hexagon Series on Human and Environmental Security and Peace. Cham, Switzerland: Springer.

Shaaban, Mostafa, Jürgen Scheffran, Jürgen Böhner, and Mohamed Elsobki. 2017. "Sustainability Assessment of Electricity Generation Technologies in Egypt Using Multi-Criteria Decision Analysis." *Energies* 11 (5): 1–26.

Voinov, Alexey, David Arctur, Ilya Zaslavskiy, and Saleem Ali. 2008. "Community-Based Software Tools to Support Participatory Modelling: A Vision." In *IEMSs 2008: International Congress on Environmental Modelling and Software: Integrating Sciences and Information Technology for Environmental Assessment and Decision Making (4th Biennial Meeting of IEMSs)*, edited by M. Sànchez-Marrè, J. Béjar, J. Comas, A. Rizzoli, and G. Guariso, 766–74. Barcelona, Spain: International Environmental Modelling and Software Society.

Voinov, Alexey, Nagesh Kolagani, Michael K. McCall, Pierre D. Glynn, Marit E. Kragt, Frank O. Ostermann, Suzanne A. Pierce, et al. 2016. "Modelling with Stakeholders – Next Generation." *Environmental Modelling & Software* 77 (March): 196–220.

Walton, Mat. 2014. "Applying Complexity Theory: A Review to Inform Evaluation Design." *Evaluation and Program Planning* 45: 119–26.

Wiegand, Thorsten, Florian Jeltsch, Ilkka Hanski, and Volker Grimm. 2003. "Using Pattern-Oriented Modeling for Revealing Hidden Information: A Key for Reconciling Ecological Theory and Application." *Oikos* 100 (2): 209–22.

Index

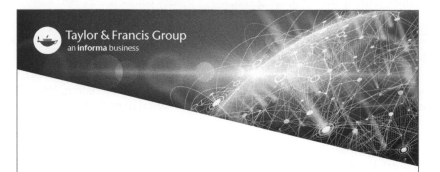

Milton Keynes UK
Ingram Content Group UK Ltd.
UKHW051946071024
449327UK00026B/2187